Steam and Gas Turbines
for Marine Propulsion

Steam and Gas
TURBINES
for Marine Propulsion
Second Edition

Maido Saarlas

Naval Institute Press · Annapolis, Maryland

Copyright © 1987
by the United States Naval Institute
Annapolis, Maryland
All rights reserved. No part of this book may be reproduced
without written permission from the publisher.

Library of Congress Cataloging-in-Publication Data
Saarlas, Maido.
    Steam and gas turbines for marine propulsion.

    Bibliography: p.
    Includes index.
    1. Steam-turbines, Marine. 2. Marine gas-turbines.
I. Title.
VM740.S2 1987   623.8'723   86-31305
ISBN 1-59114-750-6

Printed in the United States of America

Preface to the Second Edition

The last dozen years have seen the maturing of the first generation of marine gas turbines through their use and acceptance in both commercial and combatant ships. Such names as LM2500, Olympus, Spey, and Proteus have established, thanks to their versatility and proven reliability, a respectable baseline of performance to which new units will be compared.

Like the earlier units, the next generation of gas turbines will be composed of marinized aircraft derivatives, but will exhibit numerous refinements and improvements. Perhaps the most significant will be reheat and regeneration, which were tried years ago but discarded at that time for operational reasons. The vastly improved heat exchanger technology and the need for improved fuel economy will again bring about the introduction of the improved cycle and, it is hoped, a successful mating of old principles with new shipboard hardware.

In keeping with the objectives of the first edition, *Steam and Gas Turbines* has been revised to include some important topics necessary for an understanding of the current gas turbine technology and development. Additional material has been added on gas turbine cycles, turbine materials, combustion, and environmental effects. The ever-increasing scope of gas turbine applications has been duly noted by inclusion of more examples of the current ship–gas turbine configurations. Additional examples have been introduced wherever feasible.

The new material incorporates the suggestions and helpful criticism of people too numerous to thank here individually. Special gratitude, however, is due to K. L. Tuttle of the U.S. Naval Academy's Naval Systems Engineering Department for many helpful suggestions and for materially contributing to several new sections. And, as for the first edition of this book, E. P. Weinert of the Naval Ship Engineering Center, Philadelphia, who has just retired after forty-two years of service, deserves extra thanks for all the suggestions and constructive comments he has given freely during this project.

October 1986
Annapolis, Md.

M. SAARLAS

Preface to the First Edition

Steam and Gas Turbines for Marine Propulsion is partly a revision of a series of books published by the United States Naval Institute since 1946: *Naval Turbines, Naval Auxiliary Machinery,* and *Descriptive Analysis of Naval Turbine Propulsion Plants.* Partly it is a new book, describing the introduction of the gas turbine in marine power plants and providing a common and comparative treatment of steam and gas turbines used in both commercial and naval ships.

Emphasis has been placed on understanding the basic principles of the turbine as a propulsive device. Auxiliary machinery and overall power plant analysis have yielded to more detailed and practical aspects of turbine-oriented topics.

This book is intended mainly for those who are interested in finding out how and why turbines operate and are more concerned with their operation than their design: ships' officers, student engineers, and those who wish to have a general refresher on the topic.

The author is grateful to Captain M. J. Gross, USMS, Kings Point, and to E. P. Weinert, Naval Ship Engineering Center, Philadelphia, for truly constructive criticism and many helpful suggestions.

October 1977　　　　　　　　　　　　　　M. SAARLAS
Annapolis, Md.

Contents

Steam and Gas Turbines
for Marine Propulsion

Chapter One

Introduction

1.1 General

For many years the steam turbine and the diesel engine were the principal marine prime movers. Following its advent in the early forties and subsequent extensive usage in aviation, the gas turbine eventually found its way into marine applications. It is anticipated that these three types of prime movers will coexist for some time for ship propulsion.

The purpose of this book is to discuss and compare steam and gas turbines as work-producing turbomachinery in the marine environment. The diesel, being a reciprocating machine, will not be discussed here; though a worthy subject, it has received adequate treatment elsewhere. Nor are nuclear power plants included. Like boilers, they are important components of certain propulsion plants as devices that provide the thermal energy necessary for turbine operation; but they are outside the scope of this discussion.

Much has been written separately about various aspects of steam and gas turbines, and the literature is rapidly expanding in the developing areas of gas turbine technology. Voluminous operating and maintenance handbooks are issued for each new power plant and vessel, each with differing boiler, valve, pipe, and pump configurations. But less has been written about the relatively simple fundamentals of these two prime movers—material which would be useful for people who are not active in design or analysis, e.g., officers, crew, operators, and so on. Thus, the following three purposes guide this book:

1. To provide fundamentals and background for understanding the operation of steam and gas turbines based on principles of thermodynamics and fluid mechanics.

2. To study the operation of the turbine in the context of the overall power plant.

3. To provide understanding of the performance potentials and differentials of steam and gas turbines.

The emphasis herein is placed on principles, essential applications, and performance as related to marine applications, rather than construction, hardware, and design variation. Primary attention is focused on the turbine as a device that converts thermal energy into kinetic and mechanical shaft energy. Other components of the power plant are discussed peripherally only as they relate to serving the turbine needs or are necessary for transmission of power to move the vessel.

This distinction is made out of necessity rather than by choice. Other shipboard equipment, auxiliaries, and hardware are not less important; but they either serve the turbine or other functions, or are too specialized and are already adequately treated in books or manufacturers' literature (valves, pumps, piping, and so forth). Moreover, an adequate treatment of principles and developing technology under the same cover would result in excessive volume, and some parts of it would soon be obsolete.

The necessary principles of thermodynamics and fluid mechanics will be stated briefly, without derivations or amplifications, as they are required for understanding the material and for providing the background for simple calculations. References for further study and for verification of results presented here will be provided. Wherever convenient and applicable, typical examples will be included to illustrate use of the equations and to provide a feel for the magnitudes involved.

Before reviewing thermodynamics and fluid mechanics, a brief review of the basic marine power plant configuration is useful for further delineation of the role of the turbine in present practice, and to underline general features, similarities, and differences of steam and gas propulsion units.

1.2 Marine Steam and Gas Turbine Propulsion Units

A propulsion unit is generally considered to consist of the machinery and equipment, including controls, that are mechanically, electrically, or hydraulically connected to a propulsion shaft. A propulsion plant is made up of the propulsion unit, boiler, propulsion auxiliaries, piping, and electrical equipment that are required to drive one propeller. Propulsion auxiliaries (condenser, pumps, and so on) and independent auxiliaries (refrigeration, distilling, air conditioning, and the like) serve either the propulsion plant or ship services.

As will be seen, the foregoing general definitions, used mainly for steam plants, do not apply well to the gas turbine, which comes as a unit integrated with most of the auxiliaries that otherwise are external to the typical steam turbine unit and make up the bulk of the steam plant. On the one hand, the gas turbine unit can be considered as a complete propulsion plant, perhaps with the exception of the gearing and the fuel storage and delivery system. On the other hand, the gas turbine imposes new and more extensive requirements on inlet and exhaust systems, not found in steam plants.

The essential principles of both steam and gas turbine operation are the same: heat is added to a working fluid, which is expanded by turbine nozzles to convert the thermal energy into kinetic energy, which is then put through successive turbine stages where the kinetic energy is converted into mechanical (rotating shaft) energy. The net result is represented by the work (output) and the fuel consumption (input) which occurs during the propulsion cycle. A cycle is used here to mean a series of events and processes that the working fluid undergoes from the time heat is added and work is extracted until the fluid is discharged or returned for new duty. The cycle is the key factor that leads to differences between steam and gas turbine units, their operation, and construction. Although to the uninitiated eye the "innards" of the two types of turbines seem not to be very different, there are essential variations, which are dictated by the differences in the cycle as well as by the working fluid.

A steam turbine main steam cycle is represented in Figure 1.1, which indicates the four basic parts: generation, expansion, condensation, and the boiler feed systems. Only the general arrangement is shown here; a more detailed discussion is provided in Chapter 2. The working fluid is water or steam, depending on the location in the cycle. This is a *closed* cycle, where the working fluid is recirculated and reused. With a few, relatively minor variations, this arrangement may be found in any modern steam-propelled ship.

Figure 1.2 shows a typical gas turbine cycle. In addition to the usually diagrammed compressor-burner-turbine combination, the intake and exhaust systems are shown separately. In diagrams of aircraft or stationary plant applications, the latter are usually omitted as being part of airframe and jet propulsion systems, or simply insignificant for cycle analysis purposes. When a gas turbine is altered for marine

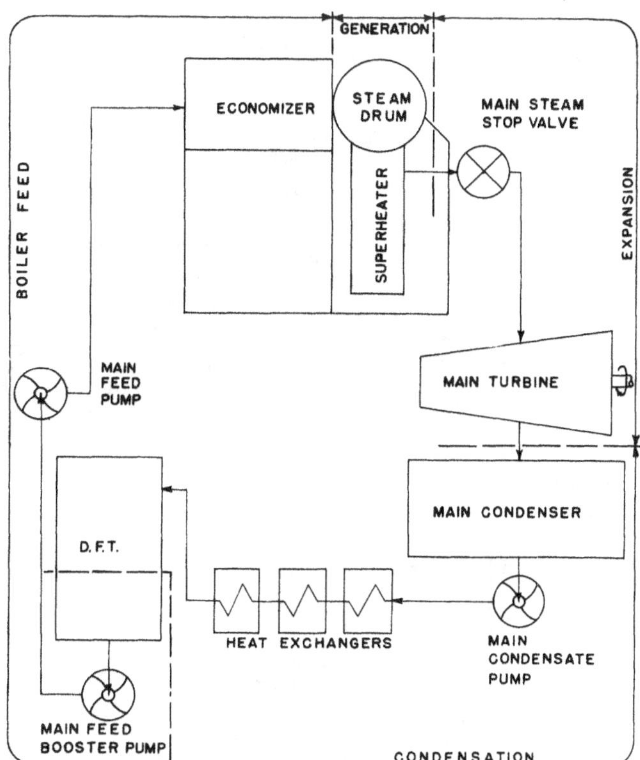

Figure 1.1 Main steam cycle

Figure 1.2 Simple gas turbine cycle

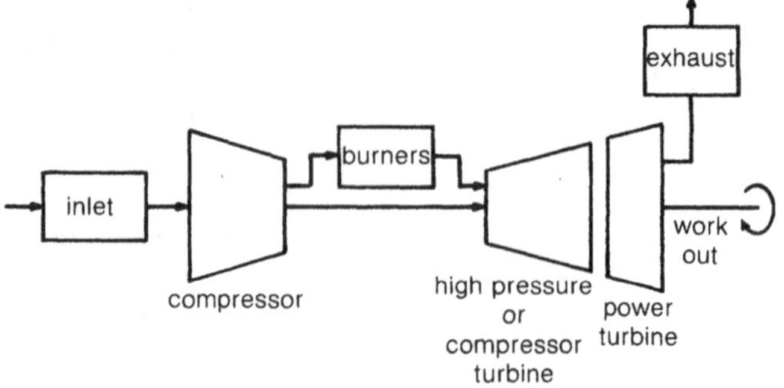

use—a process called marinization (to be discussed later)—the intake and exhaust systems can seriously affect the cycle performance.

Because the working fluid here is an air and air–combustion products mixture, which is discharged overboard, the gas turbine cycle is called an *open* cycle.

A comparison of steam and gas turbine plants is provided in Figure 1.3. A brief consideration and comparison with Figure 2.3 in the next chapter should easily show that great savings of weight and complexity are realized in the case of a gas turbine propulsion plant. Although these benefits are often somewhat offset by requirements for an auxiliary gas turbine or diesel unit to supply ship services, which are provided as an integral part of the steam plant, the gains can nevertheless be considerable.

Figure 1.3 Comparison of steam and gas turbine plants

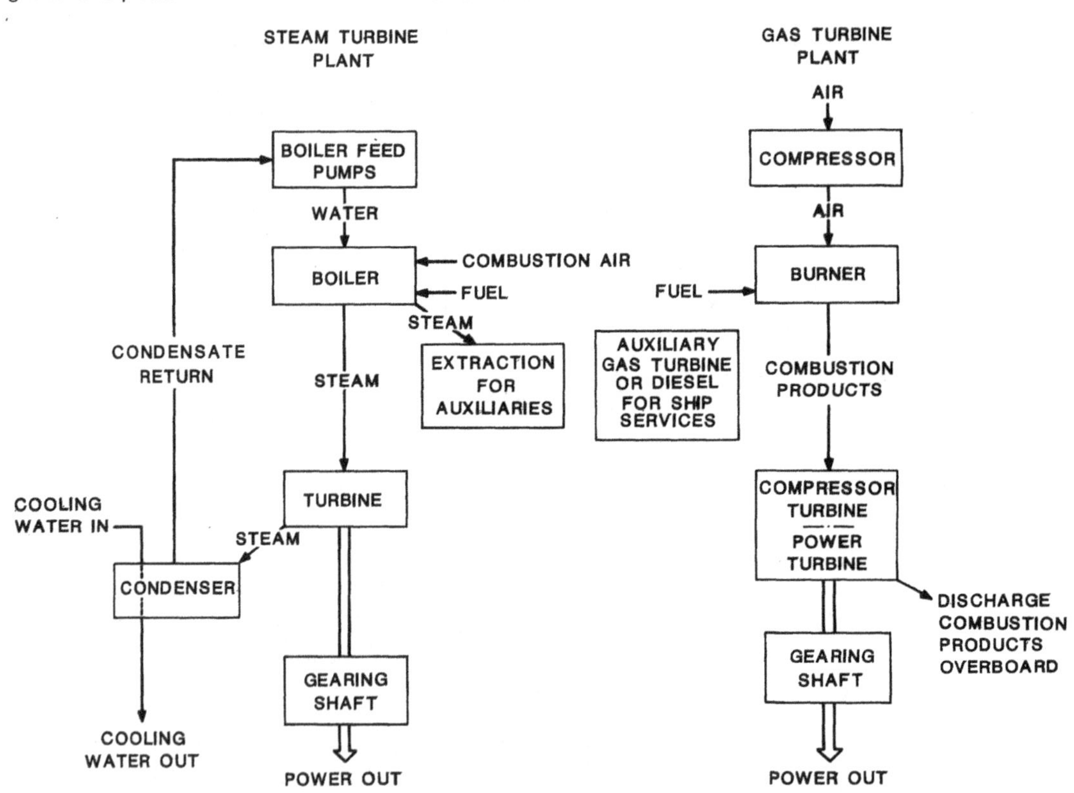

Chapter Two

Steam and Gas Turbine Cycles and Applications

2.1 Introduction

In this chapter the complete marine propulsion plant will be examined. This approach should provide a better picture of the overall operation of the plant prior to the detailed study of the turbine in the following chapters, and lay groundwork for a better understanding of the thermodynamic cycles and why their modification is necessary and useful. Both steam and gas turbines will be considered. To round off the general description, typical applications of the plants will be presented. The steam turbine arrangement described below is that of a destroyer, but is typical of many modern ships. The gas turbine plant, because of its flexibility and its rapid development, makes possible a great deal of variety and innovation in its applications. Thus, a selection of both present and possible future applications will be examined.

2.2 The Main Steam Cycle

The main steam cycle is here considered to consist of four parts or systems, namely, the *expansion system*, the *condensate system*, the *boiler feed system*, and the *generation system*. The condensate and boiler feed systems, grouped together, are often referred to as the *main feed system*.

Figure 1.1 showed a diagram of the main steam cycle, indicating these systems. The systems, which are discussed in detail in the following paragraphs, may be briefly outlined as follows:

1. Expansion system
   a. Main steam piping, valves, and fittings
   b. Turbine control valves
   c. Main propulsion turbines
2. Main feed system
   a. Condensate system
      (1) Main condenser
      (2) Main condensate pump
      (3) Various heat exchangers
      (4) Heating and deaerating portions of the DFT (deaerating feed tank)
   b. Boiler feed system
      (1) Storage portion of the DFT
      (2) Main feed booster pumps
      (3) Main and emergency feed pumps
      (4) Economizer

c. Piping and valves to reserve feed tanks (connected at various places to both the condensate and boiler feed systems)
3. Generation system
    a. Boiler
    b. Superheater
    c. Desuperheater

The expansion system includes that part of the main steam cycle beginning at the main steam boiler outlet, and ending at the interconnection between the turbine unit and the condenser. This includes the main steam piping (sometimes referred to as the main steam system) and the valves therein, the turbine control valves, and the turbines themselves. In this part of the cycle, steam is delivered to the turbine unit, its flow is controlled, and the thermal energy therein is converted into mechanical energy.

The main feed system is that portion of the main steam cycle beginning at the steam inlet to the condenser, including the condenser, condensate pump, various heat exchangers (to be described later), the DFT, the feed booster pump, and the main and emergency feed pumps. The emergency and reserve feed tanks and their interconnection to the above units will be considered later. The main feed system is subdivided into two principal systems, the condensate system and the boiler feed system (see Figure 1.1.).

The condensate system is that portion of the main feed system which begins at the condenser steam inlet and ends with the portion of the DFT that accomplishes heating and deaerating. In this system, the exhaust steam is converted to liquid condensate by rejection of heat to circulating sea water; the pressure is raised by the condensate pump; heat flows into the condensate in the various heat exchangers mentioned above and in the DFT; and dissolved air is removed from the condensate. The final deaerating process converts the condensate to boiler feed water. The boiler feed system consists of the storage part of the DFT, the feed booster pump and the main and emergency feed pumps, and the economizer. In this portion of the system the pressure of the boiler feed is raised sufficiently to inject it into the boiler drum. Before its entry into the boiler, thermal energy is added in the economizer from the boiler stack gases.

The generation system in the main steam cycle includes the boiler, the superheater, and the desuperheater, all of which are ordinarily in one integral unit. It is in this portion of the system that the largest quantity of heat is added to the working fluid.

2.3 The Auxiliary Steam System

All of the propulsion auxiliaries named above, and many other auxiliaries, require a source of power or a prime mover to drive them, or a source of heat to enable them to function. The prime movers are either small steam turbines or electric motors. To distribute the steam for driving these turbines, and to furnish energy to heat exchangers, an auxiliary steam system is provided.

In the 600 psig main steam systems that predominated until the 1960s, it was the practice to provide saturated steam at 600 psig for the principal auxiliary loops. Subsidiary systems, requiring lower pressures, were supplied from this principal loop through reducing valves. The saturated steam for the system was supplied by routing a portion of the steam from the superheater outlet of the boiler back through desuperheater coils located in the boiler steam drum. This scheme provides a continuous steam flow through the superheater tubes, even when the turbines are shut down.

In the first 1200 psig system, the principal auxiliary steam system was still at 600 psig. In this system, the superheated steam at boiler pressure is desuperheated as before, then led to a reducing station where its pressure is reduced to 600 psig.

In more recent 1200 psig installations, superheated steam at boiler pressure is used to drive the auxiliary turbines. This steam is obtained from a branch connection to the main steam line near the boiler outlet.

In all systems, steam at lower pressures is required for various services, some of which are itemized below. In the first two types of installation, this steam is obtained by reducing valves from the auxiliary steam line. In the third type of installation, the steam is obtained by desuperheating a portion of the superheated steam in the steam drum desuperheating coils, and then reducing the pressure to that desired.

The steam system that supplies the major auxiliaries is termed the *auxiliary steam system*. Systems at lower pressures are identified by their pressures, e.g., the *150 lb auxiliary steam system*. Three of these systems, together with their uses, are:

1. 150 psig auxiliary steam system
    a. Fuel oil service heater
    b. Air ejectors
    c. Lube oil purifier heater
    d. Reducing valve for 100 psig auxiliary steam system
2. 100 psig auxiliary steam system
    a. Laundry and tailor shop
    b. Reducing valve for 50 psig auxiliary steam system
3. 50 psig auxiliary steam system
    a. Galley cooking
    b. Hot water system
    c. Ship's heating system

To summarize, the auxiliary steam system provides steam

Figure 2.1 Auxiliary steam system

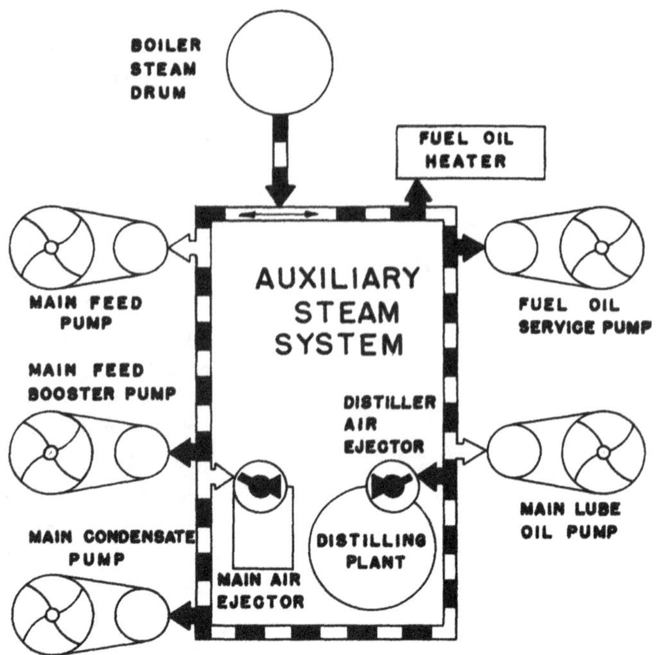

to power the driving turbines of propulsion auxiliaries, and it also provides steam through reducing valves to lower-pressure auxiliary systems. Figure 2.1 shows a skeletonized auxiliary steam system.

2.4 Auxiliary Exhaust System

The propulsion turbine exhausts into the main condenser. Similarly, the turbo-generator, being a sizable unit, usually exhausts into its own condenser (but also may exhaust into the main condenser). It is impracticable, however, to provide an individual condenser for each small turbine that drives an auxiliary pump. Consequently, a piping circuit is provided to collect the discharged steam from all of the auxiliary turbines in a propulsion plant. This circuit is called the *auxiliary exhaust system*. Pressure in this system is maintained at 15 psig by an automatic valve system.

Steam in the auxiliary exhaust system at 15 psig retains considerable thermal energy, which is utilized in two important ways. First, the steam is used in the DFT, to heat the condensate and to purge it of dissolved air. A major portion of the auxiliary exhaust steam is used in this manner. This portion combines with the feed water in the heating and de-aerating process and so returns to the boiler. Second, a smaller portion is used to provide a part of the thermal energy required to operate the distilling plant, in which fresh water is obtained from sea water. This portion is condensed, and finds its way back to the boiler through the low pressure drain system.

When more steam is delivered to the auxiliary exhaust

system than is required for these two functions, the pressure in the system commences to rise above 15 psig. This rise in pressure causes an automatic unloading valve to open, allowing steam to flow from the auxiliary exhaust system to the main condenser. This valve closes when the pressure falls back to 15 psig. The steam that flows into the main condenser loses its thermal energy to circulating sea water and is condensed. Note that the thermal energy in the portions of auxiliary steam used in the DFT and distilling plant performs useful functions. However, the steam unloaded to the condenser loses its thermal energy to the sea. Therefore it is desirable to control the supply of auxiliary exhaust steam discharged into the system to balance the sum of the requirements of the DFT and the distiller. This is done by proper choice of the number of steam-driven pumps and electrically-driven pumps in use under various plant load conditions.

Under some load conditions, there may be insufficient steam in the auxiliary exhaust system to maintain the required pressure, even when no electric pumps are in use. In such circumstances provision is made for augmenting the pressure by admitting steam directly to this system from the auxiliary steam system, through a pressure-reducing valve.

Figure 2.2 shows the outline of the auxiliary exhaust sys-

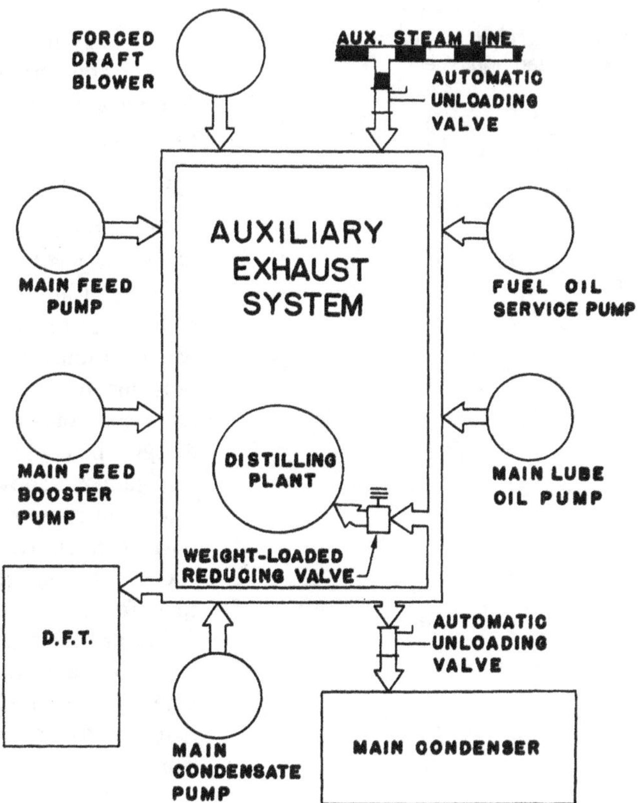

Figure 2.2 Auxiliary steam system

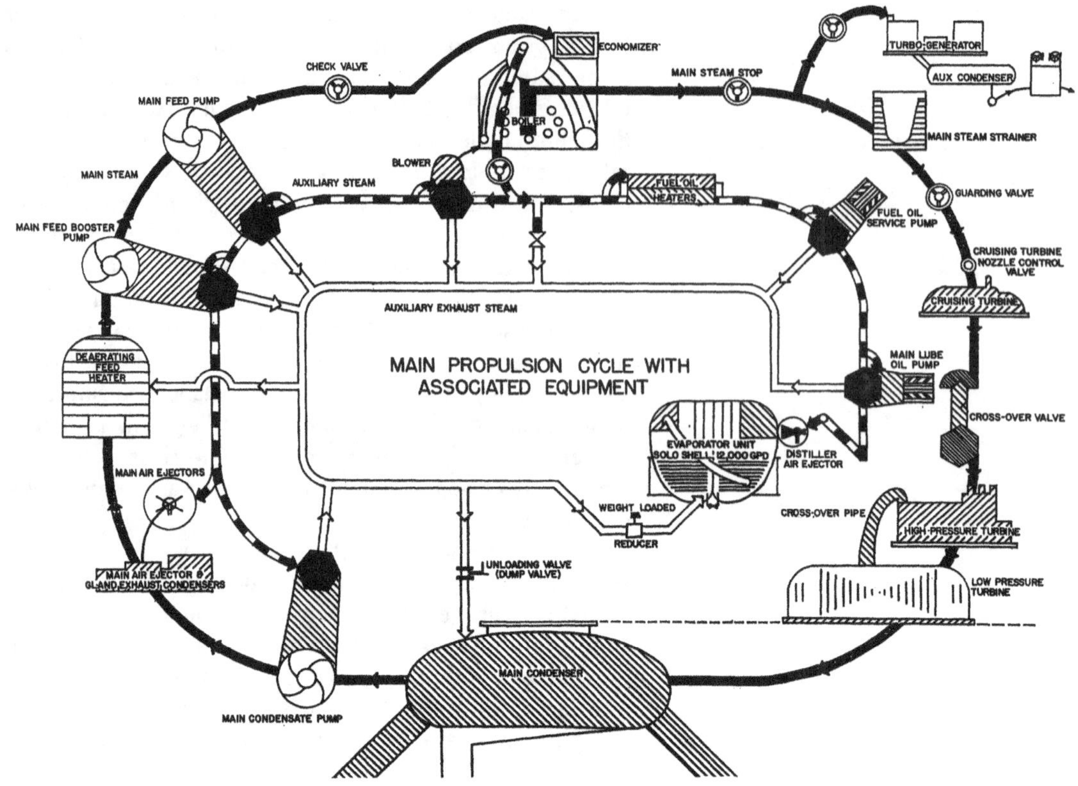

Figure 2.3 Main propulsion cycle with associated equipment

tem. The systems described above are integrated into a diagram for one propulsion plant and are shown in Figure 2.3. A typical shipboard machinery arrangement is shown in Figure 2.4.

2.5 Gas Turbine Plant

In contrast to the rather sizable and heavy steam power plant, the gas turbine unit is simplicity itself. In that simplicity and in relatively low weight lie two of the principal advantages of gas turbine plants. Because it functions as part of a unit, it is impractical and misleading to consider only the turbine as the energy conversion device. Since other components are closely matched to the turbine in the design, the entire unit must be studied (see Chapter 6).

Figure 2.5 shows a gas turbine unit with its auxiliaries. The main components of the unit are the compressor, combustors or burners, and turbines. The rotating compressor sucks in air through the intake section, compresses it, and then forces it into combustion chambers. There fuel is mixed and burned, raising the temperature of the mixture of air and combustion products. This high-energy mixture then enters the compressor turbine, where part of the energy is extracted to turn over the turbine that drives the compressor. This compressor–burner–high-pressure-turbine section is

Figure 2.4 Typical shipboard machinery
arrangement

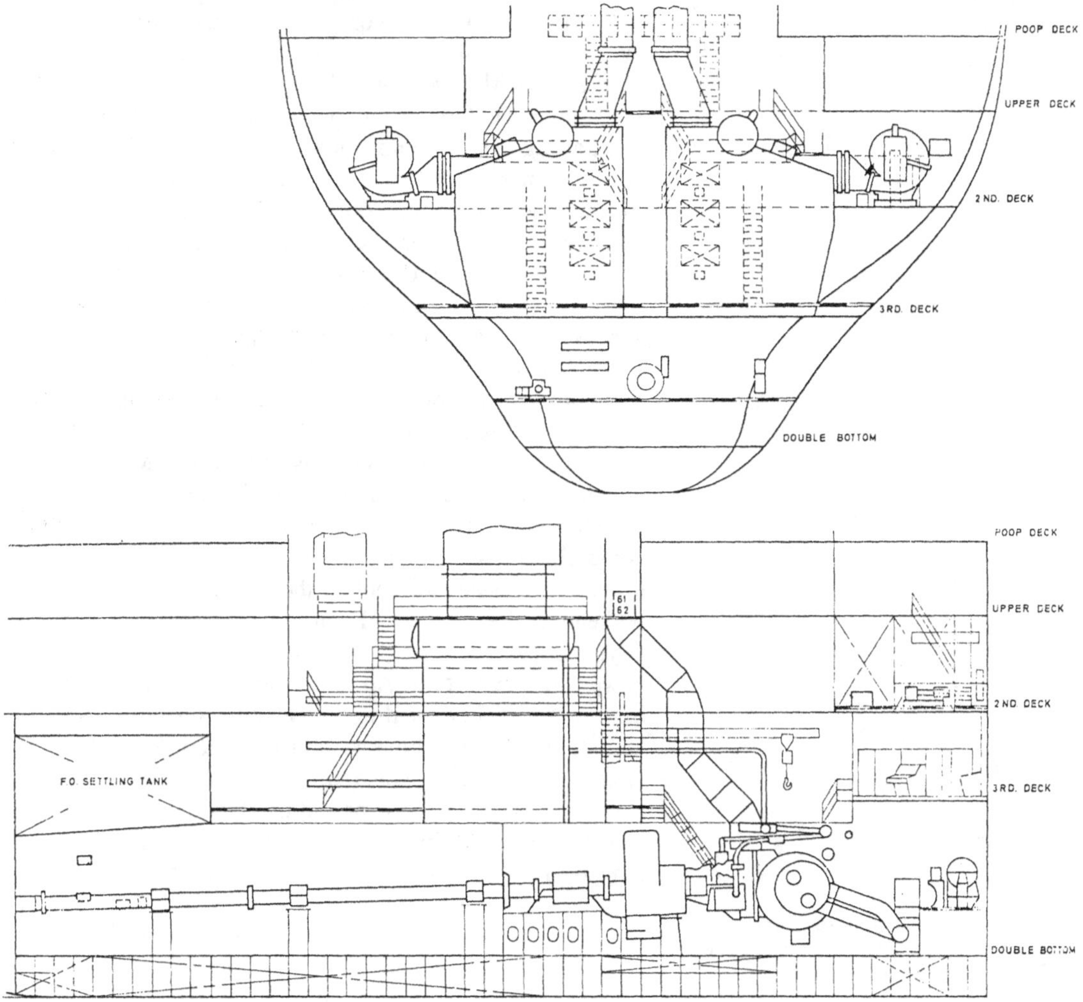

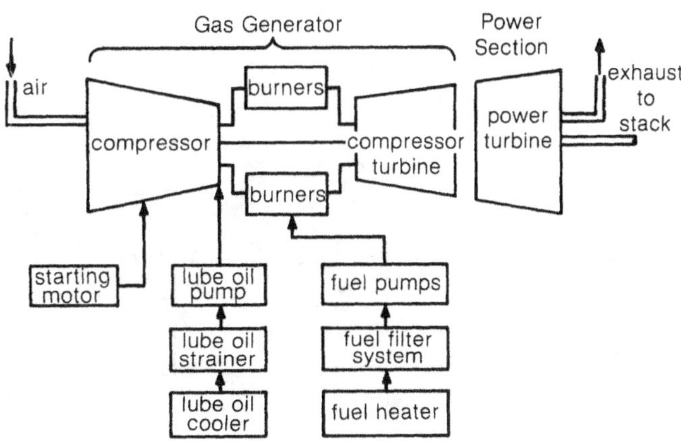

Figure 2.5 Gas turbine unit with
auxiliaries

also known as the gas generator. The essential auxiliaries are the lube oil system, the fuel heating and pumping system, and the starting motor. The starting motor is driven by an auxiliary diesel generator, or by batteries.

In some earlier versions, the power take-off shaft was directly connected to the high pressure turbine, which then rotated at high speed in unison with the entire unit. Although such turbines proved successful and are still in use, severe load and flexibility limitations are placed on the gearing and the rest of the power train in that arrangement. In most newer versions an additional turbine section, the power turbine, is added, which is free to rotate at its own various load speeds, thus permitting the gas generator to run at an optimum speed for a required unit power setting.

Figure 2.6 shows a Bristol Proteus gas turbine engine, which is a single turbine unit, but with the power take-off shaft at the compressor end and connected to a V-drive, which contains the clutch and reduction gears. Also shown are the intake and exhaust arrangements. Here the engine compartment serves also as the inlet plenum chamber, providing a means for cooling the exterior engine parts and preheating the inlet air. Figure 2.7 shows a Rolls-Royce Olympus unit arrangement in its enclosure and with the power turbine section. An interesting aspect of the gas turbine nomenclature of different manufacturers is indicated here. The power turbine section is sometimes called "the engine," with the gas generator considered only as a device, similar to the boiler, for providing the engine with proper working fluid. A General Electric LM-2500 gas turbine arranged in its enclosure is shown in Figure 2.8. Also shown

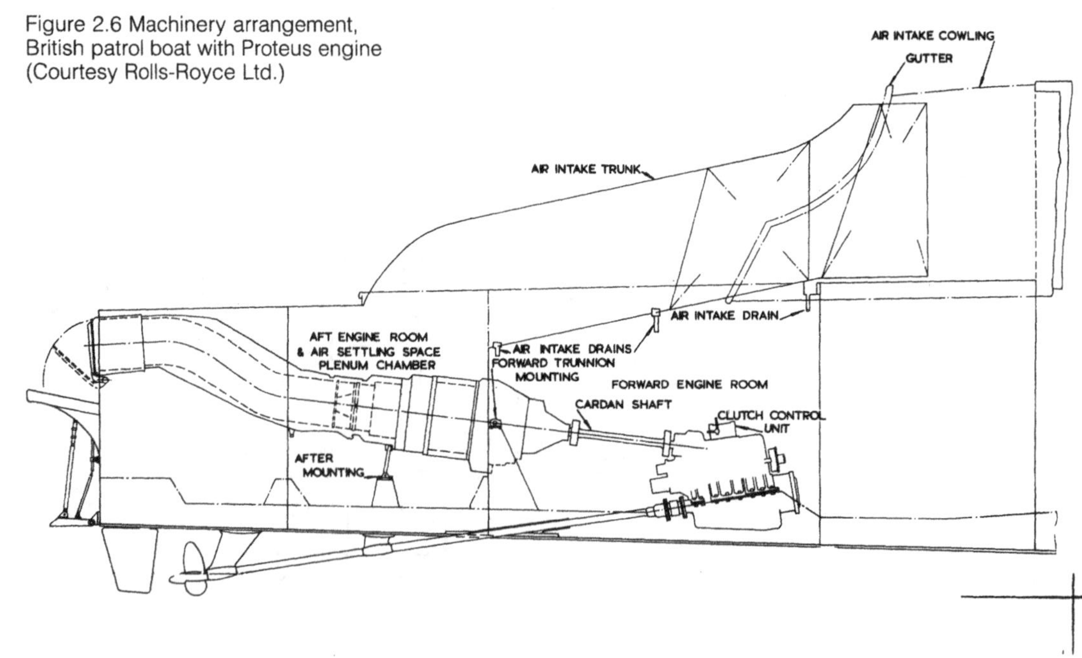

Figure 2.6 Machinery arrangement, British patrol boat with Proteus engine (Courtesy Rolls-Royce Ltd.)

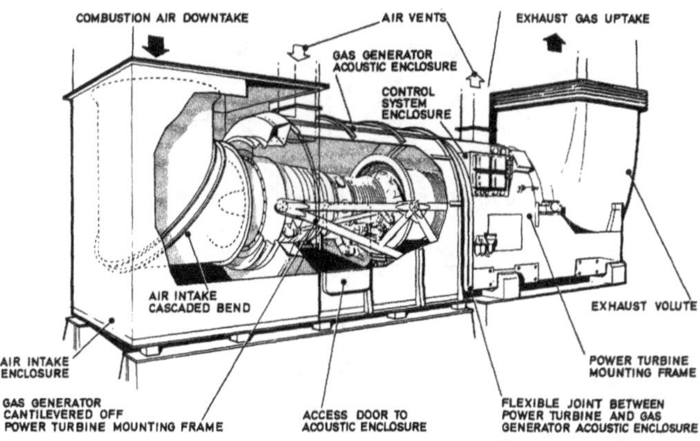

Figure 2.7 Olympus engine, showing cantilever mountings and enclosures (Courtesy Rolls-Royce Ltd.)

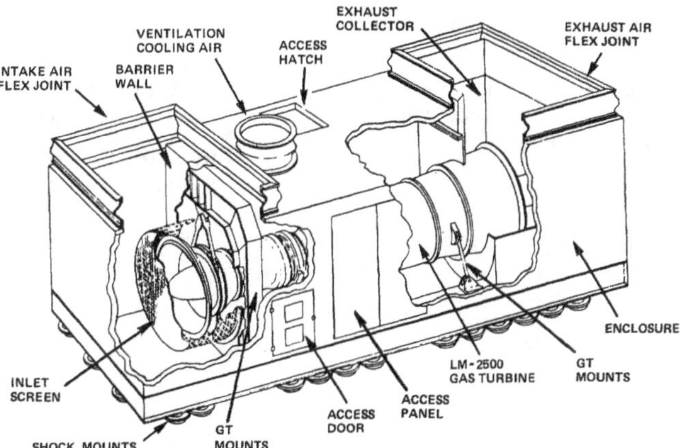

Figure 2.8 General Electric LM-2500 gas turbine unit (Courtesy General Electric Co.)

are shock mounts, which minimize the effect of ship vibration, and inlet screens, which prevent entry of potentially damaging foreign objects.

These three units represent some of the basic types of gas turbine units and the power take-off connections. There are numerous other ways of utilizing gas turbines for marine applications, either singly or in combination with steam or diesel units. Gas turbines will be considered in more detail in Chapter 7.

In summary, this chapter has discussed the basic functions and arrangements of steam and gas turbine power plants. Although detailed comparisons of these marine power plants will be provided later, the two main comparative features should be reemphasized here: the gas turbine unit is much smaller and lighter, and the gas turbine requires a bare minimum of auxiliaries for unit operation (see Figure 2.9).

The gas turbine propulsion unit, however, does not provide the power to run the auxiliaries necessary for ship's

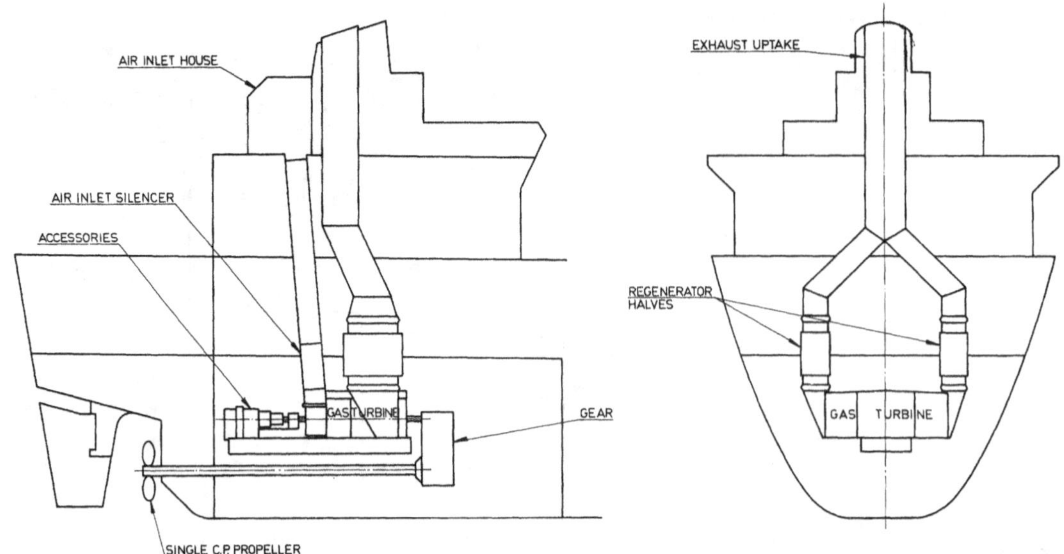

Figure 2.9 Typical heavy duty gas turbine unit in a cargo ship (Courtesy General Electric Co.)

operation as does a steam propulsion unit. Thus, other smaller gas turbine units or auxiliary diesel generators are used for generating power for a ship's electricity needs and for hotel services.

Chapter Three

Basic Considerations of Fluid Flow and Thermodynamics

3.1 General

In this chapter basic relationships of fluid flow and thermodynamics will be discussed. For one who has taken elementary courses in these subjects, the recollection of these relationships will not be difficult. No attempt is made here to develop the principles fully. The person who desires a more thorough treatment should refer to standard textbooks in the various subjects.

Three basic physical laws applying to the flow of fluids will be discussed in connection with fluid flow principles. The three governing laws are: (1) the law of conservation of mass, (2) the law of conservation of energy, and (3) the law of conservation of momentum.

3.2 Conservation of Mass, Continuity

Continuity of fluid flow in a channel is obtained when the same mass of fluid flows past every point in the channel (whether it be an open or a closed channel) in the same increment of time. That is, between any two points simultaneously under observation, there is no accumulation or depreciation of the quantity of fluid anywhere between the two points.

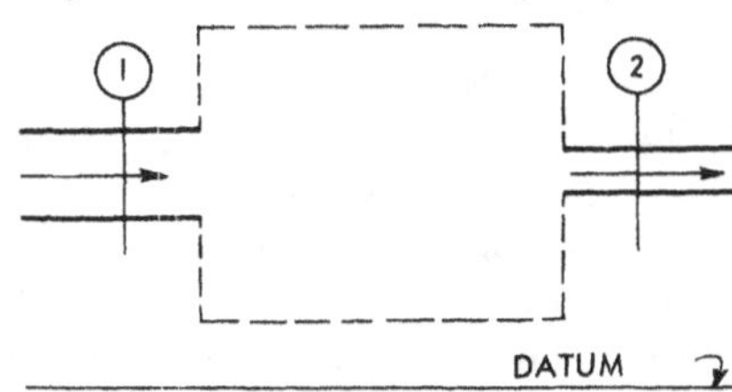

Figure 3.1 Fluid flow in a simple system

Consider the flow depicted in Figure 3.1. Fluid flows past an observation point, section 1, into a system whose configuration is unknown. It flows out through an observation point at section 2. The cross-sectional area (A) of the channel (or pipe) at sections 1 and 2 is measurable. Also the specific weight (weight per unit volume—w) of the fluid and its average velocity (V) are determinable at the two sections. The volume, Q, passing each section per unit time is found by multiplying the cross-sectional area by the velocity:

$$A \text{ ft}^2 \times V \text{ ft/sec} = Q \text{ ft}^3/\text{sec}$$
$$A \text{ m}^2 \times V \text{ m/sec} = Q \text{ m}^3/\text{sec}$$

The mass flow rate of flow, m, is determined by multiplying this volume by density, ρ (lb/ft³) (kg/m³), or by the reciprocal of the specific volume, v (ft³/lb) (m³/kg):

$$\dot{m} = \rho AV = \frac{AV}{v} \left(\frac{\text{lb}}{\text{ft}^3} \times \text{ft}^2 \times \frac{\text{ft}}{\text{sec}} = \frac{\text{lb}}{\text{sec}} \right)\left(\frac{\text{kg}}{\text{sec}} \right)$$

It should be remembered that the velocity and density or specific volume represent average values at any given section of the flow channel. Under the circumstances, the mass flow rate is equal at all sections, and the law of conservation of mass is satisfied. Thus:

$$\dot{m} = \frac{A_1 V_1}{v_1} = \frac{A_2 V_2}{v_2} \qquad (1)$$

This equation is the mathematical expression of continuity of flow. When the fluid is incompressible, there is no change in density during flow, or v_1 is equal to v_2, and the volume rate of flow is constant at all sections.

$$Q = A_1 V_1 = A_2 V_2 \qquad (2)$$

When dealing with compressible fluids, changes in density are frequently small and may be neglected. Under such circumstances equation (2) remains valid.

3.3 Conservation of Energy, Bernoulli's Equation

Ignoring thermal energy, the energy contained in a moving fluid may be of two basic types: potential and kinetic. The potential energy consists of two components, *potential head* and *pressure head*.

A potential head exists whenever a body of fluid is located above a given datum. A pound of water 10 feet above a datum plane has the capacity to perform 10 ft-lbf of work in falling to the datum plane; therefore it contains 10 ft-lbf of potential energy with respect to that datum plane. Since this energy, expressed on a unit mass basis, has the dimensions ft-lbf/lb, the pound units are conventionally canceled, and the energy is expressed in feet of *potential head,* where head refers simply to the height above the established datum plane. Thus a head, Z, of 10 feet indicates a potential energy of 10 ft-lbf/lb (Nm/kg or J/kg).

A closed container filled with liquid may be put under pressure by connecting a pipe to the container and filling it with liquid to some level above the top of the container (Figure 3.2). The pressure at section 1 of the pipe is equal to the pressure at the top of the container; else there would be flow. The volume of fluid resting on section 1 is hA, if A is the cross-sectional area of the pipe. Its weight is $\rho(g/g_c)\,hA$, and its pressure, or weight per unit area is:

$$p = \frac{\rho g h A}{g_c A} = \rho h \ \text{lbf/ft}^2 \ (\text{N/m}^2) \ \text{or Pa (Pascal)} \qquad (3)$$

where g = gravitational acceleration
   = 32.2 ft/sec² (9.81 m/sec²)

    g_c = conversion constant to yield Newton's inertial law dimensionally homogeneous, i.e.,
   $F = m(g/g_c)$.

The value of g_c may be determined using either of our two basic measurement systems:

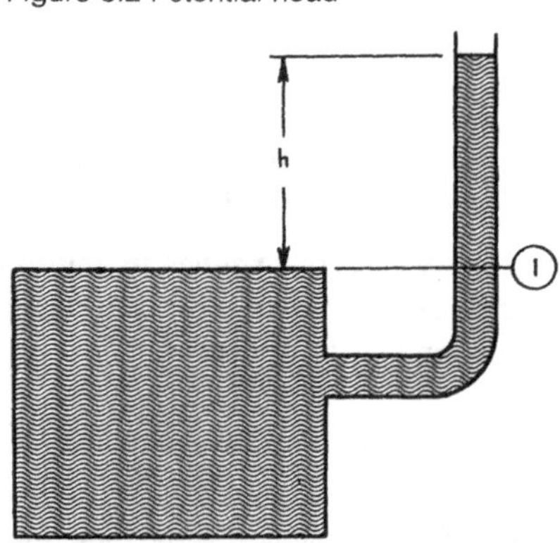

Figure 3.2 Potential head

Engineering English System

$$F = \text{(lbf)}$$
$$m = \text{(lb)}$$
$$g_c = 32.2 \; \frac{\text{lb ft}}{\text{lbf sec}^2}, \text{ and } \frac{g}{g_c} = 1 \text{ for many engineering}$$
applications

SI System (International System of Units)

$$F = \text{(N) Newtons}$$
$$m = \text{(kg)}$$
$$g_c = 1 \; \frac{\text{kg m}}{\text{N sec}^2}$$

In common practice pressure is expressed in pounds per square inch (psi), and the head, h, in feet. Then equation (3) can be written:

$$h = \frac{144p}{\rho} = 144pv \text{ ft} \qquad (4)$$

which defines h as *pressure head*. Like the potential head, this quantity could be expressed in units of ft-lbf/lb (J/kg). Thus the pressure, no matter how produced, denotes a capacity for creating a potential head, and therefore a capacity for doing work, and it is thus a form of potential energy.

Kinetic energy is a property of a body in motion. The kinetic energy of a mass, m, moving at a velocity V is:

$$KE = \frac{mV^2}{2g_c} \text{ ft-lbf (J)}$$

or on a unit mass basis:

$$KE = \frac{V^2}{2g_c} \; \frac{\text{ft-lbf}}{\text{lb}} \text{ (J/kg)} \qquad (5)$$

It follows that the total mechanical energy of a flowing fluid is the sum of the potential head and the pressure head (which together measure the potential energy) and the kinetic energy:

$$E = Z + \frac{144p}{\rho} + \frac{V^2}{2g_c} \qquad (6)$$

If no energy is added to or taken from a fluid passing through a system in steady flow, the law of conservation of energy states that the energy of each mass unit of the fluid remains constant. Thus, for any two sections in the system, denoted by the subscripts 1 and 2 (Figure 3.1):

$$Z_1 + \frac{144p_1}{\rho} + \frac{V_1^2}{2g_c} = Z_2 + \frac{144p_2}{\rho} + \frac{V_2^2}{2g_c} \qquad (7)$$

Equation (7) is applicable to incompressible flow. Therefore the density is the same on both sides and it is not identified by a subscript denoting where it was measured. Equation (7) is the well known *Bernoulli equation*.

It is important to note the difference between the terms compressible and incompressible *fluid* and compressible and incompressible *flow*. Generally, liquids are assumed to be incompressible, and all of the preceeding equations apply to them. Gases, on the other hand, are compressible. However, in many instances, processes involving gases approximate the conditions for incompressible flow, if there is no appreciable change in the density of the gas in the flow process.

3.4 Conservation of Momentum—The Momentum Equation

The momentum relation is derived from the Newtonian relationship:

$$F = ma$$

where F = force
m = mass
a = acceleration

Since acceleration is time rate of change of velocity, the relationship may be written:

$$F = m\frac{dV}{dt} = \frac{m}{\Delta t}(V_2 - V_1) \tag{8}$$

In equation (8), F is the force required to produce a change of velocity of a mass, m, of fluid from V_1 to V_2 in a time interval Δt. The ratio $m/\Delta t$ may be interpreted as the *mass rate of flow,* and the equation takes the form:

$$F = \frac{\dot{m}}{g_c}(V_2 - V_1) \tag{9}$$

where V_2 = final velocity, ft/sec (m/sec)
V_1 = initial velocity, ft/sec (m/sec)
F = force *on the stream* required to change its velocity, lbf (N)

The term F in equation (9) is the force exerted *on the stream*. The force, R, which a stream exerts on any object is equal and opposite to F. Thus:

$$R = -F = \frac{\dot{m}}{g_c}(V_1 - V_2) \tag{10}$$

It is important to note that V_1 and V_2 in equations (9) and (10) are *vector* quantities having both magnitude and direction. Their difference is a vector difference, and must be used as such in the equations. Two examples follow to illustrate this.

Example 3.1: A stream of water strikes a vane as shown in the figure, and it is deflected through an angle of 30°. The

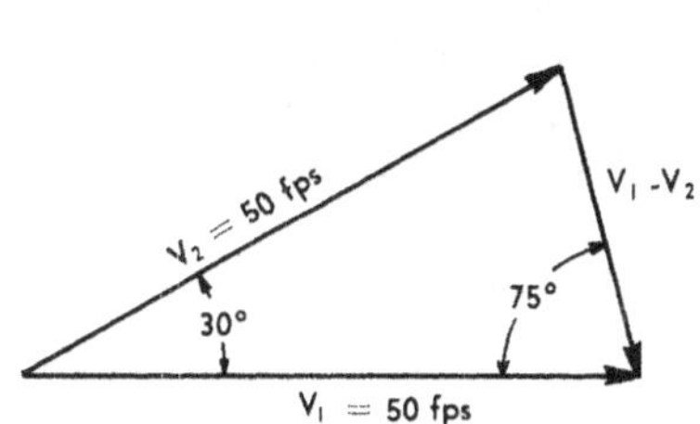

stream velocity is 50 ft/sec at both entrance and exit and its cross-sectional area is 4 in². What is the magnitude and direction of the force produced on the vane?

Solution: There is no change in the *magnitude* of the stream velocity. The difference, $V_1 - V_2$, may be found vectorially as indicated in the vector diagram. Solving the diagram either analytically or graphically gives the magnitude of $V_1 - V_2$ as 25.8 fps and its direction as 105° from the original direction of V_1. Therefore:

$$R = \frac{\dot{m}}{g_c}(V_1 - V_2)$$

$$= \frac{Q\rho}{g_c}(V_1 - V_2)$$

$$= 50 \times \frac{4}{144} \times \frac{62.4}{32.2} \times 25.8$$

$$= 69.4 \text{ lbf, in the direction of } \Delta V$$

Example 3.2: Water flows through a 3-in. fire hose at 10 fps, entering the nozzle on the end of the hose at this velocity. The nozzle exit diameter is 1¼ in. What is the magnitude and direction of the force on the nozzle due solely to the change in velocity of the fluid?

Solution: Applying the continuity equation to determine the exit velocity:

$$A_1 V_1 = A_1 V_2$$

$$\frac{\pi}{4}\frac{(3)^2}{144} \times 10 = \frac{\pi}{4}\frac{(1\frac{1}{4})^2}{144} V_2$$

or:

$$V_2 = \left(\frac{3}{1.25}\right)^2 (10) = 57.6 \text{ fps}$$

Since there is no change in the direction of the fluid between the entrance and exit sections, the resultant force is determined by use of the magnitude difference of $V_1 - V_2$, and is opposite to the direction of V_1, the direction of flow.

$$R = \frac{\dot{m}}{g_c}(V_1 - V_2)$$

$$= \frac{Q\rho}{g_c}(V_1 - V_2)$$

$$= 10 \times \frac{\pi}{4}\frac{(3)^2}{144} \times \frac{62.4}{32.2} \times (10 - 57.6)$$

$$= -45.3 \text{ lbf (opposite to the direction of flow)}$$

Example 3.3: Whenever a change in fluid velocity occurs, a change in fluid pressure also occurs, according to Bernoulli's equation. The above nozzle discharges to atmospheric pressure, 15 psia. Find the force on the nozzle caused by the pressure change across the nozzle, and find the total force on the nozzle.

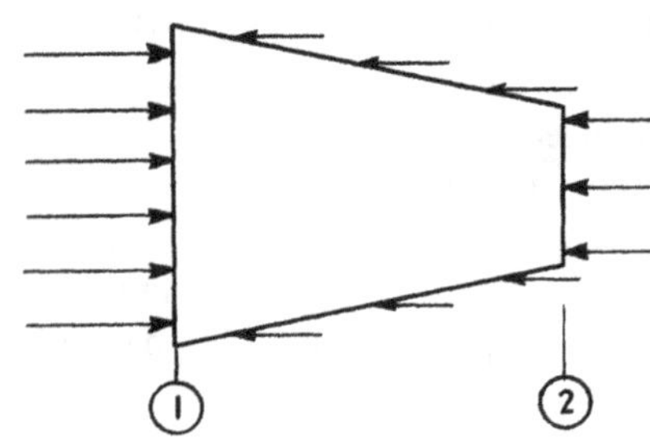

Solution: The fire nozzle pressure acts on the cross section of the nozzle at the entrance. Atmospheric pressure acts on the outside of the nozzle, including the exit area. The net pressure force is equal to the cross-sectional area at 1 multiplied by the difference between the entrance and exit pressures. Since the exit pressure is atmospheric, the entrance pressure can be found by Bernoulli's equation:

$$144\frac{p_1}{\rho} + \frac{V_1^2}{2g_c} = 144\frac{p_2}{\rho} + \frac{V_2^2}{2g_c}$$

whence:

$$p_1 = p_2 + \frac{(V_2^2 - V_1^2)\rho}{288g_c}$$

$$= 15 + \frac{(57.6^2 - 10^2)62.4}{288 \times 32.2}$$

$$= 15 + 21.6 = 36.6 \text{ psia}$$

Then the net pressure force is:

$$F = A_1(p_1 - p_2) = 21.6(3)^2\frac{\pi}{4}$$

$$= 152.7 \text{ lbf}$$

The total force on the nozzle is the algebraic sum of the above force and the force found in part (a). Choosing the plus sign in the direction of flow, the total force is:

$$F_T = 152.7 - 45.3 = 107.4 \text{ lbf}$$

in the direction of flow. This is the force exerted on the threads holding the nozzle on the end of the hose.

Example 3.4: For example 3.3 calculate the pressure p_1 at the inlet in the SI system.

Solution: At standard atmospheric pressure $p_2 = 14.7$ psi which corresponds to 101,350 N/m² or Pa. At 15 psi, $p_2 = 101,350 (15/14.7) = 103,418$ Pa. Since the water density, at normal pressure and temperature, is 1,000 kg/m³:

$$p_1 = p_2 + \frac{(V_2^2 - V_1^2)\rho}{g_c} \quad V_2 = 17.6 \text{ m/sec}$$

$$V_1 = 3.05 \text{ m/sec}$$

$$= 103,418 + \frac{(17.6^2 - 3.05^2)1000}{1}$$

$$= 103,418 + 300,458$$

$$= 403,876 \text{ N/m}^2$$

$$= 403.9 \text{ k Pa}$$

3.5 Thermodynamic Laws

The operation of all engines designed to produce mechanical work from thermal energy is based on the fundamental set of principles known as *the first and second laws of thermody-*

namics. They are briefly discussed here to provide a basis for understanding some of the broader aspects of engine performance, and consequent design arrangements. These laws are deceptively simple in their statement, but they have such extensive ramifications that their proper application is essential in nearly all fields of engineering.

The first law of thermodynamics is a form of the law of conservation of energy. It may be simply stated as follows: *When a quantity of heat energy disappears, an equivalent quantity of mechanical, electrical, or chemical energy appears*. We are concerned in this study only with the conversion of thermal to mechanical energy.

The second law of thermodynamics is concerned with the proportion of available heat energy that can be converted into mechanical energy, i.e., the largest fraction of a given quantity of heat energy which can be made to disappear and then reappear in the form of mechanical energy. Clausius's original statement of this law was as follows: The transformation of heat to work is always dependent on a temperature drop and the transfer of heat to a lower temperature level. The law has been restated in numerous forms. Perhaps the most useful form to the engineer is as follows: *It is impossible to construct an engine that will deliver mechanical work derived solely from the cooling of a single heat source if no heat is rejected to a reservoir at a lower temperature*. A moment's reflection on these statements will establish that the requirements for producing mechanical energy from thermal energy are *first*, a source of heat; *second*, a device or engine for converting a part of the thermal energy to mechanical work; *third*, a reservoir (often called a "heat sink") at a lower temperature than the source, to which heat can flow; and *fourth*, a working fluid, which provides the means of transporting the thermal energy from one part of the system to another. It should also be apparent from the second law that no matter what kind of engine is utilized, some of the heat energy can never be converted to work and must be rejected to the sink. This law has never been proved in the rigorous sense, and has been subjected to doubt and criticism, *but no one has ever demonstrated its failure*.

The science of thermodynamics is concerned primarily with the conversion of thermal energy into useful work. Certain definitions are necessary in the following discussion. These are:

(a) *Work*—mechanical energy in transition. Work is evidenced by a force moving through a distance. Work, as such, may not be stored, but may be converted into another form of transitory energy, *heat;* or into *kinetic energy;* or into a form of stored mechanical or thermal energy.

(b) *Heat*—thermal energy in transition. For heat to flow,

a temperature difference must exist. Heat may not be stored as such, but it may be converted into another form of transitory energy, *work;* or into a form of stored thermal energy.

(c) *Internal energy*—stored thermal energy. When heat or work is done on a substance, the molecular kinetic energy may be increased, as evidenced by an increase in temperature of the substance, resulting in an increase in internal energy.

3.6 The General Energy Equation

From equation (6), and the above definitions, we can include the consideration of thermal energy, and express the energy associated with a unit weight of fluid at any instant as:

$$E = \frac{Z}{J} + \frac{pv}{J} + \frac{V^2}{2g_cJ} + u \qquad (11)$$

The constant J, Joule's constant, is used to express the relationship between thermal and mechanical energy. J equals 778 ft-lbf/BTU. The term u has been added to Bernoulli's equation to account for stored thermal or internal energy.

When a fluid flows from one point to another, heat may be transferred to or from the fluid and external work may be done on or by the fluid. The energy associated with the fluid at the second point must be the net remaining after the heat and work are transferred. Denoting the heat transferred as q, and considering it a positive quantity when *added to* the fluid; denoting the external work as w, and considering it as a positive quantity when *done by* the fluid; and considering flow through some system from point 1 to point 2, the following equation can be written:

$$\frac{Z_1}{J} + \frac{p_1v_1}{J} + \frac{V_1^2}{2g_cJ} + u_1 + q_{12} - w_{12} =$$

$$\frac{Z_2}{J} + \frac{p_2v_2}{J} + \frac{V_2^2}{2g_cJ} + u_2 \qquad (12)$$

Equation (12) is called the *general energy equation*. In the absence of chemical, electrical, magnetic, and capillary forces, it expresses the energy relation in a fluid before and after a mechanical and/or thermodynamic process. It holds for any fluid.

3.7 Properties of a Fluid

A *property* of a fluid is defined as any observable characteristic. Three common and well-known thermodynamic fluid properties are pressure, temperature, and density (or specific volume). Other properties are viscosity, internal energy, enthalpy, and entropy. For a pure fluid (a chemically and mechanically homogeneous fluid) any two thermodynamic properties are sufficient to determine all other thermodynamic properties. Thus, it may be said that a pure fluid has a fixed thermodynamic *state* when any two *independent* thermodynamic properties are fixed. Note that work and heat are not properties. Potential and kinetic energy may be thought of as mechanical properties, if desired, but they are completely independent of thermodynamic properties, except as equation (12) relates them.

Since the term pv/J and the term u in equation (12) are both composed of thermodynamic properties, it is often convenient to combine them into another property, *enthalpy*, especially in processes involving flow of the thermodynamic fluid. Thus:

$$h = u + \frac{pv}{J}$$

where h = enthalpy, BTU/lb (J/kg)
$\quad u$ = internal energy, BTU/lb (J/kg)
$\quad pv/J$ = "pressure energy," BTU/lb (J/kg)

Equation (12) may be written conveniently, by combining the u and pv terms into the single h term, as:

$$\frac{Z_1}{J} + h_1 + \frac{V_1^2}{2g_cJ} + q_{12} - w_{12} = \\ \frac{Z_2}{J} + h_2 + \frac{V_2^2}{2g_cJ} \tag{13}$$

where all energy terms are expressed in thermal units on a per pound basis, BTU/lb (J/kg).

One other fluid property is important in the use of the general energy equation. This property has the name *entropy*. The reader may refer to any standard text on thermodynamics for a complete discussion of its meaning and derivation. Its usefulness in this book lies in the following fact: *when the kinetic energy of the fluid is increased in an ideal reversible process, or when work is done by the fluid in a similar process, either being accomplished at the sole expense of stored fluid energy, the entropy of the working substance remains constant.* On the other hand, a process involving only the addition or rejection of heat to or from the fluid is *always* accompanied by a change in entropy.

We will be able to analyze nozzles and turbines in the subsequent discussion by first considering the thermodynamic processes as isentropic, and then considering the factors which cause these processes to depart from the ideal.

The thermodynamic properties of working fluids are commonly given in terms of tabulated values or in graphical form. For the working fluids of interest in this book, steam and air, one can find the property values in references 1 and 2 at the end of this chapter. Most standard thermodynamic texts (e.g. references 3–5) also give an abbreviated account of these values. The two most often used graphical forms, the temperature-entropy diagram and the enthalpy-entropy diagram, also called the Mollier chart, are shown for steam

Figure 3.3 *T-s* and *h-s* diagrams for steam

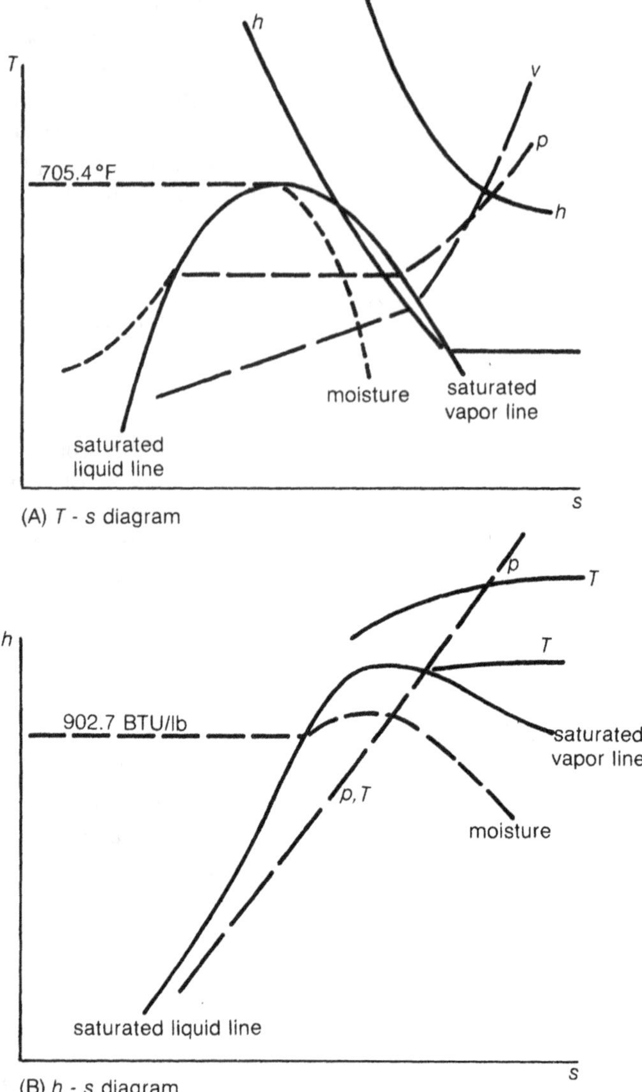

in Figure 3.3. Mollier diagrams are used in practical cycle calculations. The *T-s* diagram is useful for simple analysis of power cycles and will also be utilized in the following chapters. Moreover, in a steam cycle the working fluid may exist in a state of liquid, vapor, or their mixtures; and the graphical representation permits an easy representation and immediate overview of all processes involved in a cycle.

The thermodynamic properties, and also processes, of air can be presented advantageously on similar charts. For temperatures higher than the critical value, and for pressures lower than the critical value, in the pure gas phase, these properties can be given by simple analytical expressions using the perfect gas law. Although air is essentially a mixture of nitrogen and oxygen, its behavior is closely approximated by the equation of state for perfect gases:

$$pv = RT \tag{14}$$

The term R is the gas constant, given by:

$$R = \bar{R}/M$$

where $\bar{R}$ is the universal gas constant:

$$\bar{R} = 1545.33 \frac{\text{ft-lbf}}{\text{lb mole } °R}$$

$$8314.5 \frac{\text{J}}{\text{kg mole K}}$$

and $M = 28.97$ is the molecular weight of air. Thus, for air:

$$R = 53.34 \frac{\text{ft lbf}}{\text{lb } °R}$$

$$= 287 \frac{\text{J}}{\text{kg K}}$$

The term p is the absolute pressure (lbf/ft^2 or N/m^2), v is the specific volume (ft^3/lb or m^3/kg), and T is absolute temperature in degrees R (Rankine) or K (Kelvin). Since the internal energy u of a perfect gas is only a function of temperature, the following simple results can be derived:

$$h = u + pv = u + RT \tag{15}$$
$$= c_p T \tag{16}$$
$$u = c_v T \tag{16a}$$

where $c_p = c_v + R/J$

c_p = specific heat at constant pressure, BTU/lb $°R$ or J/kg K

c_v = specific heat at constant volume, BTU/lb $°R$ or J/kg K

For air:

$$c_p = 0.24 \frac{\text{BTU}}{\text{lb } °R}$$

$$= 1004 \frac{\text{J}}{\text{kg K}}$$

$$c_v = 0.171 \frac{\text{BTU}}{\text{lb } °R}$$

$$= 717 \frac{\text{J}}{\text{kg K}}$$

Potential for confusion is minimized if one realizes the letter J, when used in an equation like (13) or (16a), stands for the conversion factor 778 ft-lbf/BTU. When used with units, J is an abbreviation for joule(s). These values are average values of c_p and c_v at low pressures and temperatures; care should be exercised when using equations (15) or (16), as in their derivations it has been assumed that c_p and c_v are constant. Use of equations (15) and (16) vastly simplifies many calculations, although at the expense of accuracy, as direct use of temperature only is involved. For more precision, use of h and u is recommended with reference to air tables (see reference 2).

By means of the perfect gas law one obtains:

$$pv^k = \text{const}$$

and:

$$\frac{p_1}{p_2} = \left(\frac{v_2}{v_1}\right)^k = \left(\frac{T_1}{T_2}\right)^{k/(k-1)} \tag{17}$$

where $k = c_p/c_v = 1.4$ for air.

The subscripts 1 and 2 in equation (17) and in the following process descriptions refer to inlet and outlet conditions, respectively.

In many cases involving flow problems, the process discussions and the resulting equations can be simplified if the concept of *stagnation state* is introduced. More precisely, the isentropic stagnation state would be obtained if a flow were decelerated in a reversible adiabatic manner to zero velocity. Neglecting the potential head, equation (13) then yields:

$$h_2 = h_1 + \frac{V_1^2}{2g_c} \equiv h_0$$

which is the definition of stagnation enthalpy. It is sometimes also called total enthalpy, due to the fact that both static (h_1) and dynamic ($V_1^2/2g_c$) effects are included. The pressure, density, and temperature corresponding to the stagnation enthalpy are labeled with subscripts 0 or t (for total), and are called stagnation pressure, density, and temperature, respectively. In this state, $h_0 = c_p T_0$. The static and stagnation states for pressure on the h-s diagram are shown in Figure 3.4. Equation (17) is valid also for isentropic stagnation states. The above results will prove extremely useful in cycle discussions applicable to gas turbine operations.

Before proceeding to cycle discussions, a few words concerning work and heat are in order.

3.8 Evaluation of Heat and Work

The evaluation of the quantity of energy added to or extracted from a working fluid in the form of heat or work is conveniently obtained from equation (13). Since the contribution of the difference in potential energy terms $Z_1 - Z_2$ is negligible for all applications in this book, equation (13) becomes per pound of flow:

$$h_1 + \frac{V_1^2}{2g_c J} + q_{12} - w_{12} = h_2 + \frac{V_2^2}{2g_c J} \tag{18}$$

which will now be specialized for pertinent processes found in the power cycle.

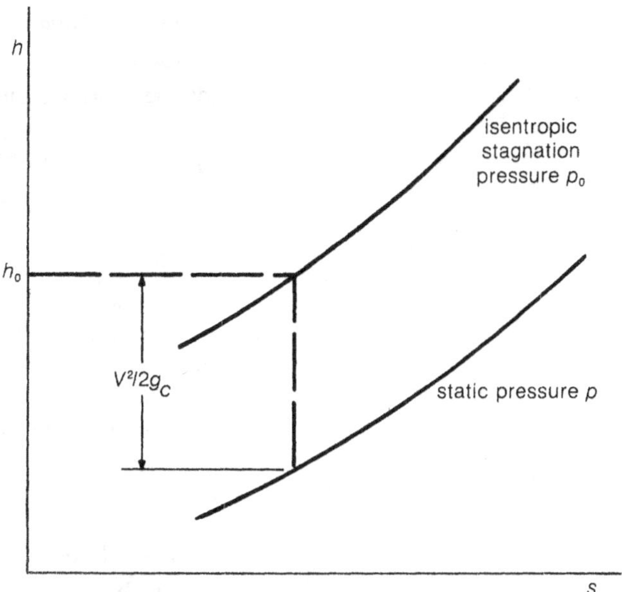

Figure 3.4 An *h-s* diagram illustrating stagnation state

Pumps, Compressors

Pumps and compressors perform essentially the same function. They increase the pressure of the working fluid and forward it into the next section, where the energy of the fluid will be raised by addition of heat. Pumps handle primarily liquids, while the term compressor implies that the working fluid is a gas. Since the difference between the inlet and exit kinetic energy terms is negligible for most pump and compressor applications, the energy equation above yields the work performed on the fluid as:

$$w_{12} = q_{12} + h_1 - h_2 \text{ BTU/lb (J/kg)} \qquad (19)$$

In pump applications the heat transfer q_{12} is negligible, and the pump work can be further developed as:

$$w_{12} = h_1 - h_2 = \frac{p_1 v_1 - p_2 v_2}{J} + u_1 - u_2$$

In liquid pumps $u_1 - u_2$ is very small, and so is the change in specific volume. Hence, unless the pressure change is very large, pump work can be calculated from

$$w_{12} = \frac{v_1(p_1 - p_2)}{J} \text{ BTU/lb (J/kg)}$$

For compressors, it is not immediately obvious that the heat term in equation (19) can be eliminated, since intercoolers are often employed for cooling the working fluid. In what follows, however, it is assumed for simplicity that both compressor and pump work are given by:

$$w_{12} = h_1 - h_2 \text{ BTU/lb (J/kg)} \qquad (20)$$

and w_{12} is a negative term, since h_2 is larger than h_1.

Boilers, Combustors

Boilers and combustors are used to raise the temperature of the working fluid. In the absence of any work in the boiler and ignoring again the kinetic energy terms, the required heat addition is given by:

$$q_{12} = h_2 - h_1 \text{ BTU/lb (J/kg)} \qquad (21)$$

Nozzles

In power plant applications a nozzle is a device for converting thermal energy into kinetic energy. Mechanical work or heat transfer does not take place in the nozzle process. Since inlet kinetic energy is usually very small due to low entrance velocity, one obtains from equation (18):

$$V_2^2 = 2g_c J(h_1 - h_2)\frac{\text{ft}^2}{\text{sec}^2}\left(\frac{\text{m}^2}{\text{sec}^2}\right) \qquad (22)$$

which determines the entrance velocity to a turbine rotating stage.

Turbines

In the context of the energy equation (18), a turbine is considered to be a steady flow device regardless of the number of stages and individual processes occurring in the various stages. Furthermore, for overall turbine analysis the nozzle is considered to be an integral part of the turbine; thus the inlet and exhaust kinetic energy difference $V_1^2 - V_2^2$ can be assumed to be negligible. (It should be noted, however, that this is not always a valid assumption.) For the steam turbine the heat transfer term is neglected on the basis that the turbine is usually well insulated, an assumption which is also useful for gas turbines, considering the large amounts of energy transferred inside as compared with the small exterior unit skin surface area available for heat loss. Using the above assumptions the turbine work is given by:

$$w_{12} = h_1 - h_2 \text{ BTU/lb (J/kg)} \qquad (23)$$

It is apparent from this expression that the turbine produces mechanical work by virtue of the enthalpy difference of the working fluid.

3.9 The Carnot Cycle

An important consequence of the laws of thermodynamics was developed by Sadi Carnot, who showed that no engine that produces mechanical work from heat can be more efficient than the ideal engine operating on the Carnot cycle, whose efficiency is given as:

$$\eta = \frac{\text{work output}}{\text{heat input}} = \frac{T_1 - T_2}{T_1} = 1 - \frac{T_2}{T_1}$$

In the equation, T_1 is the mean absolute temperature of heat addition to the working fluid from the source; T_2 is the mean absolute temperature of heat rejection from the working fluid to the sink. Note that this expression is independent of *the type of engine, the kind of working fluid, and the source of heat* (whether from chemical combustion or nuclear fission). The term η then represents the maximum fraction of heat generated that can be converted to work in any cycle for which T_1 and T_2 have been established.

Figure 3.5 shows the Carnot cycle on a *T-s* diagram as a rectangle where the working substance is compressed isentropically from *a* to *b*, then heat is added reversibly at constant temperature T_1, isentropic expansion takes place between *c* and *d,* and the cycle is completed by a reversible heat transfer at constant temperature T_2 between *d* and *a.* The rectangular area (A) a'-*b*-*c*-d' represents the heat added. The area (B) a'-*a*-*d*-d' represents the heat rejected, and the area $(C = A - B)$ *a*-*b*-*c*-*d* gives the net work. The efficiency can be calculated as $\eta = C/A$.

As an example, let us compute the ideal amount of work that can be obtained from one pound of fuel oil in a steam plant whose mean temperature of heat addition is 625°F and which rejects heat to a condenser at 85°F. Each pound of fuel contains about 18,000 BTU of thermal energy. Assume that all of this energy is transferred to the working fluid. Converting the temperature to absolute (degrees Rankine), and substituting in the Carnot equation:

$$T_1 = 625 + 460 = 1085°R$$
$$T_2 = 85 + 460 = 545°R$$
$$\eta = 1 - \frac{545}{1085} = 1 - .502 = .498$$

Therefore, the maximum amount of heat which can be extracted from each pound of fuel *and converted to work* is $.498 \times 18,000 = 8964$ BTU. The remainder of the heat energy in each pound of fuel is lost to the sea through the condenser, the sea constituting the low temperature reservoir, or sink.

It must be realized that this performance is *ideal,* and assumes no energy losses of any kind except that intrinsic in the second law. In actuality, many losses occur. Some of the causes of these losses are incomplete fuel combustion; waste heat in the boiler stack gases; radiation, conduction, and convection losses from the boiler, piping, and turbine; friction and other mechanical losses in the turbine; and friction in the reduction gearing and shafting. These losses combine to yield a considerably smaller amount of useful work than that of the ideal Carnot cycle engine postulated above. The magnitudes of these losses will be discussed later.

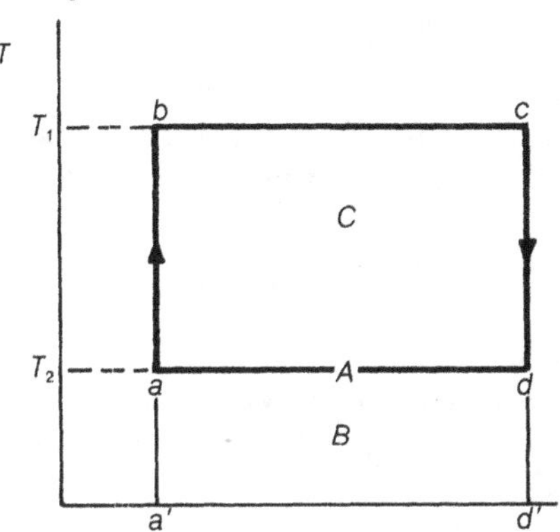

Figure 3.5 Carnot cycle on a *T-s* diagram

In practice, marine power plant performance is usually expressed as the ratio of the pounds of fuel (supplied to the cycle) to the shaft work available from the cycle, i.e., pounds of fuel per shaft horsepower hour, abbreviated as lb/SHP-hr. This is known as the specific fuel consumption:

$$SFC = \frac{\dot{m}_f}{P} \frac{\text{lb}}{\text{HPhr}} \left(\frac{\text{kg}}{\text{HPhr}} \right)$$

where $\dot{m}_f$ is fuel consumption, lb/hr (kg/hr), and P is power output, HP, with

$$P = \dot{m} w_{12}$$

where $\dot{m}$ is flow rate of the working fluid, lb/hr (kg/hr). For the preceding example, with a fuel flow of 1 lb/hr, the performance of the above cycle may be expressed as:

$$SFC = \frac{1 \times 2545}{8964} = .284 \frac{\text{lb}}{\text{HPhr}}$$

where 1 HP = 2545 BTU/hr.

Other useful conversion factors to be noted are:

$$1 \text{ HP} = 550 \frac{\text{ft-lbf}}{\text{sec}}$$
$$= 42.42 \frac{\text{BTU}}{\text{min}}$$
$$= .746 \text{ kW}$$

3.10 Rankine Steam Cycle

Many cycle arrangements have been used since the inception of the steam engine. A few of the important ones will be discussed below to give a clear understanding of why the arrangement presently used has been adopted.

As pointed out above, the Carnot cycle is an ideal cycle whose efficiency can never be matched by an actual cycle. However, it does provide a measure of maximum efficiency by which practical cycles can be compared. One such practical cycle is the Rankine cycle, which represents the events taking place in a typical conventional steam plant.

The simplest cycle, or *open cycle,* meets the four requirements discussed in Section 3.5. It consists of a *boiler* (source of heat) to liberate heat by the combustion of the fuel and transfer it to the working fluid in the form of stored thermal energy, an *engine* in which the thermal energy of the *working fluid* is converted into mechanical work, and a *heat sink* at a lower temperature than the boiler. This sink, in an open cycle, is usually the atmosphere (see Figure 3.6). Since the temperature, T_1, may be increased by boiling the water under pressure, and since a pressure difference is needed to transfer the working fluid from the boiler through the engine, the boiler pressure is maintained above that of the atmosphere.

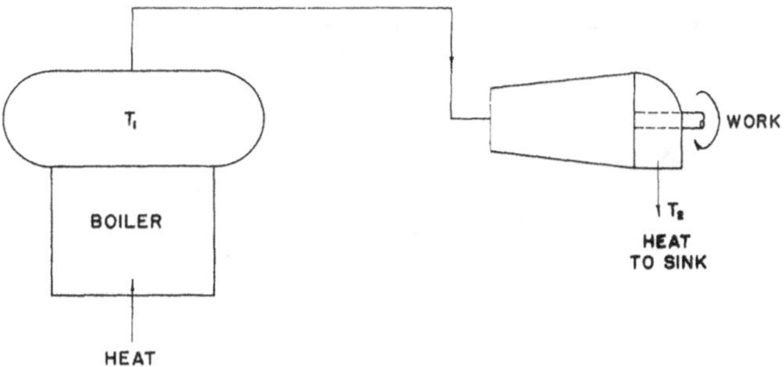

Figure 3.6 Simple open steam cycle

To avoid shutting the engine down should the boiler run dry, the working fluid must be continually supplied to the boiler. For this purpose, a feed pump must be added. Raising water from a low pressure to the high pressure necessary to introduce it into a boiler requires work. Since the power required to drive the feed pump is equivalent to a fraction of that which would otherwise be developed by the engine, the net work available from the cycle is less than the work available from the steam generated in the boiler. The work necessary to drive the feed pump is commonly developed by a separate engine, which obtains steam from the same boiler as the main engine. This engine is known as an *auxiliary* engine (see Figure 3.7). Other auxiliary engines drive fuel oil pumps, forced draft blowers, lube oil pumps, and so forth.

In the open steam cycle, the exhaust temperature, T_2, can only be as low as the boiling point of water corresponding to atmospheric pressure. Furthermore, the open cycle is wasteful of water, since the exhaust steam goes to the atmosphere. In order to conserve fresh water, a condenser is employed to condense the steam as it leaves the engine. The latent heat is removed from the steam by means of a circulating sea-water system, condensing the working fluid, which is then returned to the boiler. Thus, though the heat is still rejected, the working fluid is not lost. The addition of the condenser, and its connection back to the boiler, forms the simple *closed cycle*.

Figure 3.7 Open steam cycle with auxiliary engine

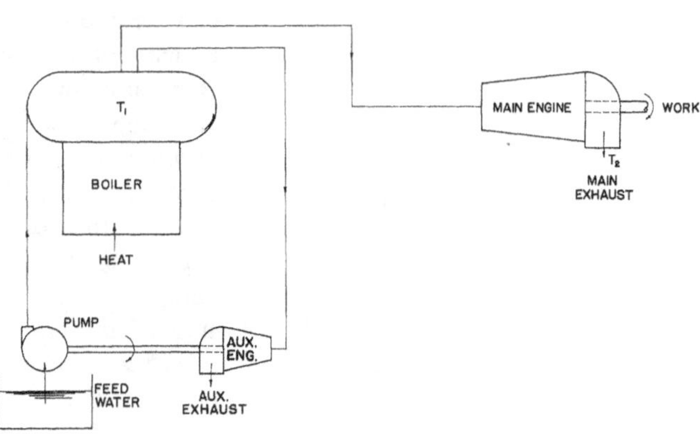

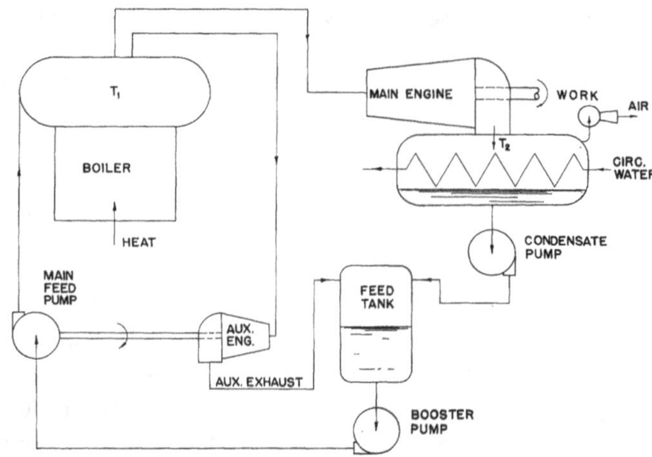

One other major advantage of a condenser is realized by placing it under a vacuum, which is easily done. Since the temperature at which steam condenses (equal to the temperature at which water boils) decreases considerably with decreased pressure, the utilization of a condenser under a vacuum lowers T_2. This increases the Carnot efficiency and makes it possible to extract more work from the steam. Figure 3.8 is a schematic diagram of a simple closed cycle.

In addition to the feed pump and condenser, which are the bare necessities for a closed cycle, practical considerations require the use of several other components in the cycle.

When the power output of the plant is varied (commonly by varying the mass flow of the working fluid through the turbine), the total quantity of working fluid circulating in the cycle changes. To accommodate this changing quantity of working fluid in circulation, a working fluid reservoir is required. This need is satisfied by a feed tank, as shown in Figure 3.8. Note that the auxiliary exhaust from the feed pump is also led to this tank, to be returned to the boiler. The feed tank has other functions as well, which will be discussed below.

Certain practical problems are engendered by placing the condenser under vacuum. Air leakage must be minimized in all parts of the system under vacuum to prevent air from accumulating in the condenser and destroying the vacuum. An air pump, or air ejector, must be provided to maintain the vacuum, and to remove any air or other noncondensable gases that do leak into the system. To help minimize air leakage, the portion of the system under vacuum must be kept as small as possible; to accomplish this a condensate pump is installed at the condenser outlet to raise the pressure of the condensed exhaust steam to higher than atmospheric pressure. This arrangement characterizes the pressure closed feed system used exclusively in modern naval ships.

The attainment of greater thermal efficiencies by increasing T_1 is limited by the temperature to which the materials in the plant can be subjected and still retain adequate strength and resistance to corrosion. Increased chemical activity at higher boiler temperatures necessitates removal from the feed water of oxygen and other dissolved gases, which tend to react chemically with the materials used in the boiler to the detriment of the strength of these materials. This removal is accomplished by a process of deaeration in the feed tank, which is accordingly called the deaerating feed tank (DFT).

The function of the main feed pump has been discussed. In addition, in many shipboard installations another pump is necessary. This pump transfers feed water from the DFT to the main feed pump, and is known as the main feed booster pump or, commonly, as the booster pump.

In commercial shipping (where there is no need to enclose all vital machinery in spaces low in the ship, or under an armored deck for battle damage protection) it is common practice to place the DFT high in the vessel, thereby eliminating the need for a booster pump. Furthermore, the DFT is often called the DC (direct contact) heater because the heating steam actually mixes with the feed water.

A final piece of apparatus found in naval pressure closed feed systems is a heat exchanger, called the economizer. It is located in the boiler uptakes just above the boiler tube array. Feed water flows through the economizer tubes before entering the boiler drum, thus absorbing waste heat from the stack gases. The economizer is usually considered a part of the boiler.

To summarize, the important components in a modern steam cycle are listed below in order of steam or feed water flow (see Figures 3.8 and 2.3):

1. Boiler (with its economizer)
2. Engine
3. Condenser (with its air ejector)
4. Condensate pump
5. DFT
6. Main feed booster pump
7. Main feed pump

Various additional processes and arrangements have been used in steam cycles in order to improve efficiency, improve operating characteristics, and decrease weight. Some of the more common of these are discussed below.

When heat is added to water to produce steam, the steam will be produced at a constant temperature as long as the steam remains in contact with the water during the heat addition process. The steam so produced is said to be saturated. As soon as this steam begins to expand in the turbine, the conversion of heat to work begins and the temperature drops, causing partial condensation. The droplets of condensed water so formed, carried through the turbine in increasing quantity at high velocity, would have a considerable erosive effect, and affect the efficiency of the turbine. Accordingly, in modern plants the saturated steam from the boiler is superheated to a sufficiently high temperature, before being introduced into the turbine, that the moisture content at the turbine outlet is not greater than 12–15%. (Superheat is an increase in temperature above the boiling temperature for a given pressure.)

Other advantages result from the use of superheat. First, a portion of the heat absorbed by the steam is at a temperature above the boiling or saturation temperature, thus increasing the mean temperature of heat addition, T_1, and resulting in some increase in the overall thermal efficiency of the cycle. Second, more energy is contained in a pound of superheated steam than in saturated steam; therefore, more work can be performed during expansion. Thus, for a given required power, less superheated steam than saturated steam is needed, and the engine size can be decreased.

The comparison below will illustrate these points. Computations are omitted for clarity:

Compare the ideal thermal efficiency and theoretical work available in each pound of steam for two engines, one operating at 600 psi saturated, and one operating at 600 psi with 350°F of superheat. Both engines exhaust saturated steam to a condenser pressure of 1 psia.

	Saturated	*Superheated*
Heat added in boiler, BTU	1133.6	1355.7
Heat rejected in condenser, BTU	737.1	851.9
Theoretical heat available for conversion to work, BTU	396.5	503.8
Theoretical efficiency, %	34.9	37.2
Moisture at exhaust, %	25	12.5

Turbine efficiency of 90% has been assumed.

For ideal cycle efficiency, thermodynamic considerations require that the heat to the cycle be received by the working fluid at a temperature as near as possible to the source of the heat. Thus it is uneconomical to inject relatively cool condensate directly into the boiler, where the temperature level is the highest in the cycle. If the condensate temperature can be raised in steps, from sources of heat lower than the boiler temperature, the cycle efficiency will increase. This is accomplished by regenerative feed heating, or extraction, as it is commonly called. In this scheme, a part of the steam is extracted or bled from the turbine at one or more points, after a part of its thermal energy has been converted into useful work, and led to heat exchangers in the feed circuit, where it gives up thermal energy to raise the feed temperature. After the bled steam is cooled (and condensed) it is

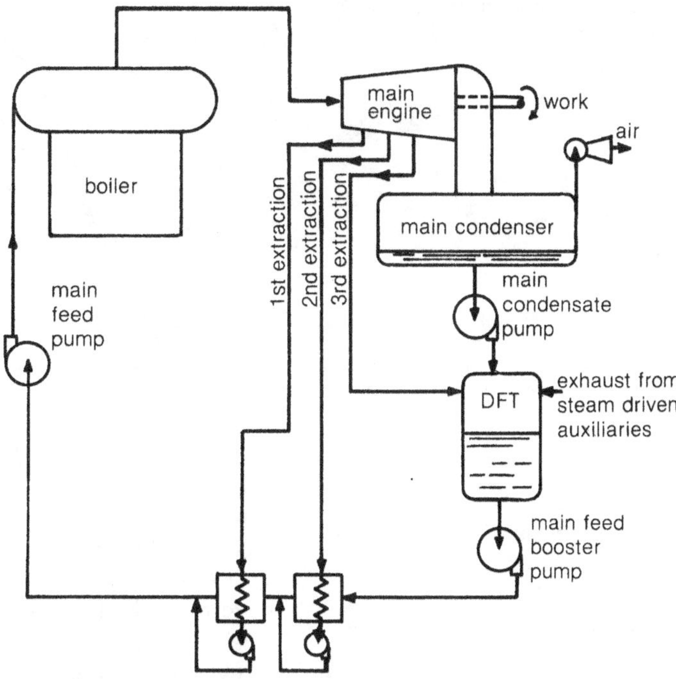

mixed with the feed water and returned to the boiler. Figure 3.9 indicates a scheme for three-stage extraction. Six or seven stages is about the practical limit in any steam plant. In modern naval practice, only one stage is used: the DFT serves as the heat exchanger for single stage extraction, with the two feed heaters indicated in Figure 3.9 not used. Even in naval installations where no steam is bled from the main turbine, the condensation in the DFT of exhaust steam from the main feed pump and other auxiliaries heats the feed, and is similar in effect to extraction.

Another scheme for improving efficiency involves adding heat to the steam after it has done some work in the turbine. This requires extracting the steam at some point in the turbine, piping it back to the boiler where more heat is added, and then returning it to the turbine immediately downstream from the extraction point and allowing it to complete its expansion. This process, called reheat, offers two advantages when properly used. First, the average temperature of heat addition to the cycle is increased, and this increases theoretical cycle efficiency. Second, more work per pound of working fluid can be extracted from the steam, while maintaining final moisture content at an acceptable level. Reheat plants involve increased plant complexity and cost, and are complicated to operate at varying turbine loads. They are not used in modern naval practice.

Before considering another means in current use of improving the cycle efficiency, the basic Rankine cycle processes are summarized in Figure 3.10. The condensate

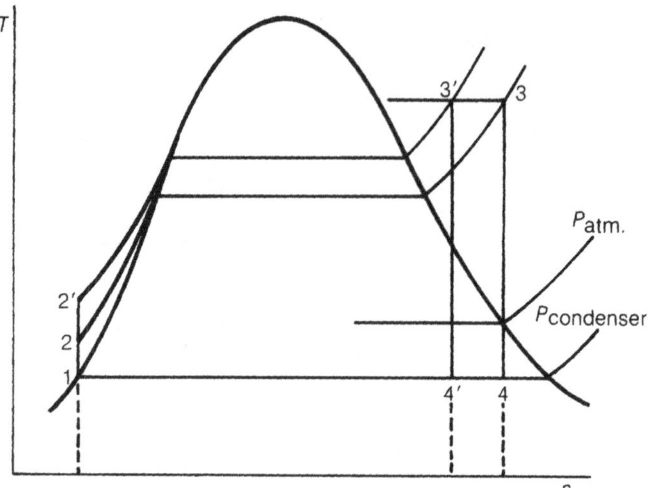

Figure 3.10 Rankine cycle in a *T-s* diagram

is brought to maximum pressure by the condensate feed booster and main feed pumps from 1 to 2. Heat is transferred to boil, evaporate, and superheat to point 3. Expansion through the turbine takes place from 3 to 4, and the cycle conditions return to initial condensate from 4 to 1 in the condenser.

The thermal efficiency of the Rankine cycle is given by (per pound basis):

$$\eta = \frac{\text{net work}}{\text{heat added}}$$

$$= \frac{w_{\text{TURBINE}} - w_{\text{(PUMP AND OTHER AUXILIARIES)}}}{q_{\text{BOILER}}}$$

From Section 3.8, corresponding to the respective points in Figure 3.10:

heat added in the boiler	$q_B = h_3 - h_2$
turbine work	$w_T = h_3 - h_4$
pump and other auxiliaries work	$w_P = h_2 - h_1$

Thus:

$$\eta = \frac{(h_3 - h_4) - (h_2 - h_1)}{h_3 - h_2}$$

$$= 1 - \frac{h_4 - h_1}{h_3 - h_2}$$

One of the ways to improve the cycle efficiency is to increase the cycle maximum pressure. This is shown as point 2' in Figure 3.10. Assuming that the maximum cycle temperature remains the same (limited by the boiler superheater metal deformation temperature), then the result is less superheat at point 3'. The expansion takes place to the same condenser pressure at 4'. As a result, the net cycle work is about the same as for the original 1-2-3-4 cycle, but less heat is rejected, and therefore the cycle efficiency is increased.

Because of improved cycle efficiency, and resulting improved specific fuel consumption, maximum naval steam plant operating pressure is now 1200 psig. An undesirable side effect is increased moisture at the last turbine stages, which increases the erosion potential and rotational losses. In addition to added erosion potential, the 1200-lb plant tends to be heavier, owing to increased turbine and piping weights necessary to contain higher-pressure steam.

In summary, the Rankine cycle efficiency can be improved by:

Increasing the maximum pressure
Increasing the superheat
Lowering the condenser pressure

The moisture content of the exhaust steam is decreased by:

Increasing the superheat
Increasing the condenser pressure
Decreasing the maximum pressure
   (at constant maximum temperature)
Introducing reheat

3.11 Gas Turbine or Brayton Cycle

Although there are numerous exceptions, the Brayton cycle represents processes of an open cycle. Gas turbine engines are practical examples of the open Brayton cycle shown in Figure 3.11. The working fluid is here assumed to be air.

The air is compressed (isentropically) from atmospheric pressure at 1 to p_2 between points 1 and 2. At 2 (exit from the compressor) the air flows into the combustion chamber, where fuel is added and burned at constant pressure to represent heat addition between points 2 and 3. An isentropic

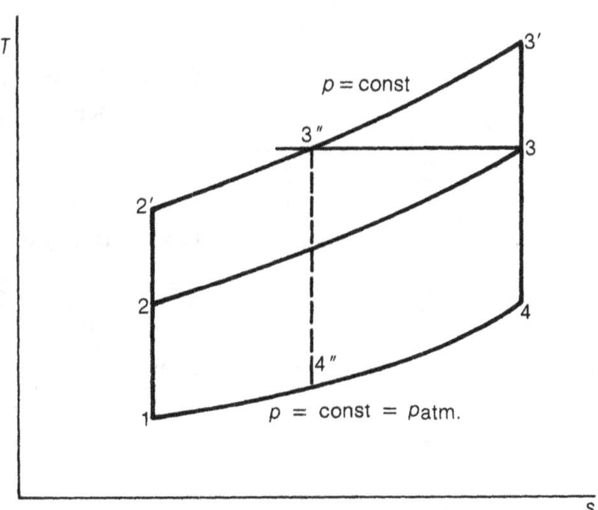

Figure 3.11 Brayton cycle on a *T-s* diagram

expansion takes place through the turbine from 3 to 4. At 4 the working fluid (mixture of air and combustion products) is discharged into the atmosphere, where it returns to normal atmospheric conditions. In a closed Brayton cycle the heat is rejected at constant pressure via a cooler between points 4 and 1.

The efficiency can be calculated, similarly to that in the Carnot cycle, as area ratios, or calculated as for the Rankine cycle:

$$\eta = \frac{\text{net work}}{\text{heat added}} = \frac{w_T - w_C}{q}$$

Now, from Section 3.8 and Figure 3.11:

$$
\begin{aligned}
\text{compressor work} \quad & w_C = h_2 - h_1 \\
\text{turbine work} \quad & w_T = h_3 - h_4 \\
\text{heat added} \quad & q_{23} = h_3 - h_2
\end{aligned}
$$

and the following obtains:

$$\eta = \frac{(h_3 - h_4) - (h_2 - h_1)}{h_3 - h_2}$$

Assuming that the perfect gas results hold, and $h = c_P T$:

$$\eta = \frac{(T_3 - T_4) - (T_2 - T_1)}{T_3 - T_2}$$

$$= 1 - \frac{T_4 - T_1}{T_3 - T_2}$$

This result immediately shows that increasing the turbine inlet temperature T_3 leads to an increase of cycle efficiency; it implies also that the cycle pressure ratio will increase, as will be shown shortly.

Rewriting now:

$$\eta_B = 1 - \frac{T_1}{T_2}\left[\frac{T_4/T_1 - 1}{T_3/T_2 - 1}\right]$$

It will be shown that the quantity in the square brackets is unity. Since $p_3 = p_2$ and $p_4 = p_1$:

$$\frac{p_3}{p_4} = \frac{p_2}{p_1}$$

From perfect gas results:

$$\frac{p_2}{p_1} = \left(\frac{T_2}{T_1}\right)^{k/(k-1)} = \frac{p_3}{p_4} = \left(\frac{T_3}{T_4}\right)^{k/(k-1)}$$

Thus:

$$\frac{T_2}{T_1} = \frac{T_3}{T_4} \quad \text{or} \quad \frac{T_4}{T_1} = \frac{T_3}{T_2}$$

and

$$T_4/T_1 - 1 = T_2/T_1 - 1$$

Then one obtains:

$$\eta_B = 1 - \frac{T_1}{T_2}$$

$$= 1 - \frac{1}{\left(\dfrac{p_2}{p_1}\right)^{(k-1)/k}} \tag{24}$$

which shows that the efficiency of the Brayton cycle depends only on the pressure ratio of the unit and increases as the pressure ratio increases. The thermal efficiency of the Brayton cycle is much lower than that of a Carnot cycle operating between the same temperature limits, as the Carnot cycle efficiency would be:

$$\eta_C = 1 - \frac{T_1}{T_3}$$

The effect of increasing the pressure ratio is shown by points 1-2'-3'-4 in Figure 3.11. The increase in the pressure ratio also brings with it a sizable increase in temperature T_3 (turbine inlet temperature). This temperature is limited by metallurgical considerations to about 2000°F or, if cooling techniques are used, to about 2400°F. If the maximum cycle temperature were held constant at 3, then an increase of cycle pressure ratio would change the cycle to 1-2'-3"-4". This cycle would have a higher efficiency than the original cycle, but decreased work output.

Example 3.5: An ideal gas turbine, as shown by the simple cycle in Figure 3.12, has the following specifications:

sea level intake air	$T_1 = 60°F = 520°R$
compressor pressure ratio	$P_2/P_1 = 12$
turbine inlet temperature	$T_3 = 2200°R$
atmospheric exhaust	$p_4 = p_a = 15$ psi
constant specific heat	$c_p = .24$ BTU/lb/°R
constant ratio of specific heats	$k = 1.4$

Calculate the work output and the cycle efficiency.

Solution: For this ideal air cycle the kinetic effects will be ignored. Thus, total and static pressures are assumed to be the same, i.e. $p_0 = p$ and $T_0 = T$.

The compressor work can be calculated by using the perfect gas results above.

$$w_c = h_2 - h_1 = c_p(T_2 - T_1)$$

$$= c_p T_1 \left(\frac{T_2}{T_1} - 1\right)$$

$$= c_p T_1 \left[\left(\frac{P_2}{P_1}\right)^{(k-1)/k} - 1\right]$$

$$= .24(520)[12^{.286} - 1]$$

$$= 129 \text{ BTU/lb}$$

Since for a simple ideal cycle the single turbine work w_t

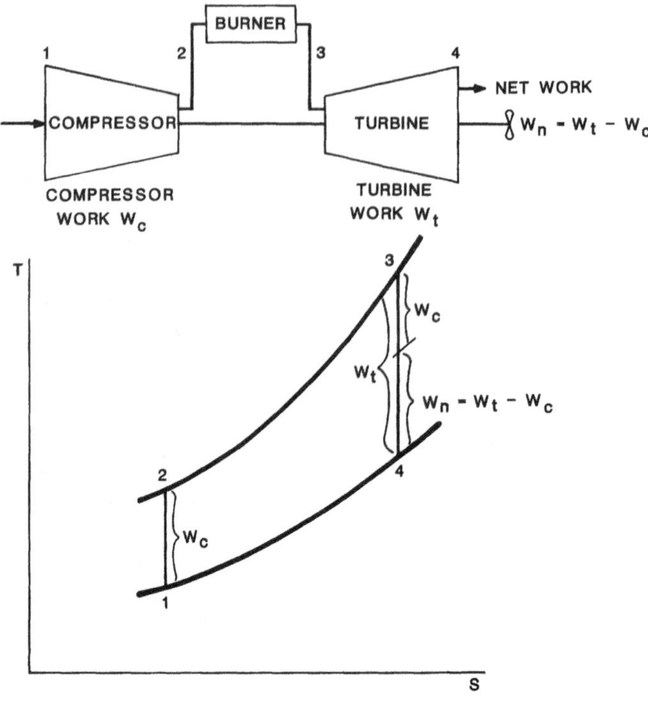

exceeds the work necessary to drive the compressor, the net work is

$$w_n = w_t - w_c$$

The turbine work is found from

$$w_t = c_p(T_3 - T_4)$$

$$= c_p T_3\left(1 - \frac{T_4}{T_3}\right)$$

$$= c_p T_3\left[1 - \left(\frac{P_4}{P_3}\right)^{(k-1)/k}\right]$$

$$= c_p T_3\left[1 - \frac{1}{\left(\frac{P_2}{P_1}\right)^{(k-1)/k}}\right]$$

$$= .24(2200)\left[1 - \frac{1}{12^{.286}}\right]$$

$$= 269 \text{ BTU/lb.}$$

The net work is then

$$w_n = 269 - 129 = 140 \text{ BTU/lb}$$

If this machine had a mass flow of 50 lb/sec, it would produce a net work of

$$w_n = \frac{140 \times 50}{.707} = 9900 \text{ HP}$$

At the same time, the compressor requires

$$w_c = \frac{129 \times 50}{.707} = 9123 \text{ HP}$$

to drive it, and the total turbine output is $w_t = 9900 + 9123 = 19023$ HP. The efficiency of this ideal machine is (equation 24)

$$\eta = 1 - \frac{1}{12^{.286}} = .51$$

which, of course, is an unrealistically high value.

If, following the discussion above, the maximum cycle temperature was held constant at 2200°R and the pressure ratio increased to, say, 20, the cycle efficiency becomes

$$\eta = .58$$

but the net work is reduced to

$$w_n = 304 - 169 = 135 \text{ BTU/lb.}$$

The preceeding example shows that the output depends on both the pressure ratio and the maximum cycle temperature, T_3. For the simple ideal cycle in Figure 3.12

$$w_n = c_p(T_1 - T_2 + T_3 - T_4)$$

becomes with the expressions for $T_1 - T_2$ and $T_3 - T_4$ developed in Example 3.5

$$w_n = c_p T_1 \left[1 - \left(\frac{P_2}{P_1} \right)^{(k-1)/k} + \frac{T_3}{T_1} \left(1 - \frac{1}{(P_2/P_2)^{(k-1)/k}} \right) \right]$$

or in terms of net specific work

$$\frac{w_n}{c_p T_1} = \frac{T_3}{T_1} \left(1 - \frac{1}{(P_2/P_1)^{(k-1)/k}} \right) - \left[\left(\frac{P_2}{P_1} \right)^{(k-1)/k} - 1 \right]$$

$$(25)$$

Figure 3.13 shows several cycles' net work values for a given maximum cycle temperature, T_3. From this figure and equation (25), it is easily seen that w_n approaches zero as the pressure ratio approaches unity, and also for the highest pressure ratio $P_2/P_1 = (T_3/T_1)^{k/(k-1)}$ where the expansion process approaches the compression line. Clearly, for some intermediate pressure ratio there will have to exist a maximum value of w_n, the net cycle work. The specific net work $w_n/c_p T_1$ is shown plotted in Figure 3.14. It is seen that there is indeed an optimum pressure ratio for maximum output for a given turbine inlet temperature T_3.

This optimum pressure ratio can be determined by the standard maximization procedure (differentiating w_n with respect to the pressure ratio and setting the product equal to zero), which yields

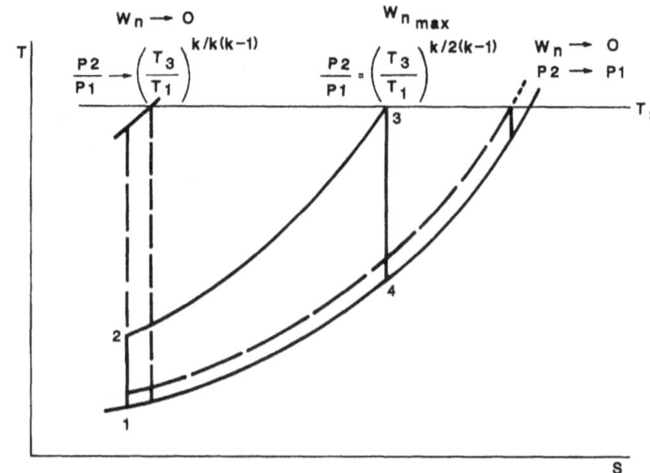

Figure 3.13 Ideal cycle maximum net work

$$\frac{p_2}{p_1} = \left(\frac{T_3}{T_1}\right)^{k/2(k-1)} \tag{26}$$

Since

$$\frac{p_2}{p_1} = \left(\frac{T_2}{T_1}\right)^{k/(k-1)}$$

the following obtains also

$$\frac{T_2}{T_1} = \sqrt{\frac{T_3}{T_1}} = \frac{T_3}{T_4} \tag{27}$$

These equations give the ideal pressure ratio required to achieve maximum work, for a given T_3, and show that the temperature ratio across the compressor should be the square root of the ratio of the cycle maximum and minimum temperatures.

The advantage of a high turbine inlet temperature is evi-

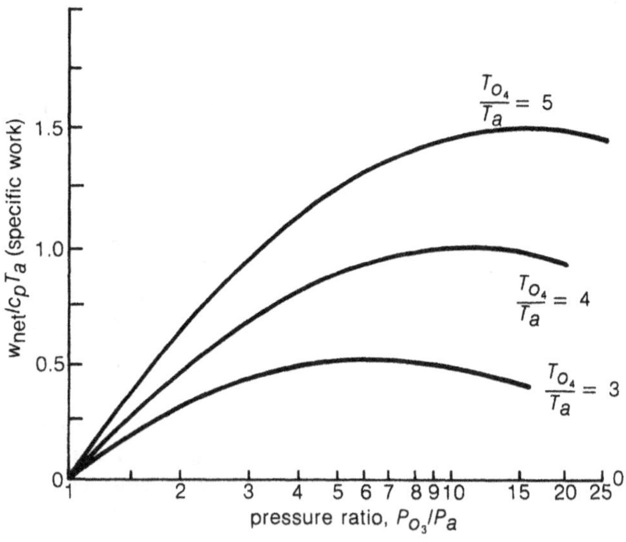

Figure 3.14 Ideal Brayton cycle specific work

dent by inspection of Figure 3.14. Although the efficiency of an ideal cycle is not dependent on the cycle temperature, as seen in equation (24), it will be shown later in the discussions on the actual cycle that the turbine inlet temperature plays a central role in proper cycle definition and optimization for a given application. Figure 3.14 indicates also that the use of higher cycle maximum temperatures requires increasing cycle pressure ratios to maintain optimum work output.

As is the case with the Rankine cycle, the basic gas turbine cycle can be improved in several ways to increase overall efficiency. Regeneration, or the heat exchange cycle, is based on the concept that the turbine outlet temperature is often higher than the compressor exit temperature, and that this temperature difference represents a source of recoverable energy which can be used to improve cycle efficiency, rather than exhausting it to the atmosphere.

For the optimum work cycle, it follows from equation (27) that $T_2 = T_4$, i.e., the compressor and turbine exit temperatures are equal. For a range of pressure ratio values between unity and $(T_3/T_1)^{k/2(k-1)}$, as shown in Figure 3.13, T_4 will be greater than T_2 and a heat exchanger can be used to preheat the air leaving the compressor. This means that less fuel needs to be burned to reach the given turbine inlet temperature T_3, thus increasing the cycle efficiency.

A flow diagram and the T-s diagram of the ideal regenerative gas turbine cycle are shown in Figure 3.15. In an

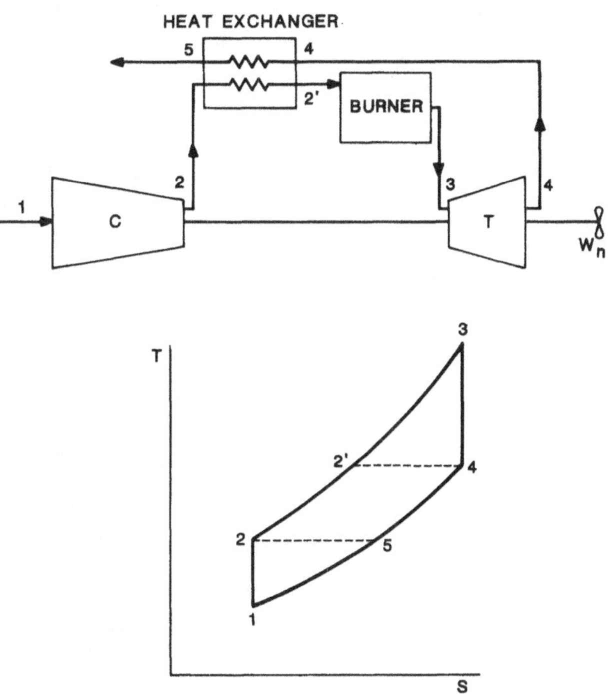

Figure 3.15 Ideal regenerative cycle

ideal heat exchanger, the turbine exhaust gas at T_4 heats the compressor outlet flow starting at T_2 until it reaches the regenerator exit temperature $T_2 = T_4$. The cycle efficiency can then be expressed as

$$\eta = \frac{T_3 - T_4 - (T_2 - T_1)}{T_3 - T_2}$$

$$= \frac{T_3 - T_4 - (T_2 - T_1)}{T_3 - T_4}$$

$$= 1 - \frac{(P_2/P_1)^{(k-1)/k}}{T_3/T_1} \tag{28}$$

The latter expression shows that the regenerative cycle efficiency depends on both the pressure ratio and the cycle maximum temperature. It increases as the maximum temperature increases, but decreases with increasing pressure ratio. Figure 3.16 shows the regenerative cycle efficiency curves decreasing with increasing pressure ratio until $P_2/P_1 = (T_3/T_1)^{k/2(k-1)}$ is reached. At this point $T_4 = T_2$. At higher pressure ratios T_4 will fall below T_2 (see also Figure 3.13), and the heat exchanger will end up cooling the compressor exhaust leading to a reduction of efficiency.

For the regenerative ideal cycle, the specific work remains unchanged, and Figure 3.14 is still valid.

Example 3.6: Calculate the cycle efficiency for the gas turbine of Example 3.5 when a regenerative heater is employed.

Solution: The increased efficiency can be obtained from equation (28)

$$\eta = 1 - \frac{T_1}{T_3}\left(\frac{P_2}{P_1}\right)^{(k-1)/k} = 1 - \frac{520}{2200}(12)^{.286} = .52$$

which is only a slight increase over the basic cycle efficiency of .51. The reason for this small change is that T_4 is barely larger than T_2. From the last example

$$T_3 - T_4 = T_3\left[1 - \frac{1}{(P_2/P_1)^{(k-1)/k}}\right]$$

$$= 2200\left[1 - \frac{1}{12^{.286}}\right] = 1119°R$$

Then

$$T_4 = T_3 - (T_3 - T_4) = 2200 - 1119 = 1081°R$$

And T_2 is calculated from

$$T_2 = T_1(P_2/P_1)^{(k-1)/k} = 520(12)^{.286} = 1058°R$$

Because of the small temperature difference, it is impractical to consider regeneration for improving the efficiency of this particular cycle.

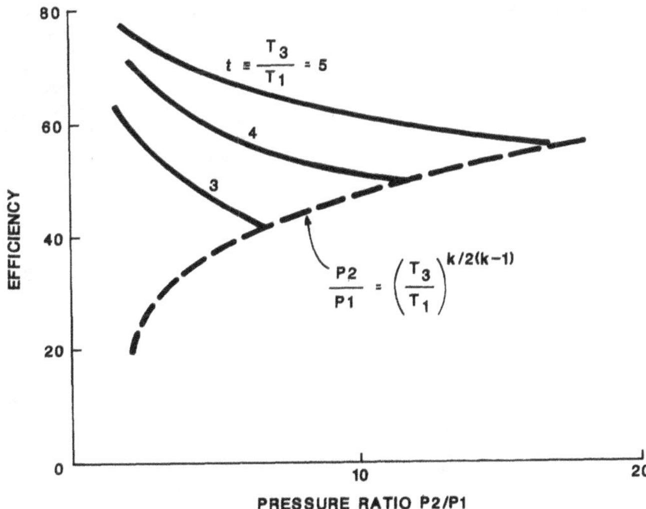

Figure 3.16 Efficiency of ideal regenerative cycle

Another method for increasing the efficiency of the gas turbine cycle is to reduce the work necessary to drive the compressor. In the steady flow situation, compression work is given by

$$w_c = c_p(T_2 - T_1) \tag{29}$$

which states that the compressor work is proportional to the temperature.

As a large pressure ratio will produce a large increase in temperature, the work required for compression is much larger in the high pressure stages than in the earlier stages of the machine. Thus, when a compression process is broken into two or more parts and cooling is introduced between the sections at essentially constant pressure, the work required to carry out the overall compression can be appreciably reduced. This is called *intercooling,* and the principle applies equally to reciprocating or rotating machinery.

Many high pressure industrial compressors take advantage of intercooling at several stages during the compression process. For effective cooling, the intercoolers require a large water flow on the cooling side of the heat exchangers. Thus, the equipment tends to be bulky and is not used for aircraft gas turbine engines. For marine applications, however, where the water supply is ever-present, intercooling offers a good method for enhancing the engine performance.

It should be pointed out that intercooling improves the cycle efficiency only when regeneration is also employed. Figure 3.17 shows the *T-s* diagram and the associated equipment for an intercooled cycle with regeneration. The compression takes place from 1 to 1′ where the intercooler will reduce the temperature (at constant pressure) to 1″. Without intercooling, the compression would proceed from 1 to 2′. The endpoint of intercooled compression is at 2, which is at

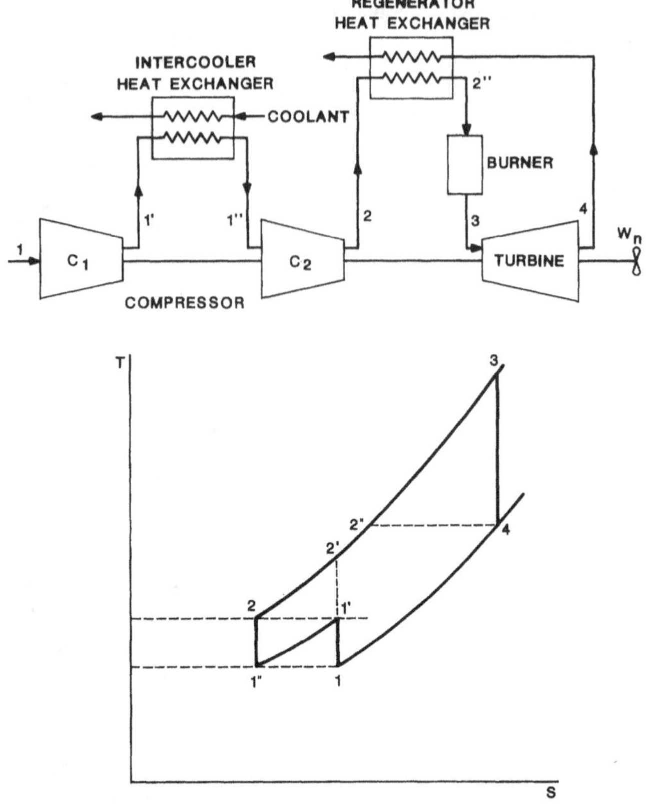

Figure 3.17 Ideal cycle with intercooling and regeneration

lower temperature than $2'$. Thus, unless regeneration is utilized between 2 and $2''$, additional heat (from fuel) must be supplied to the cycle to reach the turbine inlet temperature at 3. Without regeneration, the cycle efficiency will be decreased due to extra fuel being burned to reach a given T_3.

The compression work can be expressed by equation (29) as

$$w_c = c_p(T_1' - T_1) - c_p(T_2 - T_1'')$$
$$= c_p T_1 [(P_1'/P_1)^{(k-1)/k} - 1] - c_p T_1 [(P_2/P_1'')^{(k-1)/k} - 1]$$

where it has been assumed that the intercooler brings the temperature down to the initial temperature, i.e., $T_1'' = T_1$. Although this simplification leads to somewhat idealized calculation, it does not detract from the general conclusions.

When the above equation is optimized to determine the minimum total work of compression, one finds that

$$P_1' = \sqrt{P_1 P_2}$$

This shows that the minimum work input occurs when the total work is evenly divided by the stages. Thus

$$w_c = 2c_p T_1 [(P_2/P_1'')^{(k-1)/k} - 1] = 2c_p(T_2 - T_1)$$

which is less than the work without intercooling between T_2' and T_1.

The net work can be expressed as

$$w_n = c_p(T_3 - T_4) - 2c_p(T_2 - T_1)$$

or

$$\frac{w_n}{c_p T_1} = t(1 - 1/r) - 2(\sqrt{r} - 1) \tag{30}$$

where $t = T_3/T_1$, and $r = (P_2/P_1)^{(k-1)/k}$ have been used. Thus the ideal intercooled cycle efficiency without regeneration becomes

$$\eta = \frac{T_3 - T_4 - 2(T_2 - T_1)}{T_3 - T_2}$$
$$= \frac{t(1 - 1/r) - 2(\sqrt{r} - 1)}{t - \sqrt{r}} \tag{31}$$

and with regeneration such that $T_{2''} = T_4$

$$\eta = \frac{T_3 - T_4 - 2(T_2 - T_1)}{T_3 - T_{2''}}$$
$$= \frac{t(1 - 1/r) - 2(\sqrt{r} - 1)}{t(1 - 1/r)} \tag{32}$$

By means of a brief example calculation, it may be shown that the efficiency of an intercooled cycle with regeneration, equation (32), is higher than the one without regeneration, equation (31).

Example 3.7: An ideal intercooled cycle has turbine inlet and atmosphere temperatures $T_3 = 2000°R$ and $T_1 = 500°R$, respectively. The cycle pressure ratio is 8. Determine the cycle efficiencies without and with ideal regeneration.

Solution: Here $t = T_3/T_1 = 2000/500 = 4$, and $r = (P_2/P_1)^{(k-1)/k} = 8^{.286} = 1.81$. Thus, for the cycle without regeneration, equation (31) yields

$$\eta = \frac{4(1 - 1/1.81) - 2(\sqrt{1.81} - 1)}{4 - \sqrt{1.81}} = \frac{1.1}{2.65} = .41$$

Equation (32) yields for the regenerated cycle an efficiency of

$$\eta = \frac{1.1}{4(1 - 1/1.81)} = \frac{1.1}{1.79} = .61$$

an appreciable increase in the cycle efficiency.

A larger increase in work output (w_n) than can be obtained by intercooling can be achieved if the turbine expansion process is divided into two or more parts, and the first turbine exhaust is reheated to a higher temperature. The idea is similar to increasing net work by intercooling, as the expansion work is now increased in two stages when compared to a single stage expansion. Reheat produces more work output than intercooling because any modification of a

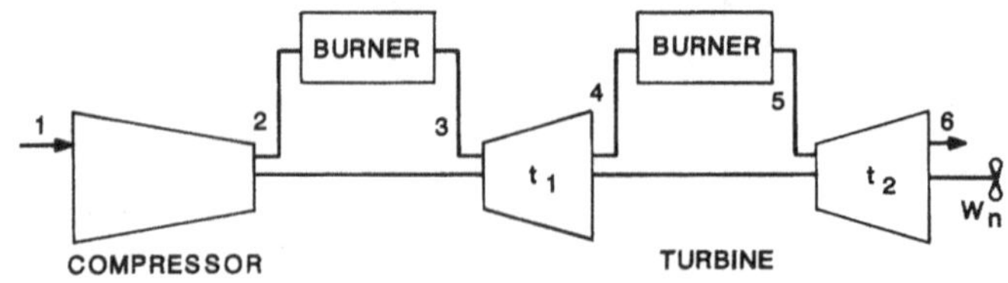

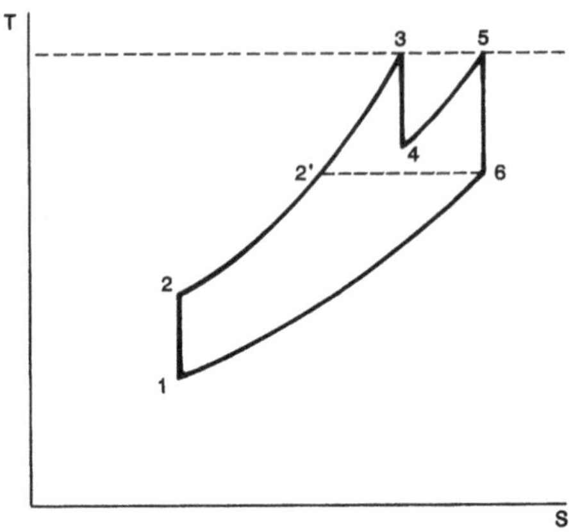

cycle at high temperatures pays more dividends than similar efforts at lower temperatures.

Figure 3.18 shows the *T-s* diagram and the configuration for the ideal cycle with reheat. Reheat increases turbine output for given amounts of compressor work and maximum cycle temperature. This process, however, requires more fuel, as two combustors are used. The final turbine temperature T_6 is higher than the exit temperature without reheat, $T_{4'}$. This, in turn, leads to very effective utilization of regeneration, which further increases the cycle efficiency.

Similarly to the intercooled cycle, it can be shown that the maximum work output can be achieved if the reheat is carried out at the point where the expansion pressure ratios (and also work outputs) are equal, or

$$P_4 = P_5 = \sqrt{P_3 P_6}$$

For ideal reheat where turbine work is given by

$$w_T = c_p(T_3 - T_4) = c_p(T_5 - T_6) = 2c_p(T_3 - T_4)$$

This becomes with the abbreviations above

$$\frac{w_n}{c_p T_1} = 2t(1 - 1/\sqrt{r}) - r + 1 \tag{33}$$

The cycle efficiency then becomes

$$\eta = \frac{w_n}{q_{added}} = \frac{2c_p(T_3 - T_4) - c_p(T_2 - T_1)}{c_p(T_3 - T_2) + c_p(T_5 - T_4)}$$

$$= \frac{2t(1 - 1/\sqrt{r}) - r + 1}{2t - r - t/\sqrt{r}} \tag{34}$$

This cycle efficiency can be appreciably improved by ideal regeneration to point $T_{2'} = T_6$. The efficiency then becomes (work is unchanged)

$$\eta = \frac{2(T_3 - T_4) - (T_2 - T_1)}{(T_3 - T_{2'}) + (T_5 - T_4)}$$

$$= \frac{2(T_3 - T_4) - (T_2 - T_1)}{(T_3 - T_6) + (T_3 - T_4)}$$

$$= \frac{2t(1 - 1/\sqrt{r}) - r + 1}{2t(1 - 1/\sqrt{r})} \tag{35}$$

Still further improvement of the ideal cycle considered here can be achieved if intercooling, reheat, and regeneration are all employed at the same time. The machine flow diagram for this situation is shown in Figure 3.19. The thermodynamic results can be combined by means of Figures 3.17 and 3.18. The work is given by

$$w_n = 2c_p(T_3 - T_4) - 2c_p(T_2 - T_1)$$

or

$$\frac{w_n}{c_p T_1} = 2t(1 - 1/\sqrt{r}) - 2(\sqrt{r} - 1) \tag{36}$$

The efficiency with regeneration (there is no point in considering this cycle without regeneration) becomes (see equation (35) for heat added)

$$\eta = \frac{w_n}{c_p(T_3 - T_6) + c_p(T_3 - T_4)} \tag{37}$$

$$= \frac{2t(1 - 1/\sqrt{r}) - 2(\sqrt{r} - 1)}{2t(1 - 1/\sqrt{r})}$$

Example 3.8: Consider the cycle between the same limits as stated in Example 3.7 but with ideal reheat and no regeneration. Then add the effect of regeneration. Lastly, consider both reheat and intercooling effects.

Solution: Without regeneration the efficiency is obtained from equation (34) with $t = 4$, $r = 1.81$ as

Figure 3.19 Flow diagram of an ideal cycle with intercooler, regenerator, and reheater.

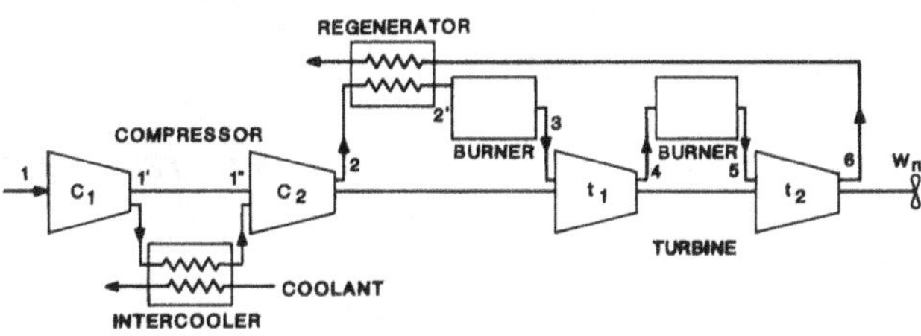

$$\eta = \frac{2 \times 4(1 - 1/\sqrt{1.81}) - 1.81 + 1}{2 \times 4 - 1.81 - 4/\sqrt{1.81}} = \frac{1.24}{3.22} = .38$$

With regeneration the efficiency is found from equation (35)

$$\eta = \frac{1.24}{2 \times 4(1 - 1/\sqrt{1.81})} = \frac{1.24}{2.05} = .6$$

The cycle efficiency with intercooling, reheat, and regeneration is calculated from equation (37)

$$\eta = \frac{2 \times 4(1 - 1/\sqrt{1.81}) - 2(\sqrt{1.81} - 1)}{2.05}$$

$$= \frac{1.36}{2.05} = .66$$

This last efficiency number approaches the Carnot cycle value between the same temperature limits, which is $\eta_c = 1 - (1/t) = .75$.

To summarize, in this last section on gas turbine cycles, it was shown that intercooling and reheat alone may be impractical, as the cycle efficiency suffers without regeneration. Effectiveness of the latter is increased by reheat and intercooling.

The design pressure ratio of the cycle depends on whether high efficiency or high work output is desired. In a later chapter this principle will be expanded to include the effect of cycle temperature ratio and component losses. For cycles without regeneration, higher pressure ratios are required with high turbine inlet temperatures.

This concludes the review of basic thermodynamics and elementary fluid mechanics necessary for understanding the power conversion processes. The discussions of cycles and concepts for improving the Brayton cycle were restricted to ideal processes with no losses, but the general conclusions are also valid for actual cycles where various component losses are included.

References

1. Keenan, J. H., Keyes, F. G., Hill, P. G., and Moore, J. G., *Steam Tables,* New York: John Wiley and Sons, 1969.

2. Keenan, J. H., and Kay, J., *Gas Tables,* New York: John Wiley and Sons, 1948.

3. Johnston, R. M., Brockett, W. A., Bock, A. E., and Keating, E. L., *Elements of Applied Thermodynamics,* Annapolis: U.S. Naval Institute, 1978.

4. Van Wylen, G. J., *Thermodynamics,* New York: John Wiley and Sons, 1959.

5. Shapiro, A. H., *The Dynamics and Thermodynamics of Compressible Fluid Flow,* Vol. 1, New York: Ronald Press, 1954.

Chapter Four

Turbomachinery Fundamentals

4.1 General

This book is essentially about turbines as work-producing devices. As seen below, there are several types of turbines that must be considered, as well as compressors, which are necessary integral parts of the gas turbine units. Thus, this chapter is devoted to understanding the fundamental aspects of turbomachinery as they apply to both turbines and compressors and their essential components: nozzles and blading.

First of all, turbomachinery can be categorized into two groups according to function: (1) work- or power-producing devices—various kinds of turbines; (2) work- or power-absorbing devices—pumps, compressors, fans. This machinery can also be classified as one of three types according to the nature of the flow path of the working fluid through the machine. When the working fluid flow direction is essentially parallel to the axis of rotation, it is called an *axial flow* machine (Figure 4.1). When the flow is mainly in a plane perpendicular to the axis of rotation, it is a *radial flow* machine (Figure 4.2.). A *mixed flow* machine exhibits considerable radial and axial velocity components at the rotor exit (Figure 4.3). Both turbines and compressors are designed in axial and radial configurations. The main propulsion units for marine applications have axial turbines, but may have axial or radial compressors, or a combination of axial-radial compressors. Auxiliary gas turbines and turbine drives are often of radial inflow type. Mixed flow machines are found in pump applications or in hydraulic turbines. Some forced draft blowers and steam turbines have also been of the mixed flow design.

There is another criterion by which turbomachinery is classified, and which also is a significant design parameter, as it affects both blading efficiency and work exchange between the fluid and the rotor. All turbomachinery, turbines and compressors, can be classified as either *impulse* or *reaction* machines. In an impulse stage (used only in turbines) there is virtually no pressure change in the rotating row, and all the expansion occurs in one or more stator rows. The Pelton water wheel is a familiar example of an impulse machine. If expansion (pressure drop) occurs in both stationary and rotating rows, then momentum exchange takes place in both rows, and the stage is referred to as a reaction stage. In actual machines combinations of impulse and re-

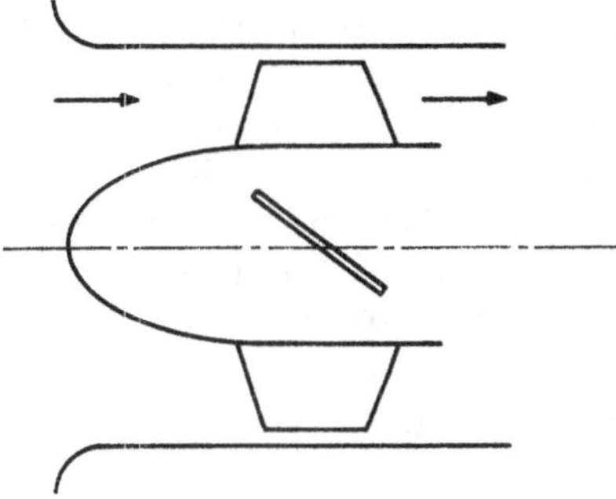

Figure 4.1 Axial flow machine

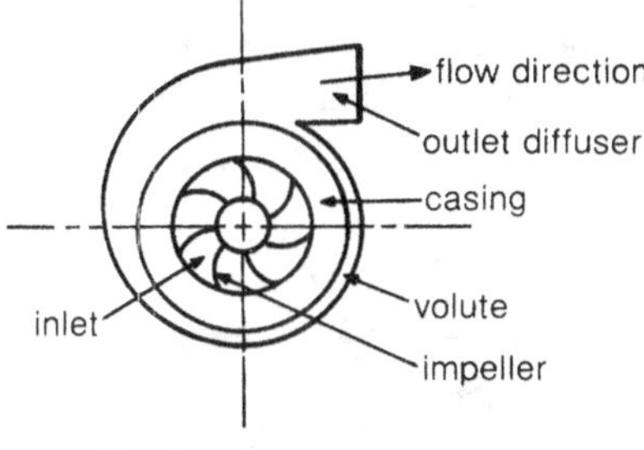

Figure 4.2 Radial flow machine

centrifugal compressor or pump

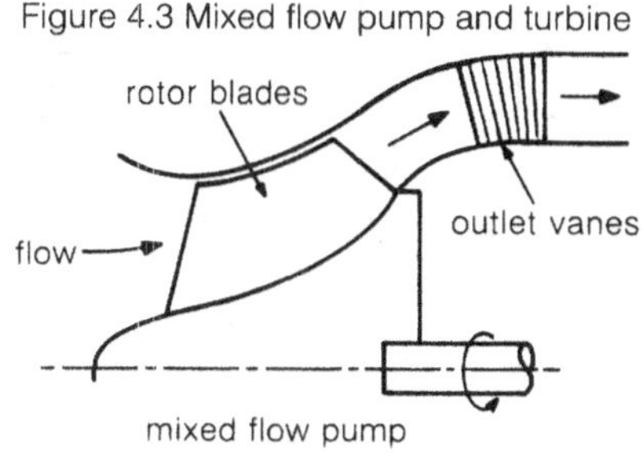

Figure 4.3 Mixed flow pump and turbine

mixed flow pump

action stages appear, as will be shown later. Moreover, in multistage steam turbines with very long twisted blades the blade root portion may behave as an impulse blade, and the tip of the blade may be designed with a high degree of reaction. Gas turbines and compressors are invariably designed as reaction stages with the degree of reaction about .5 in each type of machine, but for different reasons. They are to be discussed in more detail in the chapter on gas turbines.

4.2 Turbomachinery Principles

In Chapter 3 turbomachinery processes were analyzed in terms of the overall energy equation, and performance was evaluated in terms of the change in enthalpy. Next, momentum analysis, being unrestrictive in its approach relative to type of turbomachine, will be employed in Newton's second law form as it applies to the moments of forces. This approach leads to description of rotating machines by three distinct effects of pressure change, and yields expressions for work and power in terms of fluid and rotor velocities.

Figure 4.4 shows a rotating disc where fluid enters and leaves with velocities V_1 and V_2, respectively. For a mass m the vector sum of all external forces acting on it (about the disc axis) is equal to the time rate of change of angular momentum:

$$\tau = m\frac{d}{dt}(rV_u)$$

where τ is torque, r is distance of the mass center from the axis, and V_u is the tangential component of fluid velocity V. For steady flow through the disc and assuming that the flow is one-dimensional, the above expression becomes:

Figure 4.4 Flow through a turbomachine

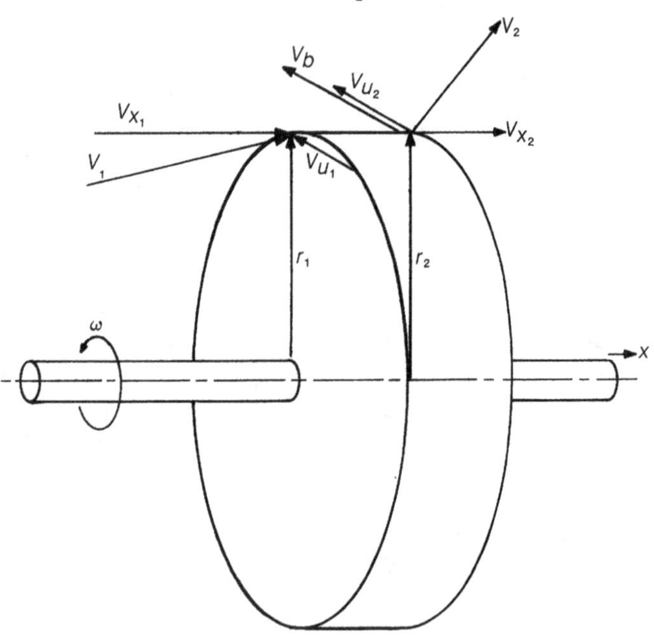

$$\tau = \dot{m}(r_2 V_{u_2} - r_1 V_{u_1}) \qquad (1)$$

For a machine rotating at constant angular velocity ω, with the blade speed $V_b = \omega r$, the work done on the fluid (power) is:

$$P_c = \tau\omega = \frac{\dot{m}}{g_c}(V_{b_2}V_{u_2} - V_{b_1}V_{u_1})\frac{\text{ft lbf}}{\text{sec}}\left(\frac{\text{Nm}}{\text{sec}}\right) \qquad (2)$$

Equation (2) is known as the Euler pump equation. If $V_{b_1}V_{u_1} > V_{b_2}V_{u_2}$, then work is done by the fluid, and:

$$P_T = \frac{\dot{m}}{g_c}(V_{b_1}V_{u_1} - V_{b_2}V_{u_2}) \qquad (3)$$

which is known as the Euler turbine equation.

Equation (2) applies to compressors, fans, and pumps, whereas equation (3) is used for turbines. To further develop these equations for practical applications, it is necessary to realize (a) that the physical process involved is that of converting kinetic energy into mechanical energy, or vice versa, and (b) that the tangential fluid velocity component V_u can be related to the blading geometry at hand. The kinetic energy–mechanical energy conversion occurs according to the momentum equation discussed in Section 3.4 (equation 8). The velocity change $(V_2 - V_1)$ causes the force on a blade, or vice versa. Since the velocity is a vector quantity having both magnitude and direction, force can be generated by (a) change in magnitude of velocity, (b) change in direction of flow, or (c) both. The purpose of blading is to effect these changes. Thus equations (2) and (3) are directly related to flow and blading geometry, i.e., turbines, compressors, pumps, and so on.

In the following development, these equations are related to velocities and pressure changes, using for the purpose a typical reaction turbine blading. The final results and conditions, however, are also valid for compressors, with an appropriate sign change between equations (2) and (3).

Figure 4.5a shows fluid flow through a turbine stage. The nozzles or vanes (stationary blades) produce a velocity V_1, which, combined with rotor blade velocity V_b, gives rotor inlet or relative velocity V_r. Rotor exit conditions are denoted with subscript 2. The angles are measured relative to the plane of rotor rotation, and α is the angle between the absolute velocity V and blade velocity V_b. Angle β is the angle between relative velocity V_r and V_b. Component V_u is the component of V in the tangential direction, or the projection of velocity V in the tangential direction, while V_x is the velocity component in the axial direction.

The velocity diagrams are the keys to the analysis of turbomachinery problems, e.g., all velocity information necessary for evaluation of equation (3) is contained here; also the angles β_1 and β_2 determine what the blade angles must be for smooth unseparated flow onto and from the blade. To

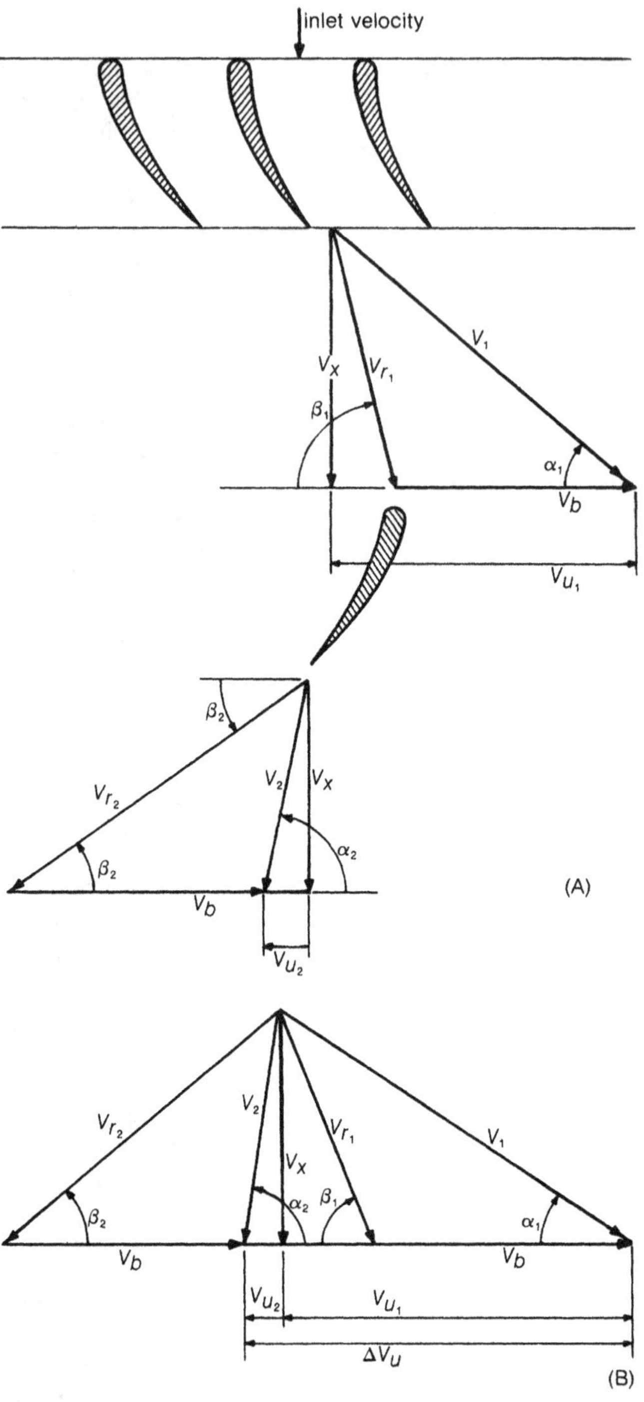

facilitate calculations and to provide an oversight, the velocity diagrams are combined into one diagram as shown in Figure 4.5b. Here the assumption is normally made that the axial velocity is constant. For the entrance and exit triangles one has, by the law of cosines:

$$V_r^2 = V^2 + V_b^2 - 2VV_b \cos \alpha$$

but:

$$V \cos \alpha = V_u$$

Thus, for entrance and exit conditions,

$$2V_{b_1}V_{u_1} = V_1^2 + V_{b_1}^2 - V_{r_1}^2$$
$$2V_{b_2}V_{u_2} = V_2^2 + V_{b_2}^2 - V_{r_2}^2$$

Subtracting the second equation from the first yields the following:

$$2(V_{b_1}V_{u_1} - V_{b_2}V_{u_2}) = (V_1^2 - V_2^2)$$
$$+ (V_{b_1}^2 - V_{b_2}^2) + (V_{r_2}^2 - V_{r_1}^2)$$

This expression can now be substituted into equation (3) to give:

$$H_T = \frac{P_{g_c}}{\dot{m}g} = \frac{1}{2g}[(V_1^2 - V_2^2)$$
$$+ (V_{b_1}^2 - V_{b_2}^2) + (V_{r_2}^2 - V_{r_1}^2)] \text{ ft (m)} \quad \textbf{(4)}$$

Equation (4), known as the turbine ideal total head equation, consists of three terms, each representing a quantity of kinetic energy per pound of fluid:

The first term, $(V_1^2 - V_2^2)$, represents the kinetic energy change between entrance and exit sections, or energy exchange due to impulse (a small pressure drop).

The second term, $(V_{b_1}^2 - V_{b_2}^2)$, represents a centrifugal effect as the fluid particles travel in radial directions through the rotor. The effect is present if the fluid changes radius between the entry and exit points, and it is the predominant term for centrifugal compressors. In both axial turbines and compressors the assumption is made that the fluid inlet radius is equal to the outlet radius (not always the case in reality), and thus this term is zero.

The internal expansion term, $(V_{r_2}^2 - V_{r_1}^2)$, results, in reaction turbines, from an increase in relative velocity as the fluid passes through the rotor. In (ideal) impulse turbines this term is zero, as $V_{r_2} = V_{r_1}$. In actual impulse wheels $V_{r_2} < V_{r_1}$ owing to friction, and the complete term represents a slight loss.

Although the total head equation is not used frequently in analysis of turbine problems, it serves a useful function in illustrating the sources of work available in the turbine. It also shows that an appreciable amount of energy leaves the rotor, as V_2 and V_{r_2} represent the kinetic energy leaving the stage. Minimizing this effect is one of the objectives during turbine design; it also depends on turbine operating conditions.

This completes the general discussion of turbomachinery; further considerations are closely tied to the various descriptive categories mentioned in the previous section, and the type of blading utilized. Thus, the remainder of this chapter is devoted to turbine nozzle and rotor descriptions.

4.3 Turbine Nozzles

In Section 3.8 the ideal nozzle process was described as obtained from the general energy equation, which gave:

$$V_2 = \sqrt{2g_c J(h_1 - h_2)} \tag{5}$$

Equation (5) is the expression for the process in an ideal nozzle or for an isentropic process. This is shown as $(h_a - h_b)$ in Figure 4.6. In an actual nozzle, there are always some friction losses, though they may be kept to a minimum by good design and careful construction. The friction on the nozzle walls results in reconversion of some of the velocity gain back into thermal energy, so that the ideal drop in enthalpy is not realized in the actual nozzle. Since the change in enthalpy is equivalent to the change in kinetic energy, somewhat less kinetic energy is obtained from a real nozzle than is indicated by equation (5) for the ideal nozzle. If h_c is the exit enthalpy (see Figure 4.6) from the real nozzle, then the efficiency of the nozzle may be defined as:

$$\eta_n = \frac{\text{actual kinetic energy obtained}}{\text{kinetic energy from an ideal nozzle}} = \frac{h_a - h_c}{h_a - h_b} \tag{6}$$

Since in steady flow the outlet pressure is fixed by the surroundings of the nozzle exit, this pressure is the same for both the ideal and actual nozzles. Figure 4.6 shows the process lines on $h\text{-}s$ coordinates.

Point a is established by the entrance pressure and temperature, point b by assuming an isentropic enthalpy drop to the exit pressure, and point c by multiplying $h_a - h_b$ by the efficiency, solving for h_c, and plotting the result on the exit pressure line. By including the efficiency term as defined by equation (6) in equation (5), the velocity for the actual nozzle can be obtained:

$$V_{\text{actual}} = \sqrt{2g_c J \eta_n (h_a - h_b)}$$

or:

$$V_{\text{actual}} = \sqrt{\eta_n}\sqrt{2g_c J(h_a - h_b)}$$
$$= C_N \sqrt{2g_c J(h_a - h_b)} \tag{7}$$

where $C_N = \sqrt{\eta_n}$ is defined as the velocity coefficient of the nozzle. For a well-designed nozzle, C_N may be taken as 0.97 or 0.98. Equation (7) is valid for any nozzle, within the limits of accuracy of C_N, which has only slight variations for well-designed nozzles operating under designed conditions.

For gas flows equation (7) can be expressed in tempera-

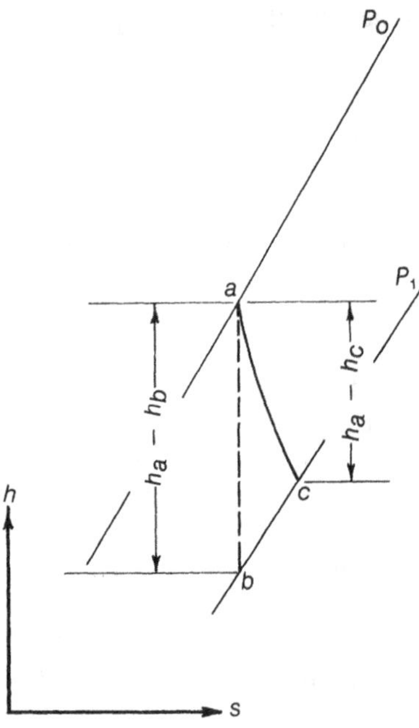

Figure 4.6 An *h-s* diagram for a real (solid line) and an ideal (broken line) nozzle.

ture or pressure ratios by the use of $h_0 = c_p T_0$ and the nomenclature shown in Figure 4.8:

$$V = C_N \sqrt{2g_c J(h_0 - h_1)}$$
$$= C_N \sqrt{2g_c J c_p (T_0 - T_1)}$$
$$= C_N \sqrt{2g_c J c_p T_0 \left(1 - \frac{T_1}{T_0}\right)}$$
$$= C_N \sqrt{2g_c J c_p T_0 \left[1 - \left(\frac{P_1}{P_0}\right)^{(k-1)/k}\right]}$$

Equation (17) of Chapter 3 was used to obtain the last result.

The shape of the nozzle is governed to an extent by the desired ratio of the outlet pressure, P_1, to the inlet pressure, P_0. A critical pressure ratio P_1/P_0 exists for all gases, which is a function of their specific heats. For dry steam the critical pressure ratio may be taken as 0.58. For air this value is 0.528. When P_1/P_0 is greater than the critical value, a converging nozzle is used (see Figure 4.7). For steam in this type nozzle, if P_0 is held constant and P_1 is set at various values from P_0 to 0.58 P_0, it will be found that the exit velocity will increase as the values of P_1 decrease. However, when P_1 drops to 0.58 P_0, the exit velocity reaches the local velocity of sound, and if P_1 is lowered further, no increase in the exit velocity results. This velocity occurs at the smallest cross section of the nozzle, or the *throat*. From equation (2) the mass flow is:

$$\dot{m} = \frac{A_T V_T}{v_T} \qquad (8)$$

where A_T = area of the throat, ft² (m²)
V_T = velocity at throat (here = V_1), fps (m/sec)
v_T = specific volume at throat (here = v_1), ft³/lb (m³/kg)

Consequently, it is seen that the convergent nozzle produces both the maximum exit velocity and the maximum mass flow at the critical pressure ratio, and that further lowering of the exit pressure does not alter this flow.

By adding a divergent portion to the nozzle when, $P_1/P_0 < 0.58$, further expansion of the steam can be accomplished with a further gain in velocity, and therefore an increase in kinetic energy is obtainable. A convergent-divergent nozzle is shown in Figure 4.8. In this nozzle when P_1/P_0 is less than 0.58, the pressure at the throat is always 0.58 P_0 regardless of the value of P_1. Therefore, the mass flow is limited by equation (8), and cannot increase beyond this. However, if the divergent section is properly designed, the velocity continues to increase as the pressure decreases from P_T at the throat to P_1 at the exit. The exit velocity is still determined by equation (7) as before.

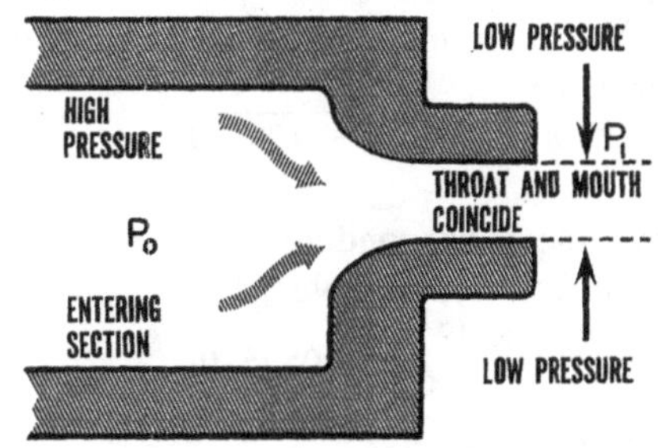

Figure 4.7 Converging nozzle

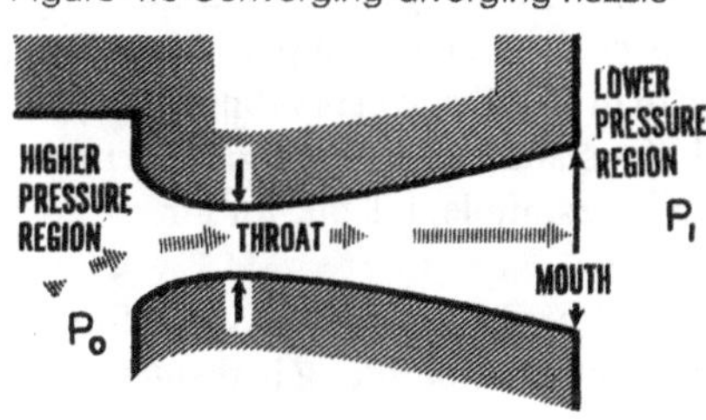

Figure 4.8 Converging-diverging nozzle

The following example will serve to illustrate the above discussion.

Example 4.1: A convergent-divergent nozzle expands steam from 350 psia and 700°F to 80 psia. $C_N = .967$. The diameter of the throat is 1 inch. Find:

(a) Outlet velocity

(b) Throat velocity

(c) The mass flow, lb/sec

(d) The kinetic energy per pound of steam.

Solution:

(a) $h_0 = 1365.5$ BTU/lb (from steam tables, for $p_0 = 350$ psia and $T_0 = 700$)

$h_1 = 1211.5$ BTU/lb (from steam tables, isentropic drop from h_0)

$$V_1 = 0.967 \ \sqrt{2g_c J(h_0 - h_1)}$$
$$= 0.967 \times 223.8 \ \sqrt{1365.5 - 1211.5}$$
$$= 2686 \text{ fps}$$

(b) $p_t = 0.58 \ p_0 = 0.58 \times 350 = 203$ psia

$h_0 = 1365.5$ BTU/lb

$h_t = 1292.0$ BTU/lb (from steam tables, isentropic drop from h_0)

$$V_t = 0.967 \times 223.8 \ \sqrt{1365.5 - 1292.0}$$
$$= 1855 \text{ fps}$$

(c) $v_t =$ specific volume corresponding to h_t
$= 2.829$ ft³/lb (from steam tables)

$$A_t = \frac{\pi d^2}{4 \times 144} = \frac{\pi \times 1^2}{4 \times 144} = .00545 \text{ ft}^2$$

$$\dot{m} = \frac{A_t V_t}{v_t} = \frac{0.00545 \times 1856}{2.829}$$

$$= 3.58 \text{ lb/sec}$$

(d) $KE = \dfrac{V^2}{2g_c J}$

$$= \frac{(2686)^2}{2 \times 32.2 \times 778} = 144.0 \text{ BTU/lb}$$

Check:

$$KE = C_N^2 (h_0 - h_1)$$
$$= (0.967)^2 (1365.5 - 1211.5)$$
$$= 144.0 \text{ BTU/lb}$$

In part (c), the specific volume at the throat used is that corresponding to conditions at the throat resulting from an isentropic drop, rather than for the actual conditions. That is, the throat conditions corresponding to point d in Figure 4.9 were used, rather than those corresponding to point c. This simplifies calculations and causes negligible error.

Example 4.2: Same as example 4.1 but for air.

Solution:

(a) The velocity can be expressed directly as a function of the pressure ratio by means of perfect gas results:

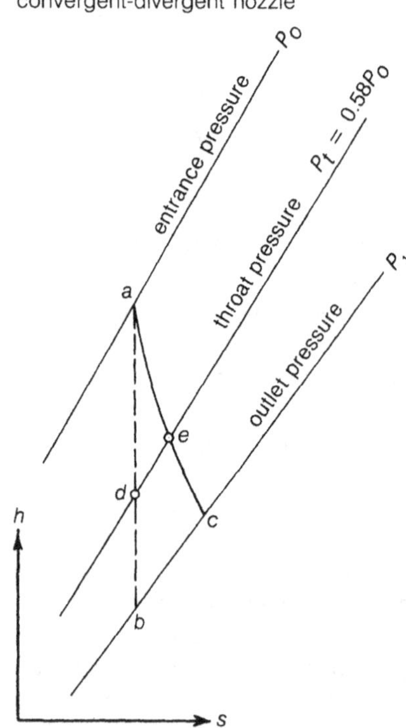

Figure 4.9 An h-s diagram for a real (solid line) and an ideal (broken line) convergent-divergent nozzle

$$V_1 = .967 \sqrt{2g_c Jc_p T_0 \left[1 - \left(\frac{P_1}{P_0}\right)^{(k-1)/k}\right]}$$

$$= .967 \times 223.8 \sqrt{.24(1160)\left[1 - \left(\frac{80}{350}\right)^{.286}\right]}$$

$$= 2119 \text{ ft/sec}$$

(b) At the throat the pressure ratio is .528, $P_t = 184.8$ psi. Thus the velocity by the above equation is (subscript t refers to nozzle throat):

$$V_t = 1475 \text{ fps}$$

(c) To calculate the mass flow, density needs to be determined first. From the equation of state,

$$\rho_t = \frac{P_t}{RT_t} \quad T_t = T_0\left(\frac{P_t}{P_0}\right)^{(k-1)/k} = 1160(.528)^{.286} = 966°F$$

$$\rho_t = \frac{184.8 \times 144}{53.3 \times 966} = .517 \text{ lb/ft}^3$$

And the mass flow rate becomes:

$$\dot{m} = \rho_t V_t A_t = .517 \times 1475 \times .00545 = 4.16 \text{ lb/sec}$$

$$\text{(d)} \quad KE = \frac{V_1^2}{2g_c J} = \frac{(2119)^2}{2 \times 32.2 \times 778} = 89.6 \text{ BTU/lb}$$

Steam supplied to the turbine flows from a line through controlling valves into a steam chest. It passes from the steam chest into groups of nozzles assembled in a circular array to direct it into the first row of moving blades or buckets. This first group of nozzles, in a ring or block, may be either convergent or convergent-divergent, depending on the designed pressure drop. For convergent-divergent nozzles, the blocks or rings are generally made from solid blocks, drilled and reamed to the required shape. For converging nozzles, the vane type construction is generally preferred. This arrangement consists of successive vanes spaced around the periphery, shaped so that the space between them gives the desired nozzle cross section. A vane type block may be either built up from individual parts or cast as a unit. Figure 4.10 illustrates the construction of these two types.

In impulse turbines, after leaving the first wheel, the steam passes through several more combinations of nozzles and wheel blades before leaving the turbine. These nozzles are of the vane type, mounted in diaphragms as shown in Figure 4.11. The buckets only serve to change the direction of the steam, and usually do not function as nozzles. The impulse turbine will be described in detail in the following pages.

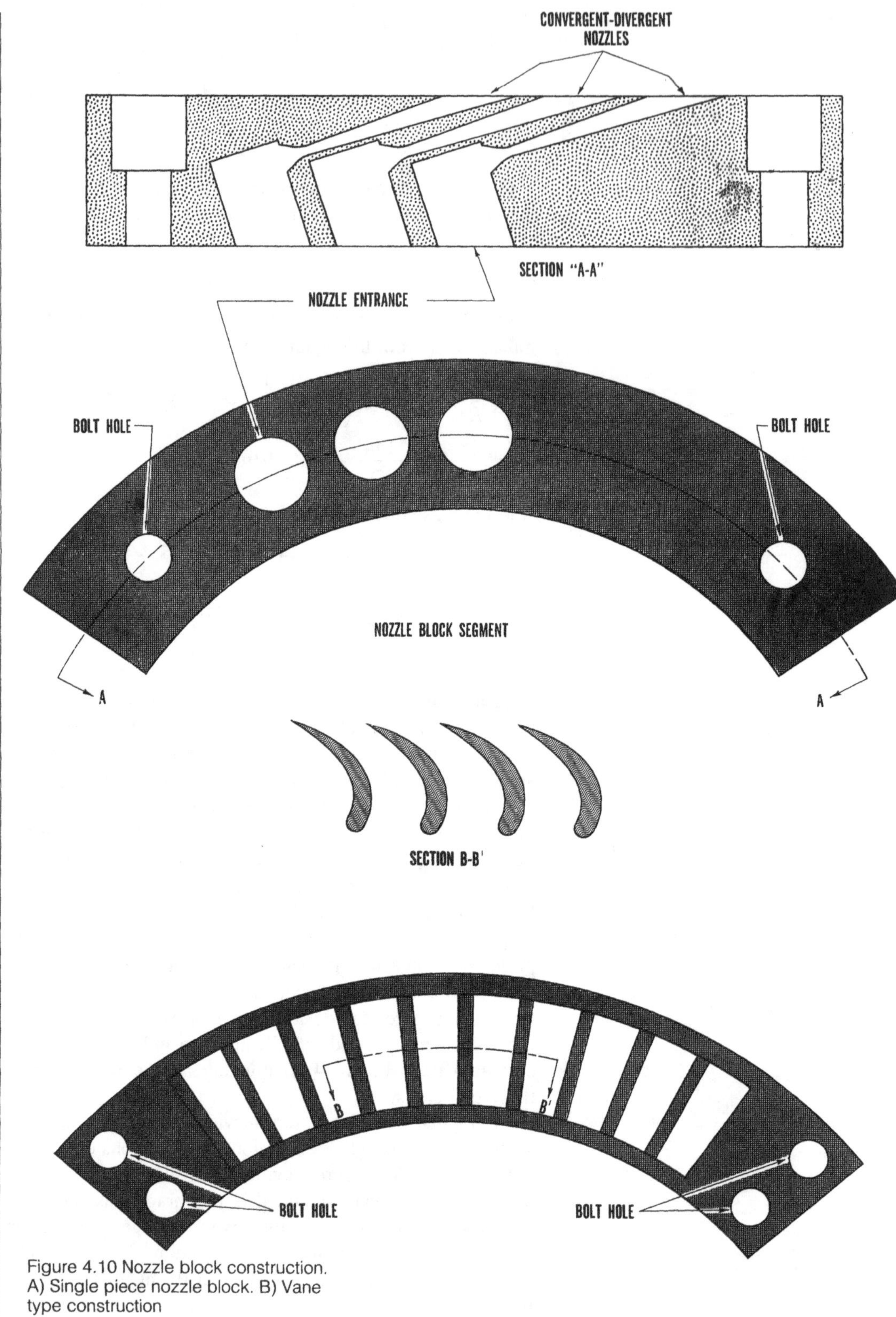

Figure 4.10 Nozzle block construction.
A) Single piece nozzle block. B) Vane
type construction

Figure 4.11 Impulse type nozzle
diaphragm

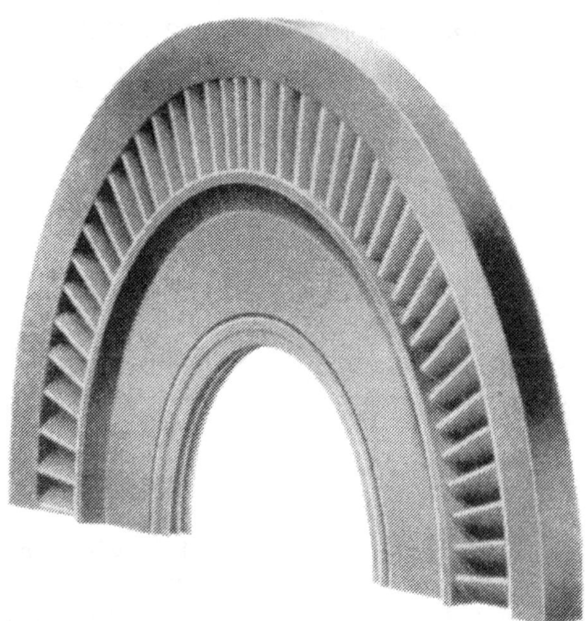

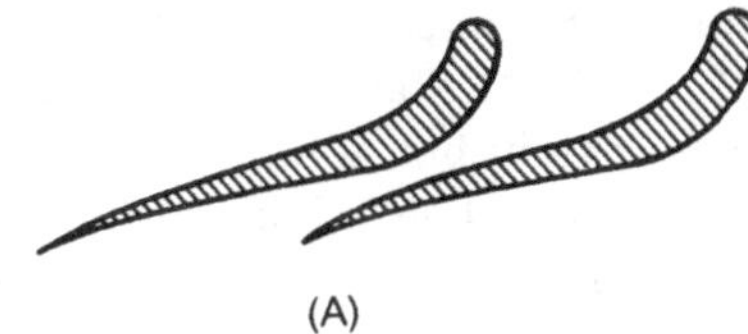

(A)

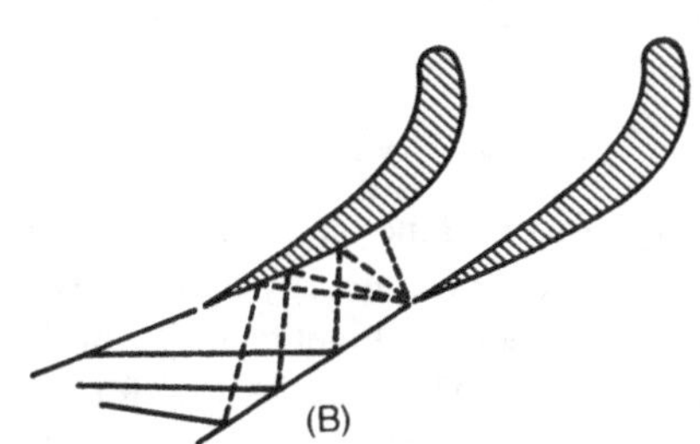

(B)

Figure 4.12 Gas turbine nozzle arrangement. A) Convergent-divergent nozzle. B) Nozzle for pressure ratio less than critical

In gas turbines the nozzles are invariably of the vane type and operate at pressures less than the critical pressure ratio. Convergent-divergent nozzles (Figure 4.12a) are not used for two reasons. First, at pressure ratios other than the design value, shocks inside the nozzle passage lead to inefficient nozzle operation. Second, such nozzles produce high exit velocity V_1, which leads to high relative velocity V_r; consequently shock waves will appear in the moving blade passages, causing additional losses. In the usual arrangement, convergent nozzles (Figure 4.12b) are used with pressure ratios that may give slightly supersonic flow due to additional expansion beyond the minimum throat area. If the exit velocity V_1 and the relative velocity V_r are not high enough to cause rotor shocks, losses are minimal. To minimize the outside blade passage expansion and flow separation from the blades, the back of the blade after the throat is straight.

4.4 Impulse Turbine

Impulse staging, the oldest form of extracting work from the fluid, is common in all high pressure steam turbines. It is not used in low pressure steam turbine stages or in gas turbines, for reasons which become obvious below.

To understand fully the processes that occur in impulse turbine blading, the ideal case will be considered first, where only a change in fluid stream occurs. This ideal system is made up of a nozzle and bucket such as is shown in Figure 4.13. The stream leaves the nozzle with velocity V_1, strikes the blade, and has a velocity of V_2 after leaving the blade. Then, from Chapter 3, the force on the stream is

Figure 4.13 Ideal zero-angle impulse stage

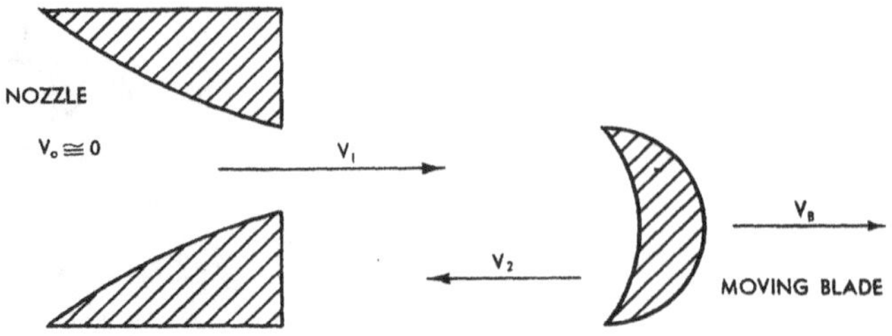

$\dot{m}(V_2 - V_1)/g_c$. This force is exerted by the moving blade, and therefore the stream exerts an equal and opposite force, R, on the blade. It is this continuous force acting on the blade which causes it to move, and to produce work and power. Its expression is

$$R = -F = \frac{\dot{m}}{g_c}(V_1 - V_2) \tag{9}$$

When the blade moves with velocity, V_b, the power is equal to RV_b.

Note that if V_b equals V_1, the stream cannot overtake the blade, and therefore it is not deflected. Also, if the blade is stationary, V_b is zero, and the power is therefore zero. The blade velocity for power production, then, lies between 0 and V_1.

At any point on the blade, the stream strikes the blade with a relative velocity $V_1 - V_b$. The blade reverses the direction of the stream relative to it; but, if we neglect friction, it does not change the magnitude of the relative velocity. *It is this fact which characterizes the blade as an impulse blade.* Therefore, the velocity of the stream leaving the blade (relative to the blade) is $-(V_1 - V_b)$. But the absolute velocity is the sum of the relative leaving velocity and the velocity of the blade, or

$$V_2 = -(V_1 - V_b) + V_b = -V_1 + 2V_b \tag{10}$$

The maximum quantity of work done on the blade in Figure 4.13 occurs when all of the kinetic energy in the stream is converted to work, that is, when the kinetic energy of the stream leaving the blade is zero. This implies that for this condition the absolute velocity of the stream leaving the blade must be zero, which occurs only when $V_2 = 0$.

From equation (10), when $V_2 = 0$,

$$V_1 = 2V_b$$

or:

$$V_b = \frac{1}{2}V_1 \tag{11}$$

Figure 4.14 Force and work for impulse stage

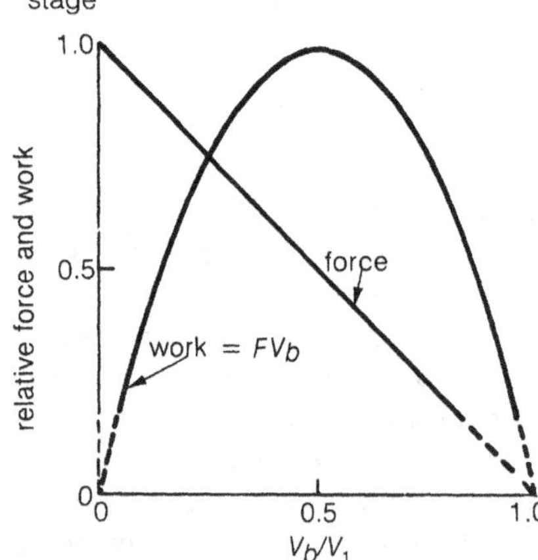

Therefore, in the zero-angle impulse stage, the maximum work occurs when the blade speed is one-half the stream speed. This result is shown also in Figure 4.14. When $V_b =$ 0, the blade is stationary, and although maximum force is exerted on the blade, no work is done. As the blade picks up speed, force diminishes but the work is increased. As the velocity ratio is increased to .5, force decreases to half of the initial value, the fluid leaves the blade at zero velocity, and maximum work is done on the blade. Further increase in V_b/V_1 to unity leads to decreasing the force and the work to zero.

In practice this ideal turbine is approached by the tangential flow turbine, used to power auxiliary machinery. It will be discussed in more detail later.

Most impulse turbines have an entrance arrangement similar to that shown in Figure 4.15. The velocity relationships are depicted in the velocity vector diagram shown in Figure 4.16. In this figure, α is the nozzle angle and V_1 is the velocity of the jet leaving the nozzle. The blade velocity, or wheel speed, is V_b, while V_{r_1} is the relative velocity entering the blade, obtained by the vectorial subtraction $V_1 - V_b$. Angle β_1 is the inlet angle; it is also fixed by the above vectorial subtraction. The flow is deflected by the blade through an angle $(\beta_1 + \beta_2)$. Angle β_2 is the exit angle and is fixed

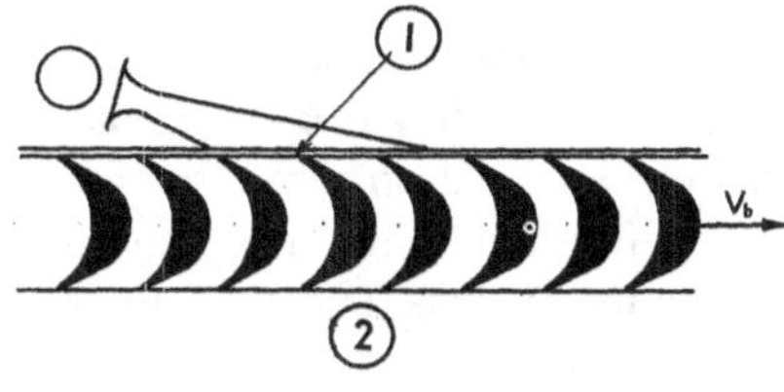

Figure 4.15 Simple impulse stage

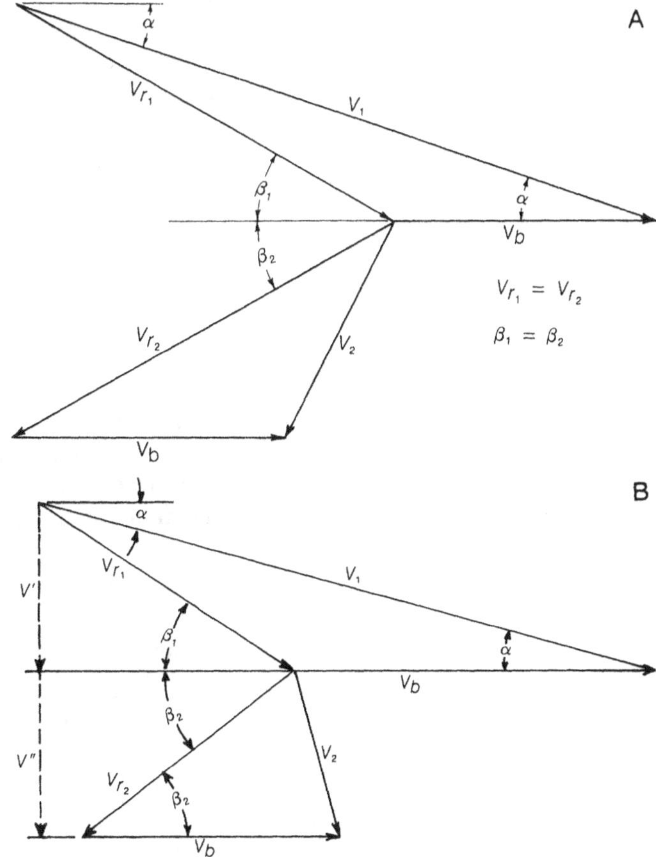

Figure 4.16 Impulse stage velocity diagram. A) Velocity diagram for frictionless stage. B) Velocity diagram for a real stage

by the blade design. In the impulse turbine, if the blade were frictionless, the relative velocity at the blade exit would be identical in magnitude to the relative entrance velocity, V_{r_1}. The effect of friction is expressed by the following relation:

$$V_{r_2} = C_b V_{r_1} \qquad (12)$$

where the constant, C_b, is the blade velocity coefficient. This coefficient is somewhat smaller and more variable than the similar (nozzle) coefficient C_N discussed previously, but a fair average value appears to be in the range of 0.8 to 0.9. Having obtained V_{r_2} from a knowledge of the angle β_2 and equation (12), the absolute exit velocity of the stream, V_2, can be found by adding the velocity of the stream relative to the blade, V_{r_2}, to the velocity of the blade, V_b.

The dotted vectors, V' and V'', represent the velocity of axial flow through the turbine. These two axial flow components are practically equal in a pure impulse stage, that is, a stage in which there is no pressure drop across the moving blades. If the stage were a frictionless impulse stage, there would be no friction loss in the buckets, and V_{r_2} would equal V_{r_1}. Also, V' would exactly equal V'', and β_1 would be identical to β_2.

The exit velocity of the flow, V_2, represents a loss to the stage, since it represents kinetic energy not utilized in the stage. This loss can be minimized by choosing an optimum turbine operating speed and then operating the turbine, conditions permitting, at or near this speed. The optimum blade speed can be obtained by defining what is known as blade efficiency:

$$\eta_b = \frac{\text{work done per pound of fluid}}{\text{energy input to blade per pound of fluid}}$$

$$= \frac{V_{b_1} V_{u_1} - V_{b_2} V_{u_2}}{V_1^2/2}$$

using equation (3) in the numerator and the total kinetic energy of the fluid in the denominator. Since $V_{b_1} = V_{b_2} = V_b$, the blade speed does not vary, and from the velocity diagram in Figure 4.16,

$$V_{u_1} = V_{r_1} \cos \beta_1 + V_b$$
$$-V_{u_2} = V_{r_2} \cos \beta_2 - V_b$$

yields:

$$\eta_b = \frac{2V_b(V_{r_1} \cos \beta_1 + V_{r_2} \cos \beta_2)}{V_1^2}$$

$$= \frac{2V_b V_{r_1} \cos \beta_1}{V_1^2}(1 + K_1 K_2) \qquad (13)$$

where:

$$K_1 = \frac{V_{r_2}}{V_{r_1}} \text{ accounts for the friction losses on the blade,}$$

and

$$K_2 = \frac{\cos \beta_2}{\cos \beta_1} \approx 1 \text{ for an impulse stage.}$$

Since $V_{r_1} \cos \beta_1 = V_1 \cos \alpha - V_b$, substitution into equation (13) yields:

$$\eta_b = \frac{2V_b}{V_1^2}(V_1 \cos \alpha - V_b)(1 + K_1 K_2)$$

$$= \frac{2V_b}{V_1}\left(\cos \alpha - \frac{V_b}{V_1}\right)(1 + K_1 K_2) \qquad \textbf{(14)}$$

If equation (14) is maximized with respect to V_b/V_1 (by plotting V_b/V_1 versus η_b, or by differentiation), for maximum η_b:

$$\frac{V_b}{V_1} = \frac{\cos \alpha}{2} \qquad \textbf{(15)}$$

and the maximum blade efficiency is:

$$\eta_{b_{max}} = \frac{\cos^2 \alpha}{2}(1 + K_1 K_2) \qquad \textbf{(16)}$$

$$\approx \cos^2 \alpha, \text{ since } K_1 \text{ and } K_2 \text{ are both near unity.}$$

Note the similarity of equation (15) to the result obtained as equation (11), for an ideal zero-angle impulse stage. Since α is seldom greater than 15°, $\cos \alpha$ is larger than or equal to .98, and the results are almost identical. Even at 20° the speed ratio turns out to be .47. Thus, the practical result is that the impulse turbine optimum speed ratio is about .5. It should also be noted that this result is independent of friction loss, but the blade efficiency is affected by it.

The velocities generated in typical high pressure turbines are so large that it is practically impossible to utilize them efficiently in a simple one-stage turbine, due to physical stress limitations on blade speed, high friction losses, and high exit losses. To overcome these difficulties, two methods of staging or compounding are used.

The first method, called velocity compounded staging, or Curtis staging, expands the fluid (steam) initially to a high velocity; but the kinetic energy from the fluid is absorbed in two or more rows of rotating blades with rows of stationary blades between each pair of moving blades. The stationary blading serves only to redirect the flow in the proper direction for entry into the next moving row. In this arrangement considerable friction losses occur, and the overall efficiency may be low. The second method is called pressure staging, pressure compounded staging, or Rateau staging. Here the large pressure drop (and high velocity) is broken up into several (two or more) parts by utilizing a series of sets of nozzles and buckets. After each nozzle a bucket wheel, running at moderate speed, utilizes efficiently a more moderate flow speed generated by the smaller pressure drops in individual stages. As velocities are lowered, so are the frictional losses. However, a larger and more complicated turbine arrangement offsets some of the gains in efficiency.

4.5 Velocity Compounded (Curtis) Staging

Consider the idealized zero angle stage in Figure 4.17. The working fluid issues from the nozzle with a velocity V_1. It approaches the first moving blade with a relative velocity $V_1 - V_b$. It is reversed, becoming $V_b - V_1$, but its absolute velocity is $V_b + V_b - V_1$ or $2V_b - V_1$. It strikes a fixed blade, which again reverses its direction but does not alter its magnitude. Its velocity is now $V_1 - 2V_b$. It approaches the second moving blade, whose speed is V_b, with a relative velocity of $(V_1 - 2V_b) - V_b$, or $V_1 - 3V_b$. It is reversed again, becoming $3V_b - V_1$ relative to the blade and $V_b + 3V_b - V_1$, or $4V_b - V_1$, in absolute magnitude. To minimize this velocity, and thus the leaving kinetic energy, we set this velocity equal to zero:

$$4V_b - V_1 = 0$$

or:

$$V_b = \frac{1}{4} V_1 \qquad (17)$$

Equation (17) shows that, for minimum leaving loss, the wheel speed is one-fourth the entering fluid speed.

It may be shown that for an ideal stage with a finite

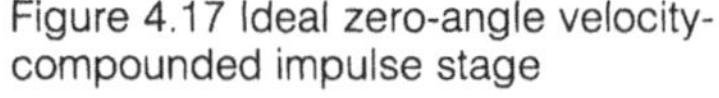

Figure 4.17 Ideal zero-angle velocity-compounded impulse stage

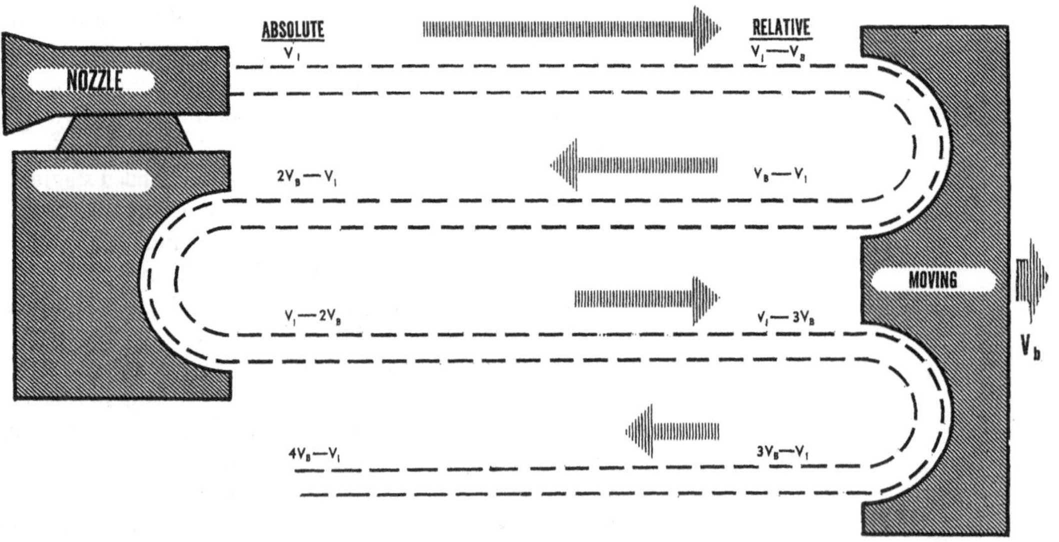

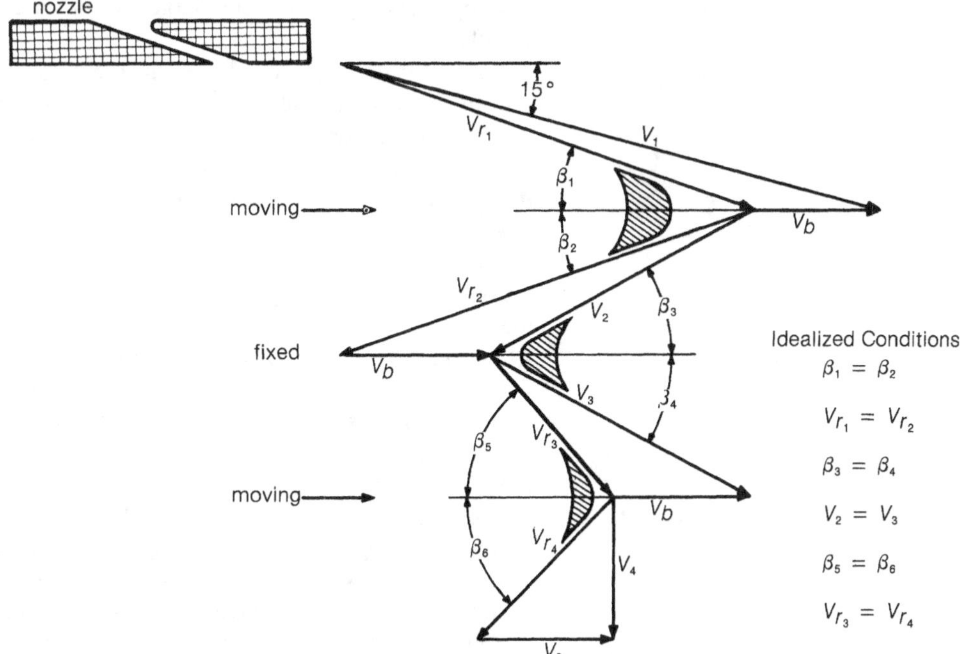

Figure 4.18 Frictionless velocity-compounded impulse stage

nozzle angle, the blade speed flow relation for minimum leaving loss is given by:

$$V_b = \frac{1}{4} V_1 \cos \alpha \qquad (18)$$

an expression similar to equation (15).

Figure 4.18 is a velocity diagram of an idealized finite angle velocity compounded stage, operating with minimum leaving loss. It is realized that in actual practice the idealized conditions are not met, but they are fairly closely approximated.

Equation (18) shows that, for a given wheel speed, a velocity compounded stage can utilize a flow velocity of twice that of the single row impulse stage, with a corresponding enthalpy drop four times as great as for a single row impulse stage. Therefore, such a stage would appear to be the energy equivalent of approximately four impulse stages. Unfortunately, the losses of such a stage are comparatively high, and so its stage efficiency is less than that of either of the two types previously described. However, it is quite advantageous for use as the first stage in a turbine for reasons discussed below.

It should be pointed out that up to this point the discussion has been a general one, since no particular working fluid has been introduced. Since the blade staging arrangements discussed in this and the next section are used primarily in steam turbines, the following examples pertain mainly to steam turbines. However, the general principles developed also apply to recent large gas turbines that utilize impulse stages.

It is highly desirable to decrease the pressure and temperature of the working fluid before its entry into the turbine casing proper. Turbine casings are made in the form of a cylinder, split and flanged into two halves along the axis of the cylinder. For very high pressures in the cylinder, the cylinder material must be stronger and heavier, and is more expensive to manufacture. Also it is more difficult with high pressures, and at the high temperatures involved, to fasten the flanges so that leakage from the joint can be eliminated. In addition, the moving blading is subjected to high stresses due to the force of the steam on the blades and to the large centrifugal forces. If high-temperature steam were also present, conditions would be conducive to creep of the metal of the blading. Instead, the high temperatures and pressures are confined to the steam chest, which is a cavity cast integrally in the upper half of the casing only. A large enthalpy drop then occurs across the nozzles at the exit from the steam chest. Thus, the steam entering the turbine shell proper is greatly reduced in pressure and temperature. This is achieved by use of a velocity compounded wheel with partial (confined to the upper half) admission in the first stage. Conditions conducive to metal creep may still exist at full power; but since full power is utilized only a small percentage of the time, and since creep is a function of time as well as temperature and stress, an acceptably long operating life is achieved before blade renewal and other overhaul work is required.

It is quite possible to build three-row and even four-row velocity compounded stages; but high losses make such designs unpopular, and they are seldom employed in main propulsion turbines. However, the large enthalpy drop thus obtainable with relatively low wheel speeds makes these designs useful in spite of inefficiency for some auxiliary turbine applications. Examples will be given later.

4.6 Pressure Compounded Staging

Figure 4.19 shows the nozzle and blade arrangement with corresponding velocity diagrams for a two-stage pressure

Figure 4.19 Pressure (Rateau) staging

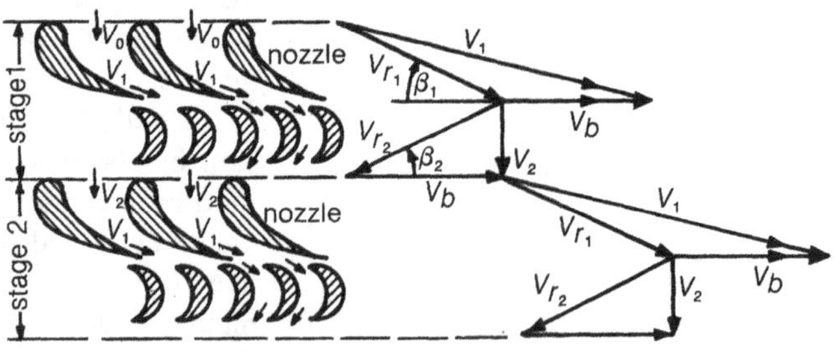

Figure 4.20 Expanding flow passage

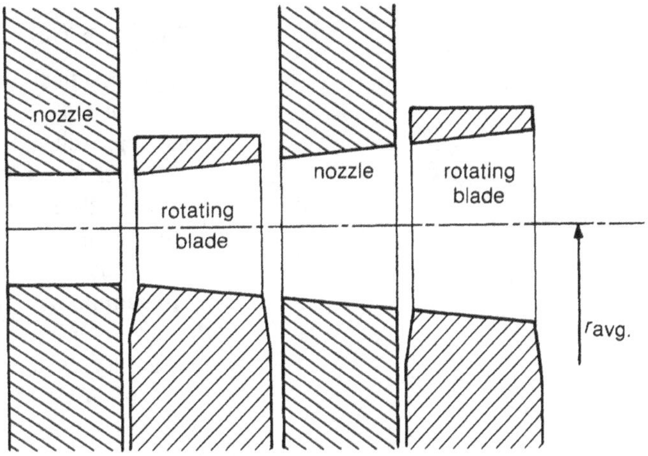

compounded (pressure-staged) impulse turbine. This arrangement (as well as the simple one-stage impulse turbine) is often referred to as Rateau staging. As is easily seen, each stage of a pressure staged impulse turbine is simply an impulse turbine rotor between two nozzle diaphragms. Each of those stages can be individually analyzed as a simple impulse stage.

The entrance velocity to each rotor is V_1 as the overall pressure-stage pressure drop is split into nearly equal amounts. The entrance velocity to consecutive nozzles is V_2, which is the absolute exit velocity from the preceding rotor. In an ideal impulse stage the relative bucket velocity V_r remains constant. In an actual turbine, due to friction, it slightly decreases from V_{r_1} to V_{r_2} with a very slight change from β_1 to β_2. The blade speed V_b remains constant even with the increase in blade height, which is necessary because of expansion through the consecutive blade rows. (See Figure 4.20.) Since the expansion and lengthening of the blade is accomplished in both outward and inward radial directions, the average radius r and blade speed V_b remain constant. The work done and the division of available energy into the proper portions for each stage are accomplished by shaping the nozzle areas.

A pressure compounded stage has three advantages over a simple impulse turbine. First, the friction losses are reduced, since the velocities are smaller than in a single impulse stage. Second, the friction losses, resulting in incomplete expansion, lead to a slight increase in the individual stage enthalpy, which is available for work in the subsequent stages. (The measure of reduction of the ideal expansion and the additional energy available in steam turbines is commonly referred to as reheat factor.) Third, the kinetic energy leaving an individual stage is not totally lost, as in a simple stage, but can be utilized in well-designed turbines with close wheel-nozzle spacing in the following stages.

4.7 Impulse Turbine Losses

In a practical impulse turbine there are five principal causes for reduction of energy available to the turbine. Various approaches to minimize these losses are discussed in Chapter 5, on steam turbines. Since impulse staging is found mostly in steam turbines, the losses described below apply mainly to steam turbines, with the exception of the packing loss, which occurs in stationary rows of all turbines.

To summarize, the shaft work of a given stage is less than the blade work produced by the stage because of additional internal losses, of which the principal ones are:

(a) *Windage loss:* The disc carrying the blades is rotating in an atmosphere of steam. Friction of the steam on the nonworking portion of the disc sets up turbulence, which consumes power.

(b) *Suction loss:* In many impulse stages the nozzles do not completely fill the entire wheel periphery. Such stages are known as partial admission stages. They are used in high pressure stages where full admission would dictate impractically short blades, due to a relatively small required flow area. When these partial admission stages are used, a pumping action takes place in the blades during the inactive portion of the arc, tending to pump steam from the upstream side of the disc to the exit side. This loss is proportional to the size of the inactive arc.

(c) *Nozzle end loss:* In a partial admission stage when the blade reaches the beginning of the admission arc, the relatively slow-moving steam therein must be accelerated again by the nozzle jet. This loss may amount to about 0.5% of the stage work.

(d) *Packing loss:* It is possible for a small amount of steam to leak from one stage to the next without passing through the nozzle. This steam passes through the diaphragm packing, which will be described later. This portion of the steam does no work.

(e) *Moisture loss:* At some point in the turbine, when steam is the working fluid, it reaches its saturation pressure and continues to expand with the accompanying formation of moisture droplets. Energy from the vapor is required to accelerate these droplets, which travel at a fraction of the steam speed, depending on the size of the droplets, the location where the droplets originate (onset of condensation), interblade spacing, and the density of the steam. Because of this slower speed, the relative velocity of the droplets to the blade actually causes a braking effect on it, resulting in an additional loss of power. Moisture effect is shown in Figure 4.21. Note that V_{r_1} strikes the face of the blade with a component in the direction of blade motion, while V_{r_m} strikes the back of the blade, opposing the motion.

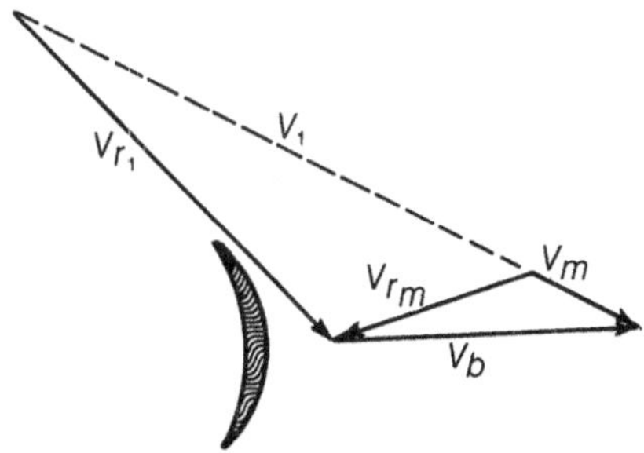

Figure 4.21 Velocity diagram with moisture effect

The total of these losses amounts to about 3% to 10% of the bucket work. Therefore the stage efficiency, defined as shaft work over energy available to the stage, is about 90% to 97% of the nozzle-bucket efficiency.

Figure 4.22 shows an exaggerated h-s diagram for the complete stage. Note from it the following relationships:

Available energy to stage:

$$h_0 - h_1$$

Entering steam velocity:

$$V_1 = C_N \sqrt{2g_c J(h_0 - h_1)}$$
$$= \sqrt{2g_c J(h_0 - h_1')}$$

Nozzle loss:

$$h_1' - h_1$$

Bucket loss:

$$h_2 - h_1' \left(= \frac{-V_{r_1}^2 + V_{r_2}^2}{2g_c J} \right)$$

Exit loss:

$$h_2' - h_2 \left(= \frac{V_2^2}{2g_c J} \right)$$

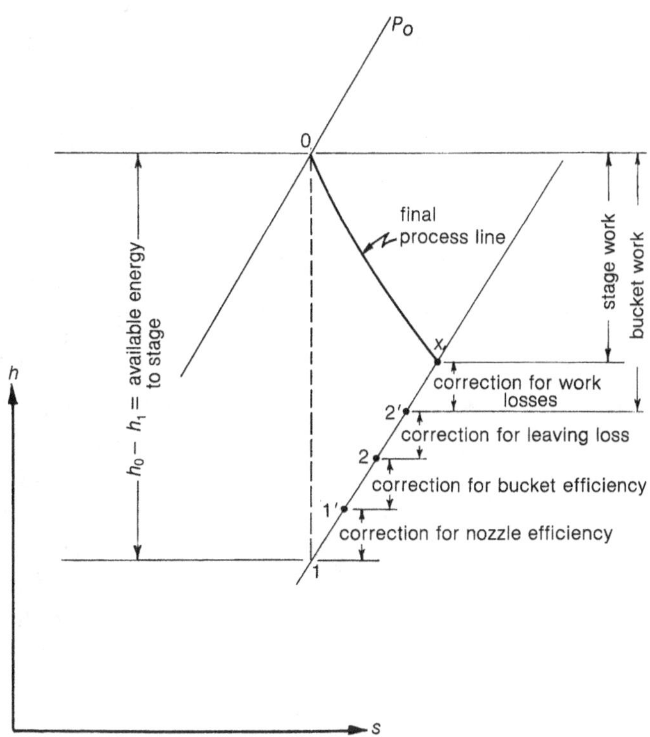

Figure 4.22 An h-s diagram for impulse stage

Nozzle-bucket efficiency:

$$\frac{h_0 - h_2'}{h_0 - h_1}$$

Bucket work:

$$h_2' - h_1 \left(= \frac{V_1^2 - V_{r_1}^2 + V_{r_2}^2 - V_2^2}{2g_c J} \right)$$

Stage work:

$$h_0 - h_x$$

Stage efficiency:

$$\frac{h_0 - h_x}{h_0 - h_1}$$

Example 4.3: The second stage inlet conditions on the HP impulse turbine at 20 knots are 68.2 psia and 392°F. The exit conditions are 55 psia and 360°F. (a) Find the stage efficiency and the stage work per pound of steam. (b) The nozzle angle is 15°, the blade velocity at 20 knots is 400 fps, the blade exit angle is 24°, $C_N = 0.975$, and $C_b = 0.85$. Construct the velocity diagram and find the bucket work. (c) Find the exit loss and the work losses. (d) Find the nozzle loss and bucket loss, and check whether the sum of all losses plus stage work equal the available energy to the stage.

Solution:

(a) $h_0 = 1228.3$ (from steam tables for entrance conditions)

$s_0 = s_1 = 1.6935$ (from steam tables for entrance conditions)

$h_1 = 1209.1$ (isentropic drop from 68.2 psia to 55 psia)

$h_x = 1214.3$ (from steam tables for exit conditions)

Stage efficiency $= \dfrac{h_0 - h_x}{h_0 - h_1} = \dfrac{14.0}{19.2} = 72.9\%$

Isentropic or ideal stage work $= 19.2$ BTU/lb

Actual stage work $= h_0 - h_x = 14.0$ BTU/lb

(b) $V_1 = C_N \sqrt{2g_c J(h_0 - h_1)}$

$\quad = 0.975 \sqrt{50100(19.2)}$

$\quad = 0.975 \times 981 = 956$ fps

$V_{r_1} = \sqrt{V_1^2 + V_b^2 - 2V_1 V_b \cos 15°}$

(See diagram)

$\quad = \sqrt{(957)^2 + (400)^2 - (2)(957)(400) \cos 15°}$

$\quad = 580$ fps

$V_{r_2} = C_b V_{r_1} = 0.85 \times 580 = 493$ fps

$V_2 = \sqrt{V_{r_2}^2 + V_b^2 - 2V_{r_2} V_b \cos 24°}$

$$= \sqrt{(493)^2 + (400)^2 - 2(493)(400)\cos 25°}$$
(See diagram)
$$= 207 \text{ fps}$$

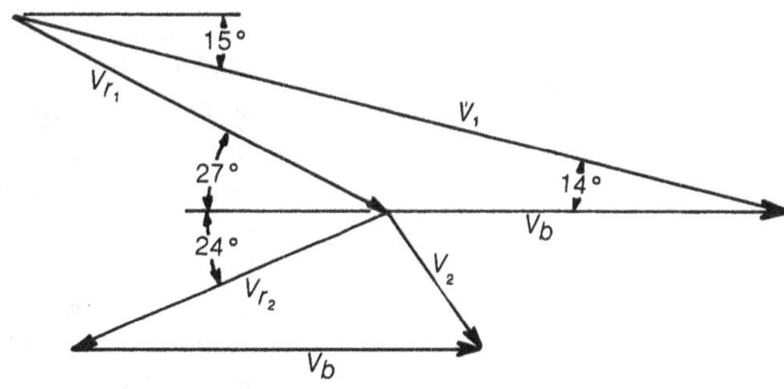

$$\text{Bucket work} = \frac{V_1^2 - V_{r_1}^2 + V_{r_2}^2 - V_2^2}{2g_cJ}$$
$$= \frac{(957)^2 - (580)^2 + (493)^2 - (207)^2}{2g_cJ}$$
$$= 15.56 \text{ BTU/lb}$$

(c) $\text{Exit loss} = \dfrac{V_2^2}{2g_cJ} = \dfrac{(207)^2}{64.4 \times 778}$
$$= .86 \text{ BTU/lb}$$

$\text{Work loss} = \text{Bucket work} - \text{stage work}$
$$= \text{Bucket work} - (h_0 - h_x)$$
$$= 15.56 - 14.0 = 1.56 \text{ BTU/lb}$$

(d) $\text{Nozzle efficiency} = C_N^2 = (0.975)^2 = 0.951$
$\text{Nozzle loss factor} = 1 - 0.951 = 0.49$
$\text{Nozzle loss} = (.049)(h_0 - h_1) = (.049)(19.2)$
$$= 0.94 \text{ BTU/lb}$$

Blade loss:

$$\text{Energy of relative velocity entering} = \frac{V_{r_1}^2}{2g_cJ}$$

$$\text{Energy of relative velocity leaving} = \frac{V_{r_2}^2}{2g_cJ}$$

$\text{Blade loss} = \dfrac{(580)^2 - (493)^2}{2g_cJ} = 1.86 \text{ BTU/lb}$

Check: From computation—

$\text{Work} + \text{work loss} + \text{leaving loss}$
$\qquad + \text{nozzle loss} + \text{blade loss}$
$$= 14.0 + 1.56 + 0.86 + 0.94 + 1.86$$
$$= 19.22 \text{ BTU/lb}$$

From h-s chart, or steam tables—
available energy to stage, $h_0 - h_1$
$$= 19.2 \text{ BTU/lb}$$

4.8 The Reaction Stage

From the foregoing description of the impulse stage, it is apparent that the process of conversion of thermal to kinetic energy takes place only in the nozzles, and that the sole function of the moving blades is to convert the kinetic energy into mechanical work. It is entirely feasible, however, to construct a stage in which the change from thermal to kinetic energy occurs in both the nozzles and the moving blades. This is done by shaping the moving blade passages so that they function as nozzles also. Such blades, in addition to changing the direction of the fluid relative to the blade, also operate to change the magnitude of its velocity relative to the blade. This scheme of operation characterizes the reaction stage. It is possible, in fact, to construct a turbine in which all of the thermal energy is changed to kinetic energy and then into mechanical energy in the moving blades alone. It is common practice, however, to so design the reaction stage that, in the normal power ranges of the turbine, about one-half of the thermal energy is converted into kinetic energy in the nozzles, or fixed blades, and one-half in the moving blades. Such a stage is called a 50% reaction stage.

Typical reaction turbine blading and the associated velocity diagram are shown in Figure 4.23. The velocity dia-

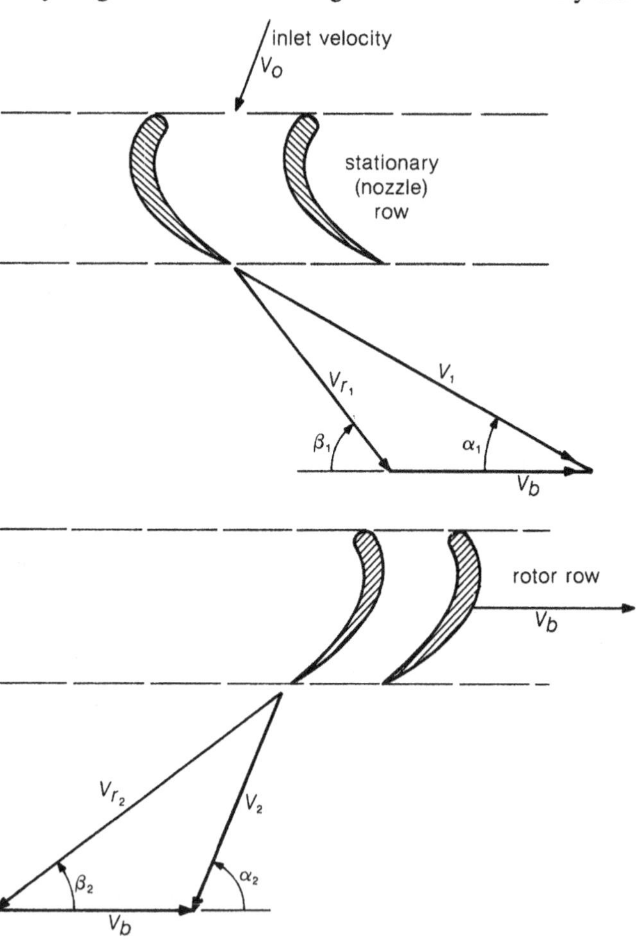

Figure 4.23 Reaction turbine velocity diagram

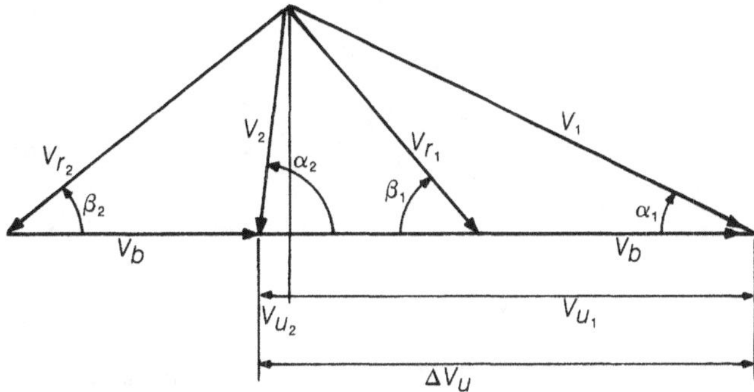

gram is very similar to that of an impulse stage shown in Figure 4.16. However, note that in this case $V_{r_2} > V_{r_1}$. It is convenient to combine the stator and rotor diagrams into one diagram as shown in Figure 4.24, since it permits a quick overview of the stage processes and simplifies calculations. Note that the stage work, $V_b \Delta V_u$, from equation (3) above,

$$P_T = \frac{\dot{m}}{g_c} V_b (V_{u_1} - V_{u_2}) = \frac{\dot{m}}{g_c} V_b \Delta V_u$$

is immediately available from the velocity diagram.

Now let the stage inlet enthalpy be h_0. Then nozzle and rotor exit enthalpies are h_1 and h_2, respectively. According to the energy equation (18), Chapter 3, without work and heat transfer in the nozzle blade rows, and omitting the conversion factors $g_c J$ for clarity, the enthalpy drops (thermal energy conversion) in the nozzle blade rows are:

$$\Delta h_n = h_0 - h_1 = \frac{1}{2}(V_1^2 - V_0^2)$$

$$= \frac{1}{2}(V_1^2 - V_2^2) \qquad (19)$$

since $V_0 \cong V_2$.

$$\text{rotor } \Delta h_r = h_1 - h_2 = \frac{1}{2}(V_{r_2}^2 - V_{r_1}^2) \qquad (20)$$

Stage reaction is defined as:

$$R = \frac{\Delta h_r}{\Delta h_r + \Delta h_s} = \frac{h_1 - h_2}{h_0 - h_2} = \frac{V_{r_2}^2 - V_{r_1}^2}{V_{r_2}^2 - V_{r_1}^2 + V_1^2 - V_2^2}$$

$$= \frac{1}{1 + \dfrac{V_1^2 - V_2^2}{V_{r_2}^2 - V_{r_1}^2}} \qquad (21)$$

Turbine stage specific work (work per pound) was given in Chapter 3 as:

$$w_2 = w_{\text{stage}} = h_0 - h_2$$

and by equation (3) above as:

$$w_{12} = w_{\text{stage}} = \frac{P_T}{m} = V_{b_1} V_{u_1} - V_{b_2} V_{u_2}$$

$$= V_b(V_{u_1} - V_{u_2}) \tag{22}$$

since in an axial turbine blade speed, V_b, can be assumed to remain constant through a stage. Combining these work expressions yields:

$$h_0 - h_2 = V_b(V_{u_1} + V_{u_2}) \tag{23}$$

where the sign on V_{u_2} has been changed to reflect the fact that on the velocity diagram, Figure 4.24, V_2 and V_1 are in opposite directions. Also it is more convenient to use absolute values in equation (23).

Before the blade efficiency and the best blade speed can be considered, two results of general interest need to be obtained. From the velocity diagram,

$$V_{u_2} + V_b = V_{r_{u_2}}$$
$$V_{u_1} - V_b = V_{r_{u_1}} \tag{24}$$

which, upon adding, give:

$$V_{u_1} + V_{u_2} = V_{r_{u_1}} + V_{r_{u_2}} \tag{24a}$$

All subscripts u refer to the components of the velocities at the base of the velocity diagram (Figure 4.24). From equation (23) the stage work is:

$$w_{\text{stage}} = V_b(V_{u_1} + V_{u_2})$$
$$= V_b(V_1 \cos \alpha_1 + V_2 \cos \alpha_2)$$

which can be expressed with (see Figure 4.24):

$$V_2 \cos \alpha_2 = V_{r_2} \cos \beta_2 - V_b$$
$$w_{\text{stage}} = V_b(V_1 \cos \alpha_1 + V_{r_2} \cos \beta_2 - V_b) \tag{25}$$

Most reaction turbines are designed to operate with reaction R at or near .5 in order to optimize stage efficiency and blade loading and to maintain sufficiently high rotor blade velocity to prevent diffusion in the rotor row. Thus, for 50% reaction ($R = .5$), equation (21) yields:

$$V_{r_2}^2 - V_{r_1}^2 = V_1^2 - V_2^2 \tag{26}$$

But the reaction can also be expressed from equations (20) and (23) as:

$$R = \frac{h_1 - h_2}{h_0 - h_2} = \frac{V_{r_2}^2 - V_{r_1}^2}{2V_b(V_{u_1} + V_{u_2})}$$

Since the axial velocity V_x can be assumed to be constant:

$$V_{r_2}^2 = V_x^2 + V_{r_{u_2}}^2$$
$$V_{r_1}^2 = V_x^2 + V_{r_{u_1}}^2$$

and

$$V_{r_2}^2 - V_{r_1}^2 = V_{r_{u_2}}^2 - V_{r_{u_1}}^2$$

From this result, and from equation (24a) R becomes:

$$R = \frac{V_{r_{u_2}}^2 - V_{r_{u_1}}^2}{2V_b\left(V_{r_{u_1}} + V_{r_{u_2}}\right)} = \frac{\left(V_{r_{u_2}} - V_{r_{u_1}}\right)\left(V_{r_{u_2}} + V_{r_{u_1}}\right)}{2V_b\left(V_{r_{u_1}} + V_{r_{u_2}}\right)}$$

$$= \frac{V_{r_{u_2}} - V_{r_{u_1}}}{2V_b} \tag{27}$$

With $R = 1/2$ this yields:

$$V_{r_{u_2}} - V_b = V_{r_{u_1}}$$

But $V_{r_{u_2}} - V_b = V_{u_2}$ from the velocity diagram. Thus:

$$V_{u_2} = V_{r_{u_1}}$$

which shows that:

$$V_2 = V_{r_1}$$

and:

$$\alpha_2 = \beta_1$$

Then equation (26) also yields:

$$V_{r_2} = V_1$$

Therefore, for 50% reaction the velocity diagram must be symmetric as shown in Figure 4.25, thus greatly simplifying all calculations.

As with the impulse turbine, the blade efficiency and best speed can now be obtained. With $V_{r_2} = V_1$ and $\beta_2 = \alpha_1$, the expression for work, equation (25) becomes:

$$w = V_b(2V_1 \cos \alpha_1 - V_b)$$

The blade efficiency is defined as the work output divided by the ideal enthalpy drop through the stage. Ideally, $\Delta h_n = 1/2\, V_1^2$. For a 50% reaction turbine stage enthalpy drop is $\Delta h_n + \Delta h_r = 2\Delta h_n = V_1^2$. Thus:

$$\eta_b = \frac{W}{\Delta h_{\text{ideal}}} = \frac{V_b}{V_1^2}(2V_1 \cos \alpha_1 - V_b)$$

$$= 2\frac{V_b}{V_1}\cos \alpha_1 - \left(\frac{V_b}{V_1}\right)^2 \tag{28}$$

Figure 4.25 Fifty % reaction diagram

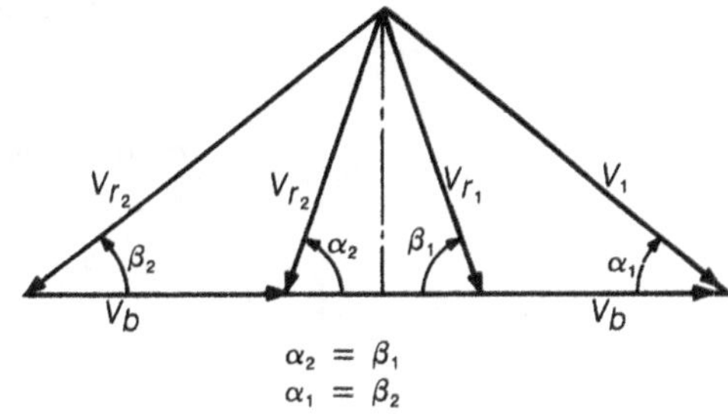

When this is maximized, it follows that for maximum η_b:

$$\frac{V_b}{V_1} = \cos \alpha_1 \tag{29}$$

and the maximum blade efficiency is

$$\eta_{b_{max}} = \cos^2 \alpha_1 \tag{30}$$

The above results have been developed by ignoring losses due to friction, exit kinetic energy loss, and so on. However, as with the results obtained for the impulse turbine, the speed ratio is independent of friction effects and the operating velocity is very close to the value given by equation (29). In jet propulsion engine applications the exit velocity V_2 is not wasted, as it is used to generate thrust. Then equation (30) will represent a fair estimate of the ideal stage efficiency.

In steam and non–jet propulsion gas turbines, the exhaust velocity V_2 is wasted and may represent a large portion of the turbine inefficiency. Minimization of this wasted kinetic energy is one of the objectives in turbine design; however, a certain amount of it must necessarily be present to assure continuity of the flow through the turbine. The turbine efficiency, called also total-to-static efficiency, must now be written as:

$$\eta_{ts} = \frac{w}{\Delta h_{ideal} + V_2^2/2} \tag{31}$$

where Δh_{ideal}, in general, should be written in terms of the total enthalpy to account for the kinetic energy, which can be appreciable (see discussion in Section 3.7), i.e., with the notation of Figure 4.23:

$$\Delta h_{ideal} = h_{0_0} - h_{0_{2_s}} \tag{32}$$

At the first glance, equation (31) implies that η_{ts} should be at its maximum when V_2 is at a minimum. For practical mass flow purposes, as mentioned above, V_2 must have a certain finite value, which is at its minimum when V_2 is axial (see also Figure 4.23). This minimum, however, does not necessarily lead to a maximum value of η_{ts}, as will be shown immediately below.

Consider an ideal (reversible) turbine where the only loss is due to exhaust kinetic energy. Then, from equation (23):

$$w = \Delta h_{ideal} = V_b \Delta V_u$$

and one obtains from equation (31):

$$\eta_{ts} = \frac{V_b \Delta V_u}{V_b \Delta V_u + V_2^2/2} = \frac{1}{1 + \dfrac{V_2^2}{2V_b(V_{u_1} + V_{u_2})}} \tag{33}$$

It may seem again that η_{ts} is at its maximum when the exhaust velocity V_2 is axial ($\alpha_2 = 90°$). But then also $V_{u_2} = 0$ and the work is diminished, which, in turn, decreases the efficiency. If the exit velocity V_2 is a small angle off the axial direction, α_2 is then slightly less than $90°$, and the work and hence the efficiency is increased for only a very small increase in the exit kinetic energy. If a rather laborious optimization process is carried out, the following expressions are obtained:

$$\alpha_2 + \alpha_1/2 = 90° \quad \text{and} \quad V_{u_2} = V_1 - V_{u_1}$$

Then the maximum (ideal) efficiency can be rewritten as:

$$\eta_{ts_{max}} = \cfrac{1}{1 + \cfrac{V_1 - V_{u_1}}{V_b}}$$

$$= \cfrac{1}{1 + \cfrac{V_x}{V_b}\left(\cfrac{1}{\sin \alpha_1} - \cot \alpha_1\right)} \tag{34}$$

The corresponding stage work is obtained as:

$$w = V_b(V_{u_1} + V_{u_2}) = V_b(V_{u_1} + V_1 - V_{u_1})$$
$$= V_b V_1$$

where $\cot \alpha_2$, or V_{u_2}, is determined by the nozzle exit angle α_1. The ratio V_x/V_b is called the flow coefficient, ϕ; its typical values are .3–.5.

The reaction of the stage can be obtained from equation (27) and by use of equation (24):

$$R = \frac{V_{r_{u_2}} - V_{r_{u_1}}}{2V_b} = \frac{V_{u_2} - V_{u_1} - 2V_b}{2V_b}$$

$$= 1 + \frac{V_{u_2} - V_{u_1}}{2V_b}$$

From the velocity diagram, Figure 4.24, $V_{u_2} = V_x \cot \alpha_2$ and $V_{u_1} = V_x \cot \alpha_1$, and the reaction becomes:

$$R = 1 + \frac{V_x}{2V_b}(\cot \alpha_2 - \cot \alpha_1)$$

$$= 1 - \phi/2 \,(\cot \alpha_1 - \cot \alpha_2)$$

$$= \frac{1}{2} + \frac{\phi}{2}(\cot \beta_2 - \cot \alpha_1) \tag{35}$$

For the maximum (ideal) efficiency case, this general expression for reaction yields:

$$R = 1 - \phi\left(\frac{V_{u_1}}{V_x} - \frac{1}{2}\frac{V_1}{V_x}\right)$$

$$= 1 - \phi \,(\cot \alpha_1 - 1/2/\sin \alpha_1 \tag{36}$$

Example 4.4: A single stage gas turbine operates at its design condition with axial inlet and exit velocities. The nozzle angle is 20°. Total temperature and pressure at nozzle inlet are 2000°R and 100 psi, respectively. Measurements show that the stage exit static pressure is 40 psi, the total-to-static efficiency is .87, and the blade speed is 1400 ft/sec. Find: (a) the specific turbine work, (b) nozzle velocity, (c) axial velocity, and (d) stage reaction. Assume that $k = 1.3$.

Solution:

(a) Denoting inlet conditions with subscript 0, the specific turbine work can be obtained from equations (31) and (32):

$$w = \eta_{ts}\left(h_{0_0} - h_{0_{2_s}} + \frac{V_2^2}{2}\right)$$

$$= \eta_{ts}(h_{0_0} - h_{2_s})$$

since $h_{0_{2_s}} = h_{2_s} + (V_2^2/2)$.

Using perfect gas relationships from Section 3.7:

$$w = \eta_{ts} c_p T_{0_0}\left(1 - \frac{T_{2_s}}{T_{0_0}}\right)$$

$$= \eta_{ts} c_p T_{0_0}\left[1 - \left(\frac{p_{2_s}}{p_{0_0}}\right)^{(k-1)/k}\right]$$

$$= .87 \times .24 \times 2000\left[1 - \left(\frac{40}{100}\right)^{(1.3-1)/1.3}\right]$$

$$= 79.6 \text{ BTU/lb}$$

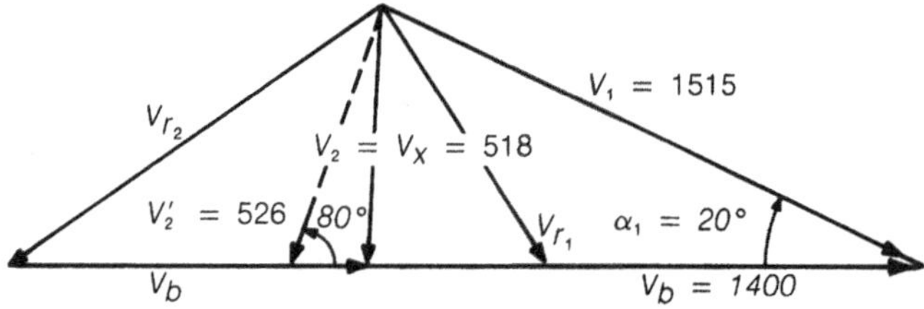

(b) Nozzle velocity V_1 is found by first calculating V_{u_1} from:

$$w = V_b \Delta V_u = V_b(V_{u_1} + V_{u_2}) = V_b V_{u_1}$$

since $V_{u_2} = 0$ by virtue of V_2 being axial. Thus:

$$V_{u_1} = \frac{w}{V_b} = \frac{79.6 \times 778 \times 32.2}{1400} = 1424 \text{ ft/sec}$$

and:

$$V_1 = \frac{V_{u_1}}{\cos \alpha_1} = \frac{1424}{.9397} = 1515 \text{ ft/sec}$$

(c) The axial velocity V_x, which in this case is also equal to V_2, is obtained from the velocity diagram as:

$$V_x = V_1 \sin \alpha_1 = V_{u_1} \tan \alpha_1$$
$$= 1515 \sin 20° = 518 \text{ ft/sec}$$

(d) The reaction is calculated from the second of equations (35):

$$R = 1 - \frac{\phi}{2}(\cot \alpha_1 - \cot \alpha_2)$$

$$= 1 - \frac{1}{2}\frac{V_x}{V_b}\frac{1}{\tan \alpha_1} \qquad \cot \alpha_2 = 0 \text{ as } \alpha_2 = 90°$$

$$= 1 - \frac{518}{2 \times 1400}\frac{1}{\tan 20°} = .492$$

Example 4.5: Consider the results of the above example and the velocity diagram to be that of an ideal reversible turbine with all losses, except those due to exhaust kinetic energy, to be neglected. Calculate: (a) the ideal efficiency and (b) the ideal maximum efficiency.

Solution:

(a) The ideal efficiency, with V_2 in the axial direction, is (equation 33):

$$\eta_{ts} = \frac{1}{1 + \dfrac{V_2^2}{2V_b V_{u_1}}}$$

since $V_{u_2} = 0$ (V_2 being axial).

$$\eta_{ts} = \frac{1}{1 + \dfrac{518^2}{2 \times 1400 \times 1424}} = .937$$

(b) For maximum ideal efficiency the rotating blading must be designed in such a manner that V_2 is in a slightly off-axial direction according to:

$$\alpha_2 = 90° - \alpha_1/2$$
$$= 90° - 20°/2 = 80°$$

Now V_2 is shown as V_2' on a dotted line in the above velocity diagram:

$$V_2' = V_x/\sin 80° = 518/.985 = 526$$

The efficiency is calculated as:

$$\eta_{ts\,max} = \frac{1}{1 + \dfrac{V_1 - V_{u_1}}{V_b}}$$

$$= \frac{1}{1 + \dfrac{1515 - 1424}{1400}} = .939$$

which is only a slight increase over the previous value of

.937. The work, however, is increased by 6.4% when calculated from:

$$w = V_b V_1 = \frac{1400 \times 1515}{g_c\,778} = 84.67 \text{ BTU/lb}$$

4.9 Radial Effects

In the discussion above it was assumed, as stated, that no variations occurred in radial direction in the flow or blade velocity. This assumption is valid in impulse turbine analysis, since the blades are short or the blade length is a small fraction of the total stage diameter. The same approach is used in reaction turbine analysis, where it represents a mean radius or pitch line approach in which the conditions at one particular blade section radius are postulated to be representative of conditions at all radii. In actual reaction turbines, however, there is considerable variation between the blade tip and hub sections due to blade height (in steam turbines reaction blade length ranges from 2 to 40 inches over a total radius ranging from 10 to 80 inches) and high rotational speeds (1500 rpm in steam turbines to over 50,000 rpm in gas turbines).

The velocity diagrams vary from hub to tip as shown in Figure 4.26. The axial velocity, V_x, varies very little along the blade and is assumed to be constant. As the blade speed varies according to $V_b = \omega r$, where ω is the angular velocity, the flow angles will also vary. Most significant is the change in the angle β_2 for both blade inlet and exit, as it represents the angle that the blade must assume in order to provide a smooth flow without separation onto and off the blade. Thus, the blade angles must vary from hub to tip, and the result is a twisted blade. A typical twisted steam blade is shown in Figure 4.27 with velocity diagrams superimposed on root and tip sections. The analyses used for steam are valid also for gas turbines with, perhaps, less drastic change of angles and little or no erosion problem.

Flow leaves the nozzle blades at the inner radius with higher speed than at the outer radius. Since bucket linear speed increases with the radius, the ratio of blade to steam speed grows steadily from root to tip. At the tip the pressure is higher than at the hub; thus there tends to be a flow along the blade also. As the flow angles change, then reaction R will also change along the blade according to equation (35). Thus, the root section is designed for almost impulse flow with extremely low reaction, while at the tip the reaction may exceed 50%. At the root the blade speed is almost half of the steam speed, and the relative velocity is changed little in the bucket. At the tip the blade speed may be twice the steam speed, with the relative speed hitting the blade almost from behind and showing a large increase in the relative

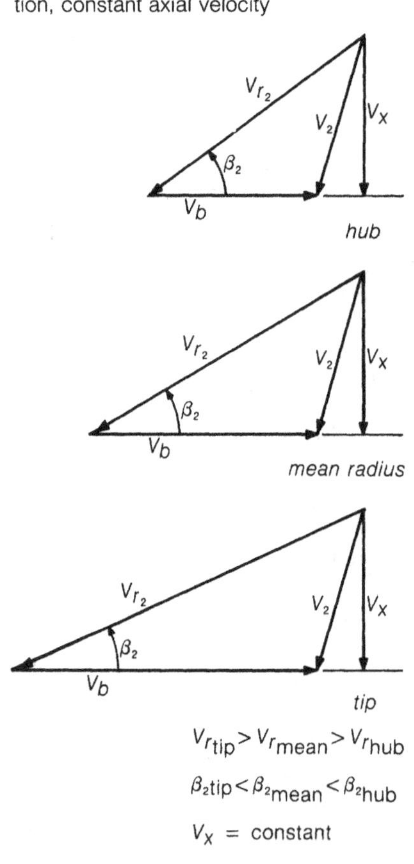

Figure 4.26 Rotor radial velocity variation, constant axial velocity

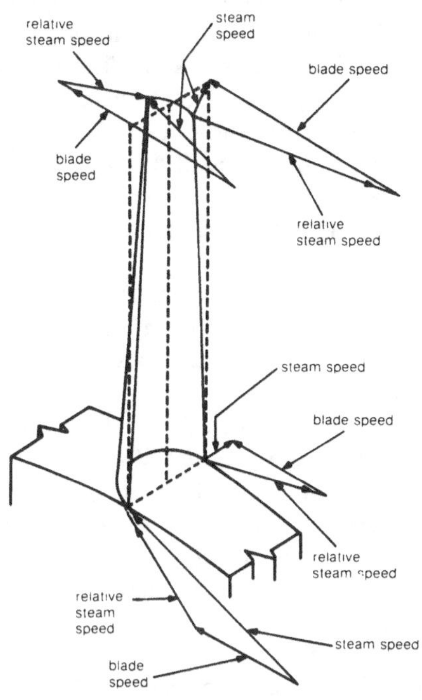
Figure 4.27 Radial velocity variation on a blade

Figure 4.28 Blade erosion pattern, stellite shield

speed exit value. This situation is typical of reaction blading and is diagrammed in Figure 4.24.

At this time Figure 4.21 should be reviewed to visualize the moisture (droplet) path and to explain the erosion pattern of Figure 4.28. At the root the droplets, being heavier than steam, have a smaller absolute velocity than the steam, and their velocity relative to the blade is almost axial. Thus, a typical erosion pattern appears well on the backside of the blade behind the leading edge. Toward the tip, the relative moisture diagram remains the same, except that the droplets tend to be smaller and the blade twisted out of the axial plane, as shown. Hence, the erosion pattern appears at the leading edge of the blade. In order to prolong the blade life, erosion shields (stellite) are attached toward the tip at the leading edge, where the blade is most vulnerable to material loss.

4.10 Reaction Stage Losses

The pressure drop across the moving blades of the reaction stage gives rise to certain differences in construction between turbines containing nearly pure impulse stages and those containing stages with a high degree of reaction. These differences will be discussed in more detail later, but it is necessary to consider their salient features briefly here.

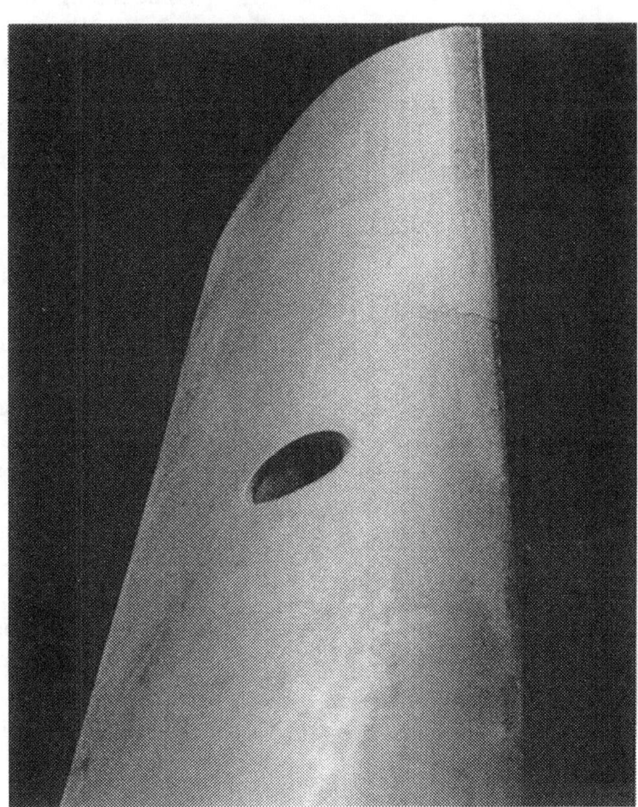

Impulse blades are carried on discs, secured to a shaft. The fixed blades, or nozzles, are mounted in diaphragms, secured to the casing. Since there is little or no pressure drop across the blades and therefore across the discs (discs often have holes in them to equalize pressure), there is little pressure on the disc area to produce end thrust. Also, there is little tendency for steam to flow around the edge of the disc instead of through the blades; therefore generous clearance is provided between the disc circumference and the casing in way of the discs.

Reaction staging, on the other hand, is characterized by a pressure drop across the blades. Drum construction is used in order to minimize the area exposed to the pressure change. That is, instead of mounting the blades on discs, the shaft is made in the form of a drum, on which the blades are directly mounted. Nevertheless, the pressure change across the blades acts upon the annular area of the blades, producing considerable end thrust. In addition, there is a strong tendency for flow to leak around the outer ends of the blades, necessitating small tip clearances between blade ends and the casing at that point.

With the above features in mind, the losses characteristic in a reaction stage (in addition to friction losses of the flow passing through the nozzles and moving blades) are:

(a) *Windage loss:* Friction of flow on nonworking portions of the moving blading and drum sets up turbulence (smaller than in a corresponding impulse stage). Gas turbine units have turbine sections of both drum and disc constructions, and windage is used to cool the disc space. In some engines, the turbine blades in the hot section are constructed with a long shank that is mounted on a smaller-diameter disc: this arrangement cuts down disc weight and improves the blade high temperature stress resistance.

(b) *Tip leakage loss:* Energy loss is experienced in that part of the flow which leaks past the blade tips without performing work. This occurs between the moving blade tips and the casing, and between the fixed blade tips and the drum. Since mechanical considerations limit the minimum clearance distance, the tip leakage area is a larger percent of total flow area in short, or high pressure, blading than in long, or low pressure, blading.

In many turbines the blades are shrouded at the tip to prevent spillage over and around the tips. The shroud forms a band around the turbine wheel in one continuous strip or, more often, ties groups of blades together, which serves to reduce blade vibration, as in Figures 4.29 and 5.23 (below). Although the shroud tends to add extra weight, it is offset (mainly in gas turbines) by use of thinner and more efficient aerodynamic blade sections, which otherwise could not be utilized because of high frequency vibration limitations. It

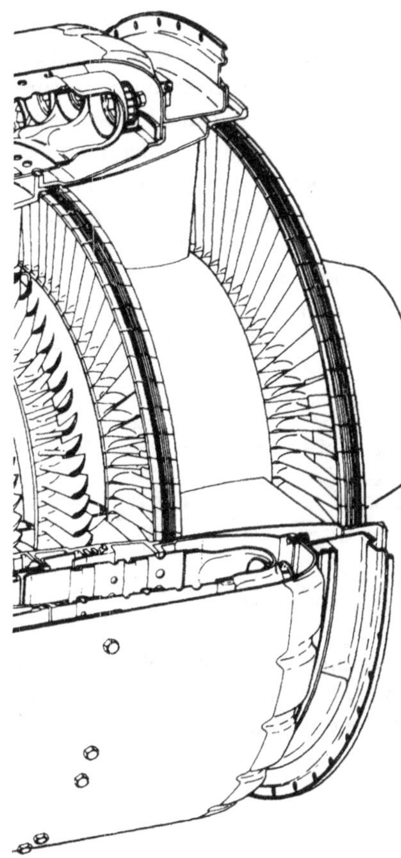

Figure 4.29 Reaction blading with shrouds

is common to use seals between the shrouds and turbine casing (more details are given in Section 5.4).

(c) *Moisture losses (in steam turbines):* Moisture losses in reaction stages occur for the same reasons as, and have similar effects to, those described for impulse stages.

In gas turbine units the turbines are not subject to moisture losses. The compressor, however, may be affected by water droplets in marine applications in severe sea conditions, or if demisters (aerosol separators) are defective.

No condensation takes place in gas turbine compressors. Any liquid there is in the form of airborne droplets—aerosols. Airborne particles are also undesirable "aerosols." The separators (ahead of the gas turbine unit) are intended to remove aerosols—droplets and dirt—and they have a certain efficiency, with each dependent upon size (of aerosol) and loading. Efficiency can be degraded to the point where the separator is defective by insufficient maintenance or poor installation.

4.11 Comparison of Impulse and Reaction Staging

Because of the pressure change across the moving blading, the partial admission arrangement used in many high pressure impulse stages is not feasible in a reaction turbine. The nozzle flow area must be kept to its proper value by using short nozzle blading (and bucket blading) and by restricting the stage diameter. Therefore, the nozzle end losses and the suction losses of a partial admission stage are not present in a reaction stage.

For impulse and reaction stages of equal enthalpy drop, the velocities produced in the reaction nozzles are considerably smaller than those in the impulse nozzles. This is obvious, because the enthalpy drop across the reaction nozzles is about one-half of that across the impulse nozzles, the remainder taking place across the buckets of the reaction stage. Thus, even though the wheel speed of such a reaction stage would be higher than in the impulse stage, the velocity relative to the wheel is somewhat less. In actual practice, the wheel speed is usually designed to the same maximum practicable value for both; this condition, then, determines the nozzle velocities and, therefore, the permissible enthalpy drop. Thus reaction stages tend to have less enthalpy drop than similar impulse stages, still further reducing local velocities. Since fluid friction losses vary approximately as the cube of the velocity of the fluid passing the surface, there is a definite advantage of the reaction stage over the impulse stage with respect to fluid friction. However, smaller enthalpy drops per stage also necessitate more stages for a given turbine, meaning larger and more expensive turbines.

In view of the last two paragraphs, tip leakage considerations favor the use of impulse blading in high pressure staging, where fluid density is high and the leakage area is a fairly large proportion of the flow area; and fluid friction losses favor the use of reaction blading in low pressure staging, where the effect of tip leakage loss is small. A multi-stage turbine can be made up of a succession of impulse (Rateau stages) turbines, but it is more usual to start with a velocity compounded (Curtis) section in order to obtain maximum work per stage and rapid lowering of pressure and temperature. This is followed by a large amount of more efficient reaction blading. Figure 4.30 summarizes velocity and pressure compounded and reaction turbine blading pressure and velocity variations.

It was shown above that the following optimum blade speed ratios should be observed:

Simple impulse, $$\frac{V_b}{V_1} = \frac{\cos \alpha_1}{2}$$

Two-stage velocity compounded impulse, $$\frac{V_b}{V_1} = \frac{\cos \alpha_1}{4}$$

Reaction, $$\frac{V_b}{V_1} = \cos \alpha_1$$

Another illuminating comparison is possible if it is assumed that the exit velocity V_2 is axial, or $V_{u_2} = 0$. Then, from equation (3), work done per pound is:

$$w = \frac{P_T}{\dot{m}} = \frac{V_b V_{u_1}}{g_c}$$

$$= \frac{V_b V_1 \cos \alpha}{g_c}$$

Figure 4.30 Blade staging arrangements

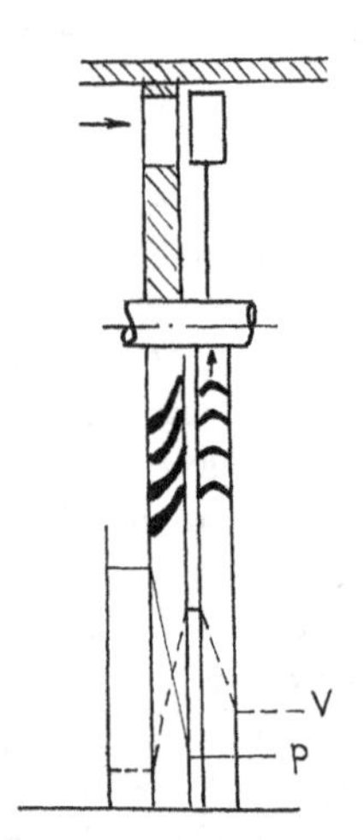

Schematic arrangement. for simple impulse (DeLaval) turbine.

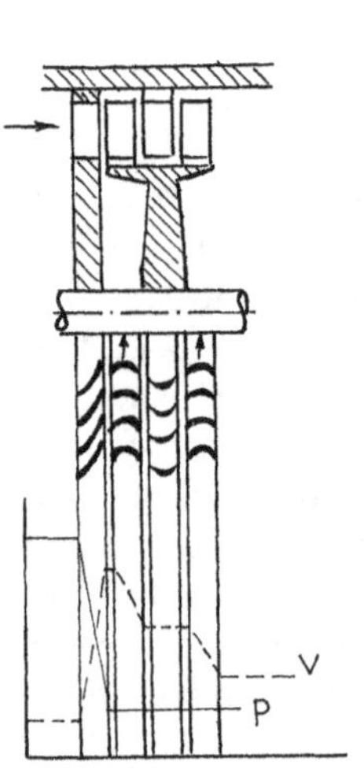

Schematic arrangement for velocity-compounded impulse (Curtis) turbine.

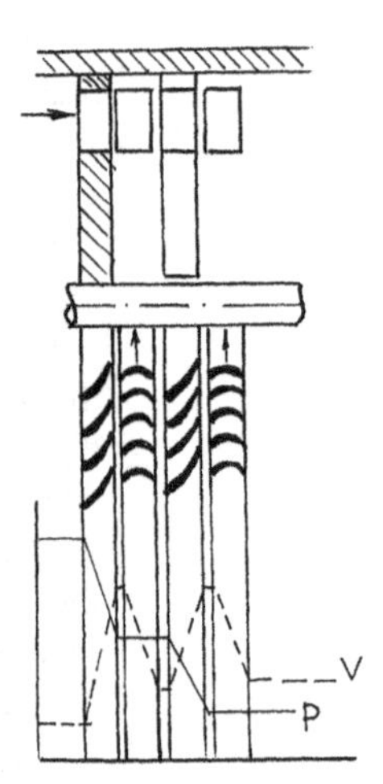

Schematic arrangement for pressure-compounded impulse (Rateau) turbine.

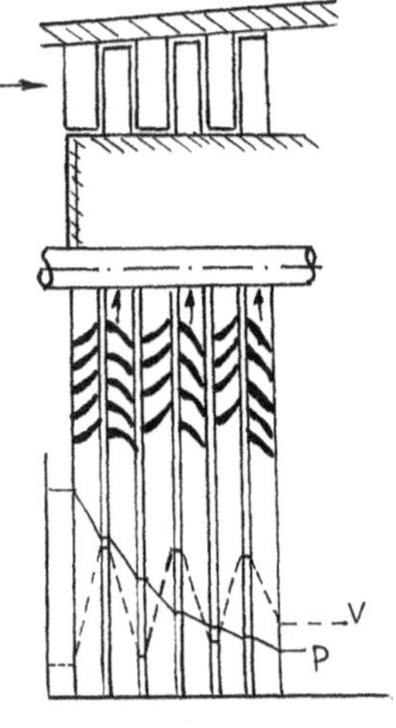

Schematic arrangemet for reaction (Parsons) turbine.

At optimum blade speed:

Impulse,
$$V_1 = \frac{2V_b}{\cos \alpha} \qquad w = \frac{2V_b^2}{g_c}$$

Two row velocity compounded,
$$V_b = \frac{4V_b}{\cos \alpha} \qquad w = \frac{4V_b^2}{g_c}$$

Reaction,
$$V_1 = \frac{V_b}{\cos \alpha} \qquad w = \frac{V_b^2}{g_c}$$

These results show why, per stage, an impulse turbine can extract more power than a reaction stage. At the same blade speed, impulse turbine work is twice that of a reaction stage. Conversely, if one were to attempt to extract the same amount of energy from both stages, then:

$$V_{b_{50\%}} = \sqrt{2}\, V_{b_{impulse}}$$

Finally, the various stage efficiencies, as obtained in preceding sections, are shown in Figure 4.31. The reaction stage produces a slightly higher peak efficiency, and retains it at a higher level over a wider range of operating conditions than the impulse stages do. This is one reason why reaction blading is used in gas turbines where a relatively wide range of blade speeds is necessary. Moreover, the re-

Figure 4.31 Comparison of stage efficiencies

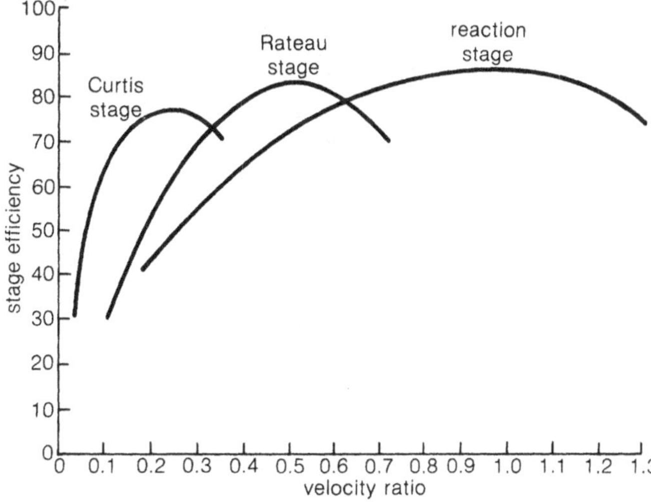

quirements for light weight and compactness will not permit the use of multiple turbines, as used in marine applications.

These comparisons conclude this chapter on basic turbine performance. In the following chapters steam and gas turbines, as used in marine applications, will be considered separately.

Chapter Five

Steam Turbines

5.1 Multistage Arrangements

After the preceeding discussion of individual types of staging, it is now possible to better understand contemporary turbine staging arrangements. From the statements previously made, it is obvious that a velocity compounded stage is a logical choice for the first stage. It is logical also to follow it with impulse staging in the high pressure end to minimize tip leakage losses, and with reaction blading where its higher efficiency can be realized in the low pressure end. Generally, this is modern practice.

Figures 5.1 and 5.2 show two arrangements of modern high pressure turbines used by most manufacturers. Figure 5.1 shows a 600 psi turbine with a velocity compounded (Curtis) stage and eleven Rateau-bladed stages. Also shown are the bypass features for efficient turbine speed control for a wide ship speed range. More will be said on this in the turbine control section. Figure 5.2 shows a 1200 psi high pressure turbine with essentially the same blading; but the bypass feature is not present.

Figure 5.3 shows a typical low pressure reaction bladed turbine of the drum construction type where the blades are attached directly to the rotor, and the fixed blading, forming the nozzles, is attached to the turbine casing. Sometimes the rotor is forged in the form of shaft and wheels, and the moving blading is then attached to the wheels; the fixed blades, in the form of diaphragms, are seated in casing grooves. The rotor and blade type arrangement is often called reaction type construction; whereas the disc and diaphragm arrangement is referred to as impulse type construction. Obviously, this nomenclature is confusing, as both types of construction are essentially reaction type; the nomenclature often reflects the user's and the manufacturer's historical usage. The configuration used for many naval turbines is the drum type which omits the wheels, but with the nozzles fixed in diaphragms, thus permitting lighter and somewhat less expensive construction.

Basic types of turbines can be divided into two main classes: condensing and noncondensing. Condensing units exhaust steam at a pressure less than atmospheric to a condenser, which vastly improves the unit's efficiency (discussed previously in section 3.10). Noncondensing units, not used in naval main propulsion applications, exhaust at pressures higher than or near atmospheric pressures, and are less expensive, as a condenser is not required. Many steam plant auxiliaries, however, such as boiler feed pumps

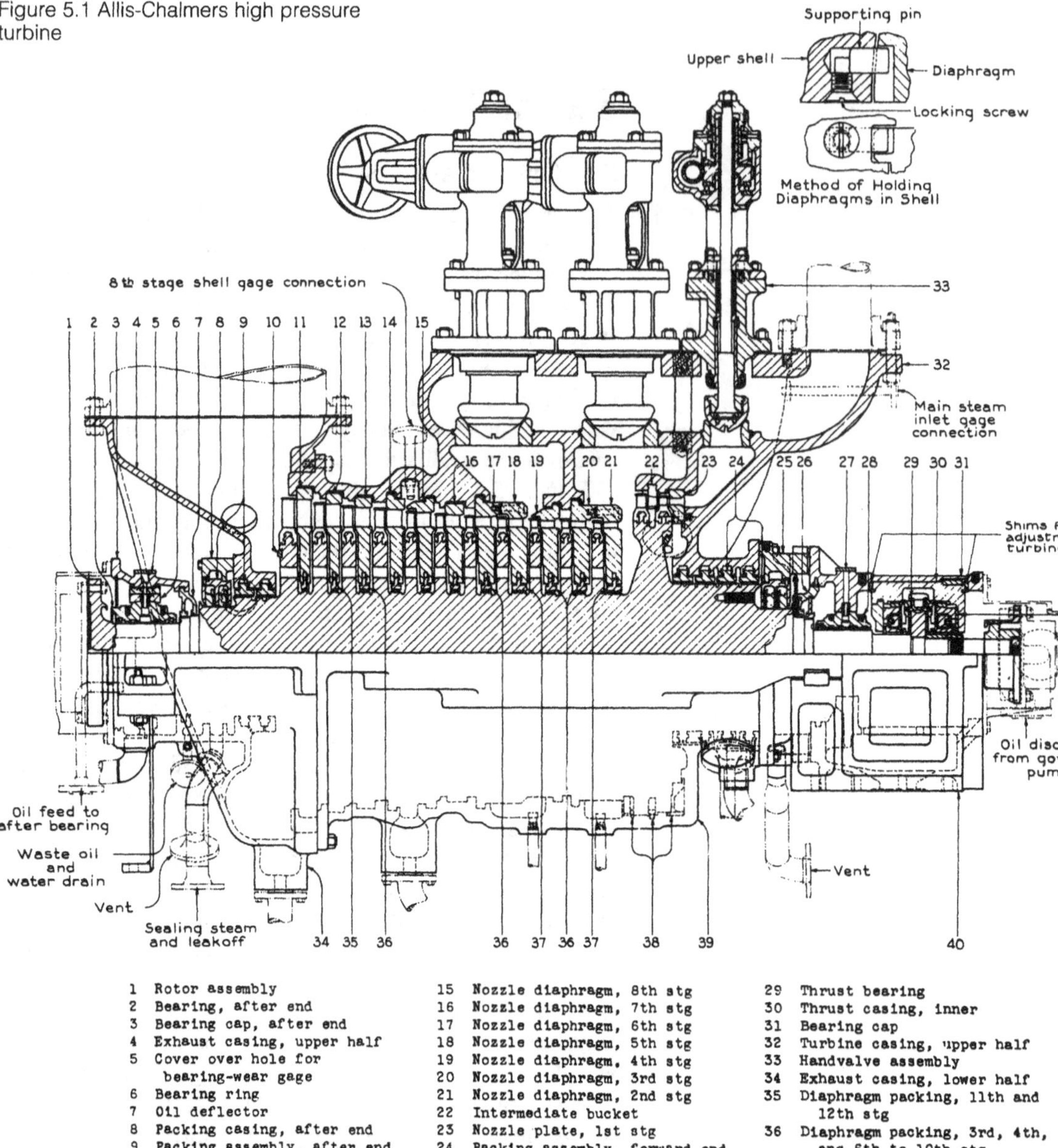

1	Rotor assembly	
2	Bearing, after end	
3	Bearing cap, after end	
4	Exhaust casing, upper half	
5	Cover over hole for bearing-wear gage	
6	Bearing ring	
7	Oil deflector	
8	Packing casing, after end	
9	Packing assembly, after end	
10	Balance plug	
11	Nozzle diaphragm, 12th stg	
12	Nozzle diaphragm, 11th stg	
13	Nozzle diaphragm, 10th stg	
14	Nozzle diaphragm, 9th stg	
15	Nozzle diaphragm, 8th stg	
16	Nozzle diaphragm, 7th stg	
17	Nozzle diaphragm, 6th stg	
18	Nozzle diaphragm, 5th stg	
19	Nozzle diaphragm, 4th stg	
20	Nozzle diaphragm, 3rd stg	
21	Nozzle diaphragm, 2nd stg	
22	Intermediate bucket	
23	Nozzle plate, 1st stg	
24	Packing assembly, forward end	
25	Packing casing	
26	Oil deflector	
27	Cover over hole for bearing-wear gage	
28	Bearing, forward end	
29	Thrust bearing	
30	Thrust casing, inner	
31	Bearing cap	
32	Turbine casing, upper half	
33	Handvalve assembly	
34	Exhaust casing, lower half	
35	Diaphragm packing, 11th and 12th stg	
36	Diaphragm packing, 3rd, 4th, and 6th to 10th stg	
37	Diaphragm packing, 2nd and 5th stg	
38	Steam shield	
39	Turbine casing, lower half	
40	Bearing standard	
41	Filler piece	

Steam Turbines

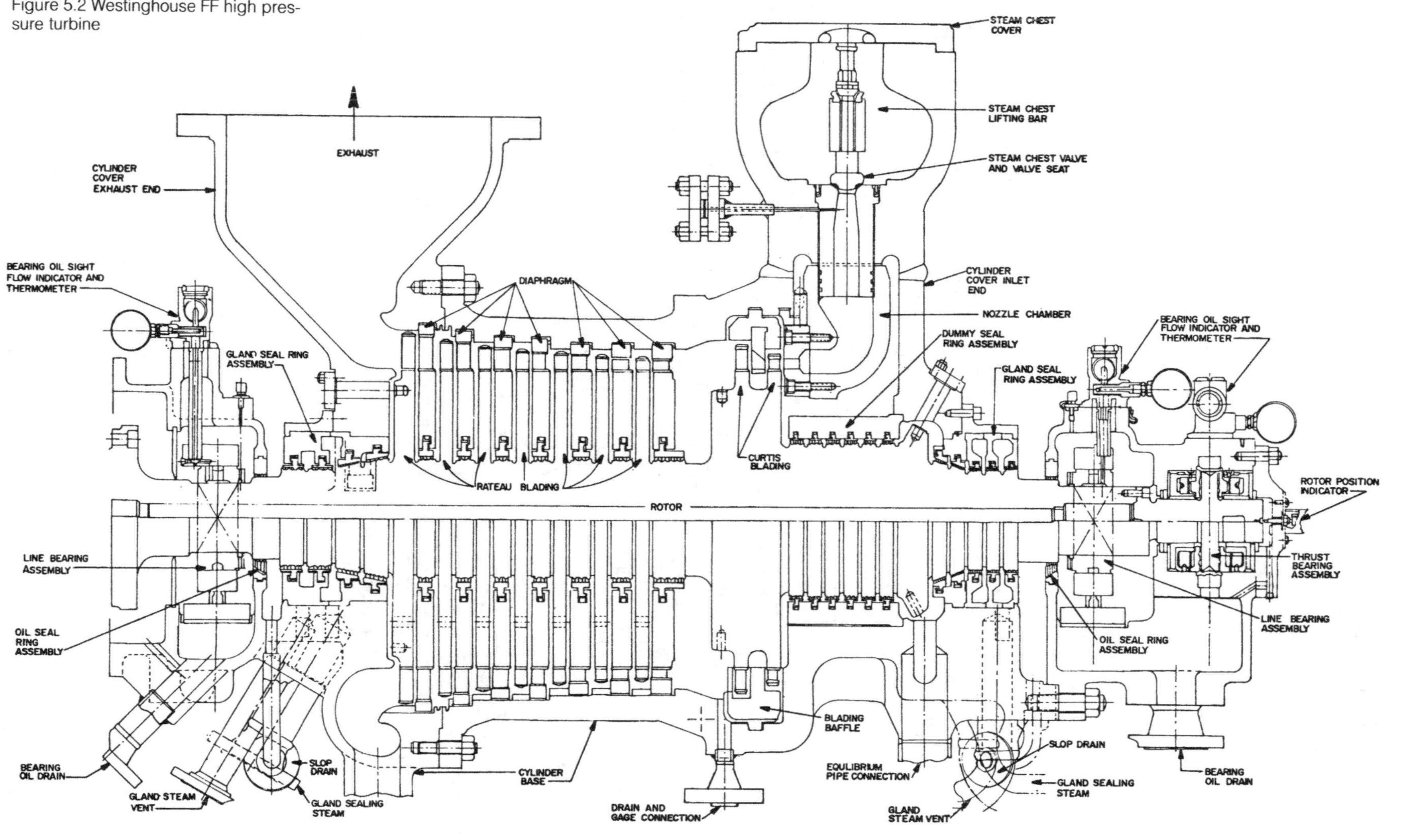

Figure 5.2 Westinghouse FF high pressure turbine

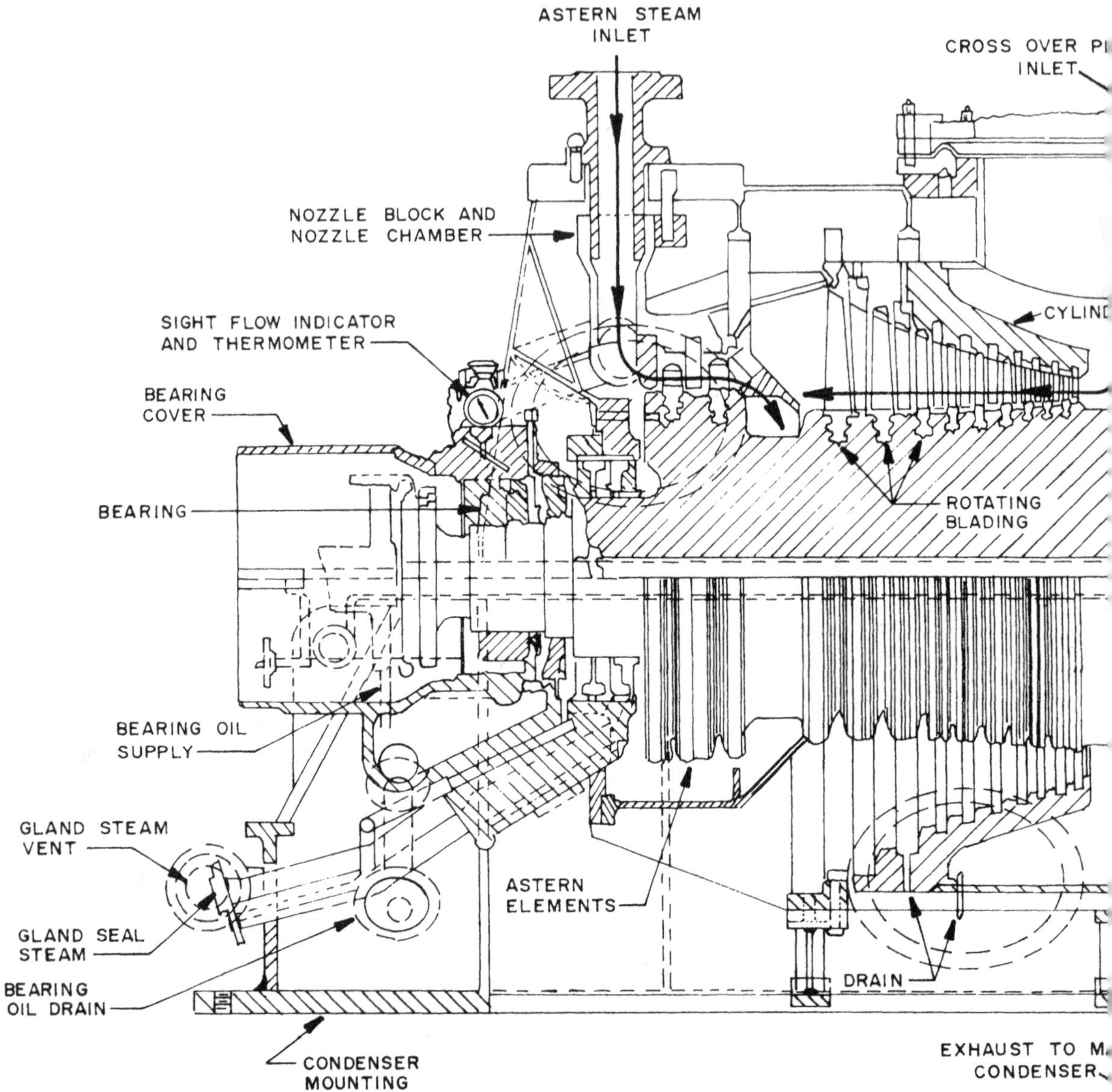

ASTERN STEAM INLET
CROSS OVER PIPE INLET
NOZZLE BLOCK AND NOZZLE CHAMBER
CYLINDER
SIGHT FLOW INDICATOR AND THERMOMETER
BEARING COVER
ROTATING BLADING
BEARING
BEARING OIL SUPPLY
GLAND STEAM VENT
GLAND SEAL STEAM
ASTERN ELEMENTS
BEARING OIL DRAIN
DRAIN
CONDENSER MOUNTING
EXHAUST TO MAIN CONDENSER

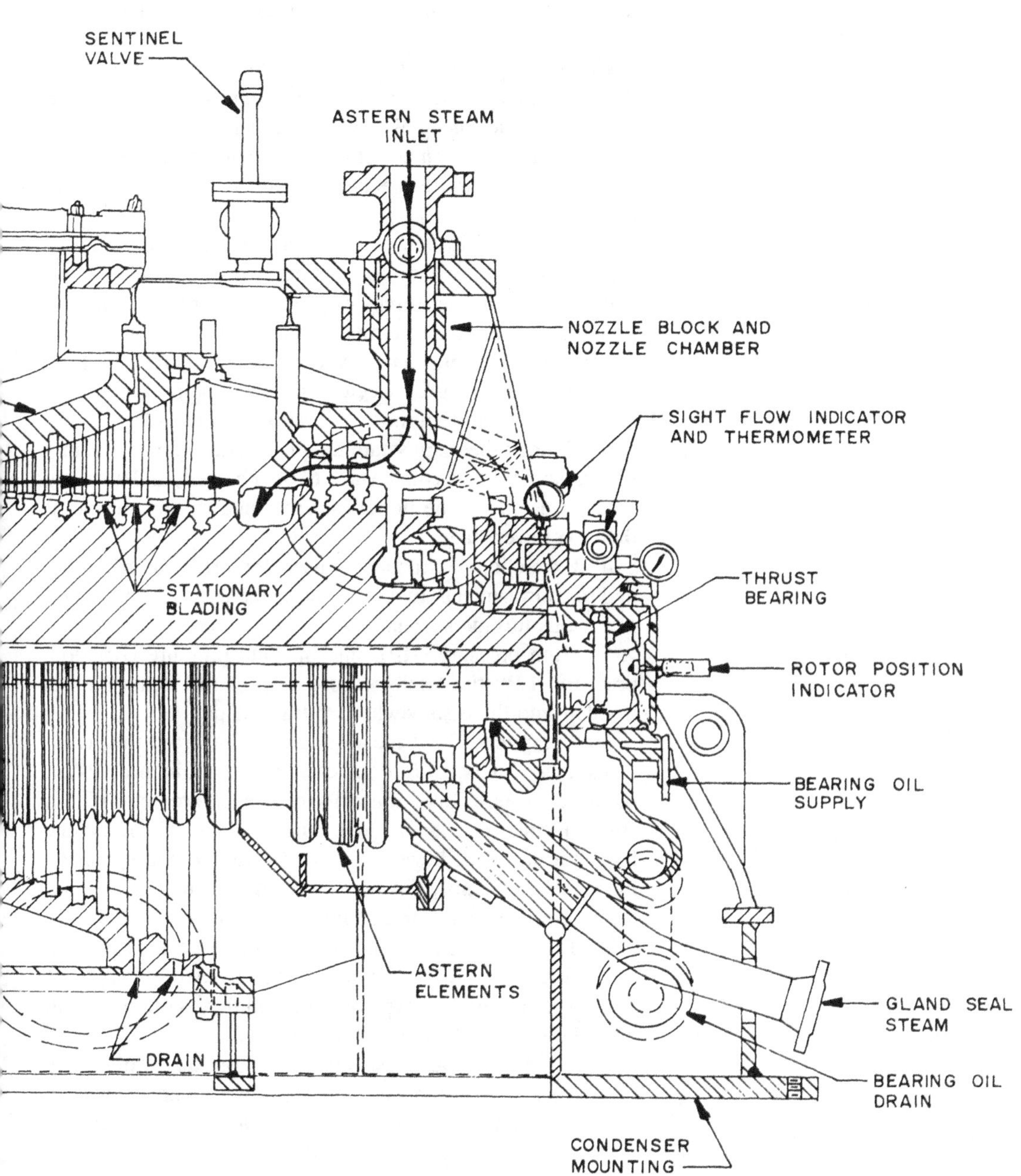

Steam Turbines

and other smaller steam-driven auxiliaries, are noncondensing type units more often than not, to save space and weight.

Condensing and noncondensing turbines can be subdivided into many variations, according to steam flow in the turbine and the casing and shaft arrangements:

1. Straight, or single, flow: complete expansion takes place from full throttle steam to exhaust in one single casing in one direction.

2. Double, or divided, flow: steam enters at the center of the turbine and flows toward the ends of the turbine to balance the axial thrust forces of the reaction blading, and to divide the otherwise extremely large exhaust area in half for best use of available space. Another benefit here is that the last-stage blade length is decreased, which helps to maintain blade strength and minimize vibration problems.

3. Reheat: the main steam flow is exhausted from the turbine at some intermediate stage to be reheated, and possibly resuperheated, at the boiler and returned to the turbine at the next lower stage.

4. Extraction: extraction is used for feed water heating and for powering auxiliary turbines; steam may be extracted at various turbine stages.

5. High pressure (HP), intermediate pressure (IP), low pressure (LP): the total pressure drop of the steam path is divided into three somewhat arbitrary steps.

6. Compound flow: the work extraction from the steam takes place in two or more casings. In main propulsion plants or where large amounts of power are required, a single turbine is not practical. The result would be a very large turbine with huge casing and rotor, providing insurmountable assembly, inspection, and overhaul problems. There are two basic compounding methods:

   a. Tandem compounding: two or more casings drive a single shaft or shaft extensions connected by means of a coupling. This type of compounding is often used in large central station power plants.

   b. Cross compound: steam expansion takes place by passing steam in series through two separate turbines on separate shafts driving separate generators or geared to a common output.

Figure 5.4 shows examples of the above classifications.

Although many earlier versions of (combatant) propulsion plants had cruising and HP turbines in tandem compound arrangement together with LP turbines in cross compound, reheat, extraction, and tandem compound turbines are not currently used in U.S. naval applications. The first two have been replaced in naval ships by the auxiliary cycle where desuperheated steam is utilized (in newer systems superheated steam at boiler pressure is used). The tandem

Figure 5.4 Turbine classifications
1. Tandem compound:
   A) Single-casing single flow
   B) Single-casing counter flow
   C) Two-casing double flow
   D) Two-casing double flow reheat
2. Cross compound:
   E) Two-casing double flow
   F) Two-casing double flow reheat

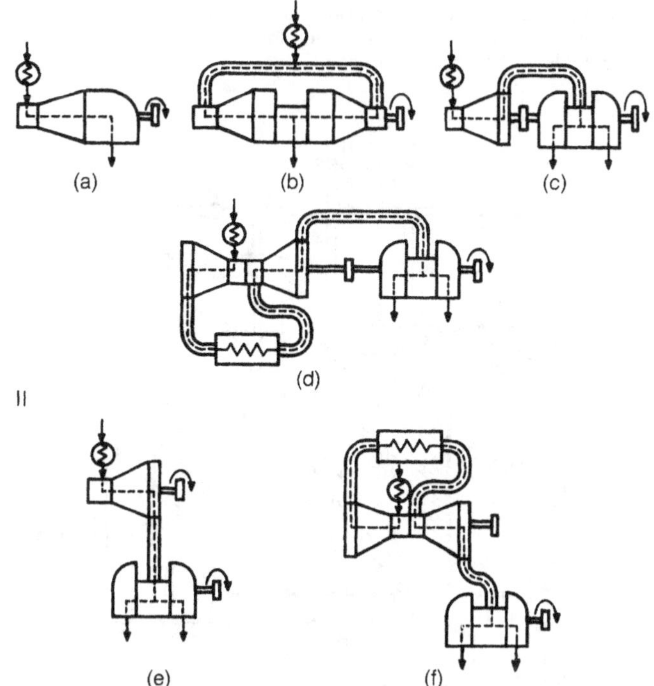

compound arrangement is seldom used because of the excessive length of such a plant. Newer naval turbine units consist of a cross compound turbine (also called a series-parallel unit for its mode of operation) with HP or HP and IP units in one casing and a divided flow LP turbine in the other casing. The steam is admitted to the HP-IP turbine and exhausted to the LP turbine through a crossover pipe. Only the HP section admits steam at low-speed, economical operation, and the IP section receives the exhaust steam of the HP turbine (series operation). At higher speeds both HP and IP sections receive inlet steam, similar to a double flow turbine (parallel operation) (see Figure 5.16, below).

Another arrangement used in earlier naval turbines is shown in Figure 5.5, where a separate cruising turbine is connected to the front end of a cross compound HP unit by a small separate reduction gear. It is used only for purposes of improving the plant efficiency at relatively low speeds, and is idled in partial vacuum at higher speeds. At these times it is supplied with cooling steam to prevent overheating. At cruising speeds, steam is admitted to the cruising turbine through separate valves and is then exhausted through the HP turbine (to cool it) into the LP turbine and to a condenser. The cruising turbine is no longer being incorporated in current applications.

In general, the cruising and HP turbine blading consists of velocity compounded (Curtis) blades for the first stage, with the remaining stages being of the Rateau type. The IP

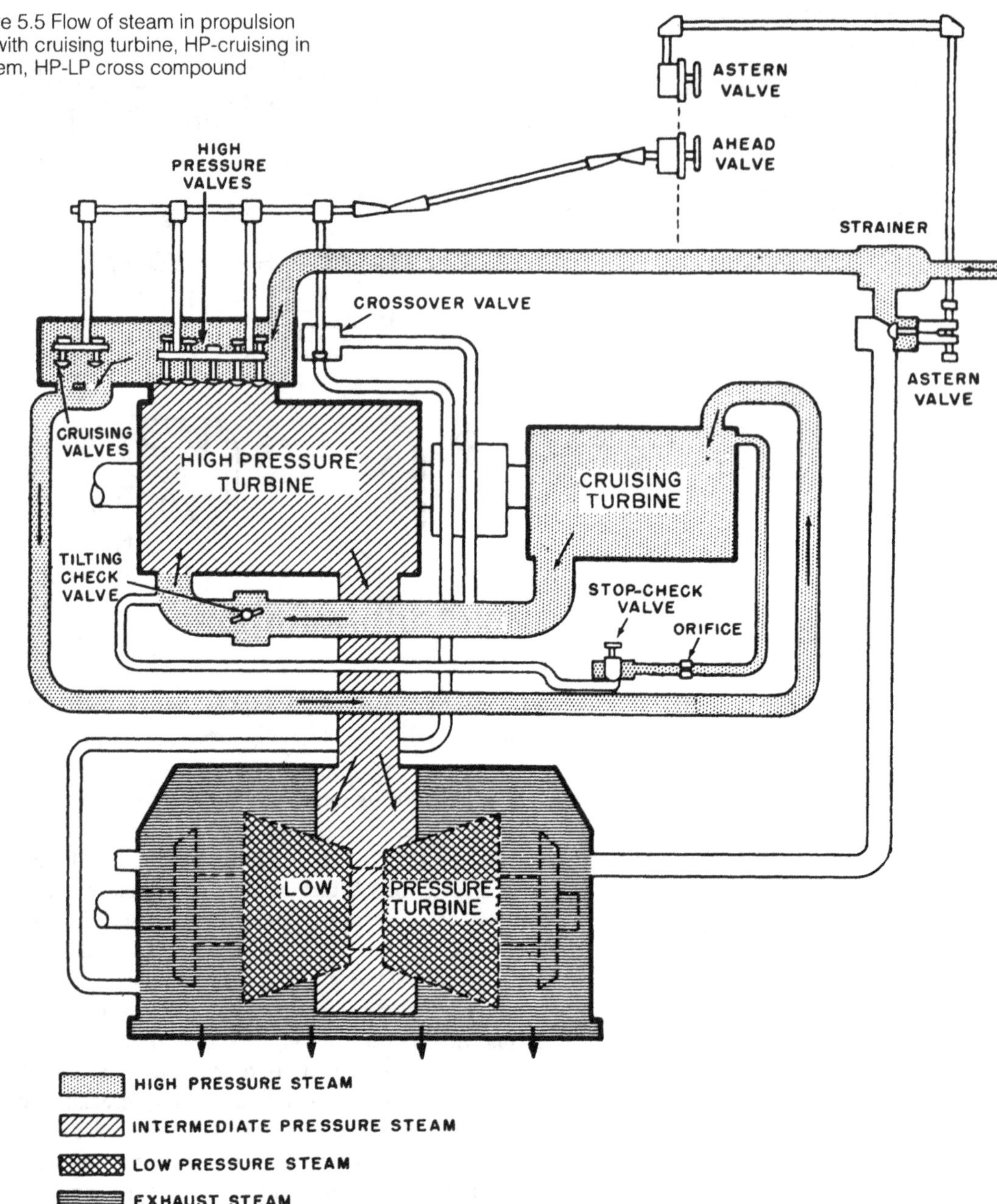

Figure 5.5 Flow of steam in propulsion unit with cruising turbine, HP-cruising in tandem, HP-LP cross compound

and LP sections are made up of reaction blading. (See Figures 5.6 and 5.7.)

It is desirable here to mention provisions for astern propulsion, to complete our discussion of the major operational characteristics of propulsion turbines. Astern operation is made possible by the inclusion of astern stages in the LP turbine. The astern turbine generally consists of one or two velocity compounded stages, sometimes in combination with an impulse stage or stages. These elements are located at each end of divided flow LP turbines or at the exhaust end

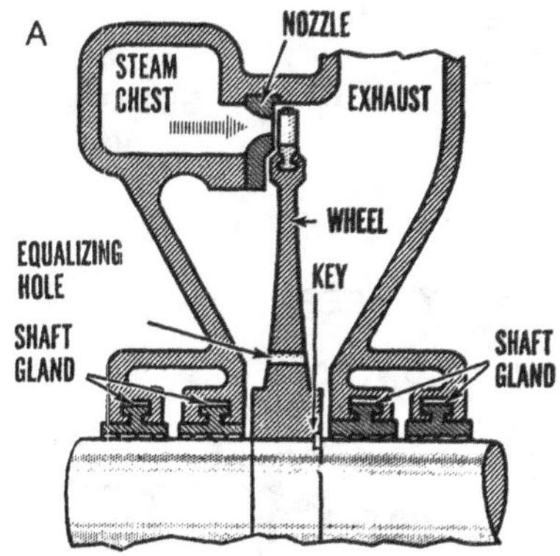

Figure 5.6 Schematic arrangement of an impulse turbine. A) Simple impulse turbine B) Velocity compounded (one Curtis) stage

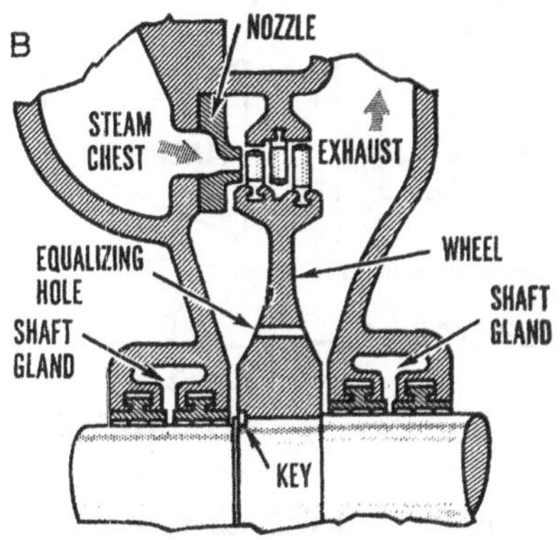

Figure 5.7 Pressure compounded impulse turbine (Rateau) stage

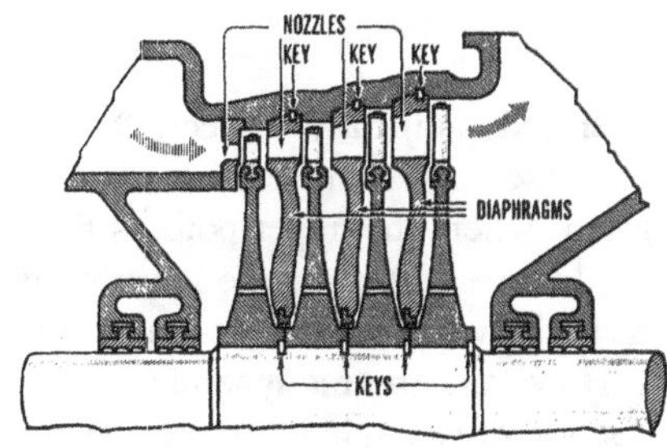

of single flow LP turbines. Steam admission is controlled by an astern throttle valve. Since these are few-stage turbines, they are characterized by high steam exit velocity, high exit losses, and generally poor efficiency. To minimize overheating caused by astern operation due to the "compressor" effect of the ahead blading working on the astern steam exhaust, a baffle is placed between the last stage of the ahead blading and the astern element (see Figure 5.3). This helps to route the astern exhaust flow directly to the condenser. Nevertheless, overheating of the ahead stages occurs during extended periods of astern operation because of friction and windage losses due to dead steam surrounding the ahead elements. No practical provision has been developed to prevent this problem, and extended time of operation at full astern power is limited for this reason.

Another type of naval steam turbine propulsion plant is the turboelectric drive, Figure 5.8, which unlike the geared turbine has a single turbine unit for each shaft. The plant schematic shows a turbine driving a generator, with a direct current generator supplying excitation for the generator, which drives the motor, which then drives the propeller. Speed change and gear reduction are accomplished electrically, as is the reverse operation, since the direction of motor rotation controls the direction of propeller rotation. Thus, there is no need for an astern element.

5.2 Turbine Performance

A brief discussion of turbine performance and losses follows now, to provide background for turbine control and other mechanical features.

Calculations involving the turbine as a whole are based on the energy balance, obtained essentially by use of the general energy equation. It is rewritten here in a form that includes losses, L_m, which occur as a result of mechanical friction. The potential energy terms, Z/J, are considered as being negligibly small, as is the kinetic energy term at the turbine inlet, $V_0^2/2g_c$. Also, in a well-insulated turbine, heat exchange with the surroundings is insignificant. The energy equation then takes the form:

$$h_a = \frac{V_e^2}{2g_c J} + h_e + \frac{w}{J} + L_m$$

or:

$$w = (h_a - h_e) - \frac{V_e^2}{2g_c J} - L_m$$

where h_a = enthalpy of steam at turbine inlet, BTU/lb
h_e = enthalpy of steam at turbine exit, BTU/lb
V_e = velocity of steam at turbine exit, ft/sec
L_m = mechanical losses, BTU/lb
w = work produced at turbine shaft, BTU/lb

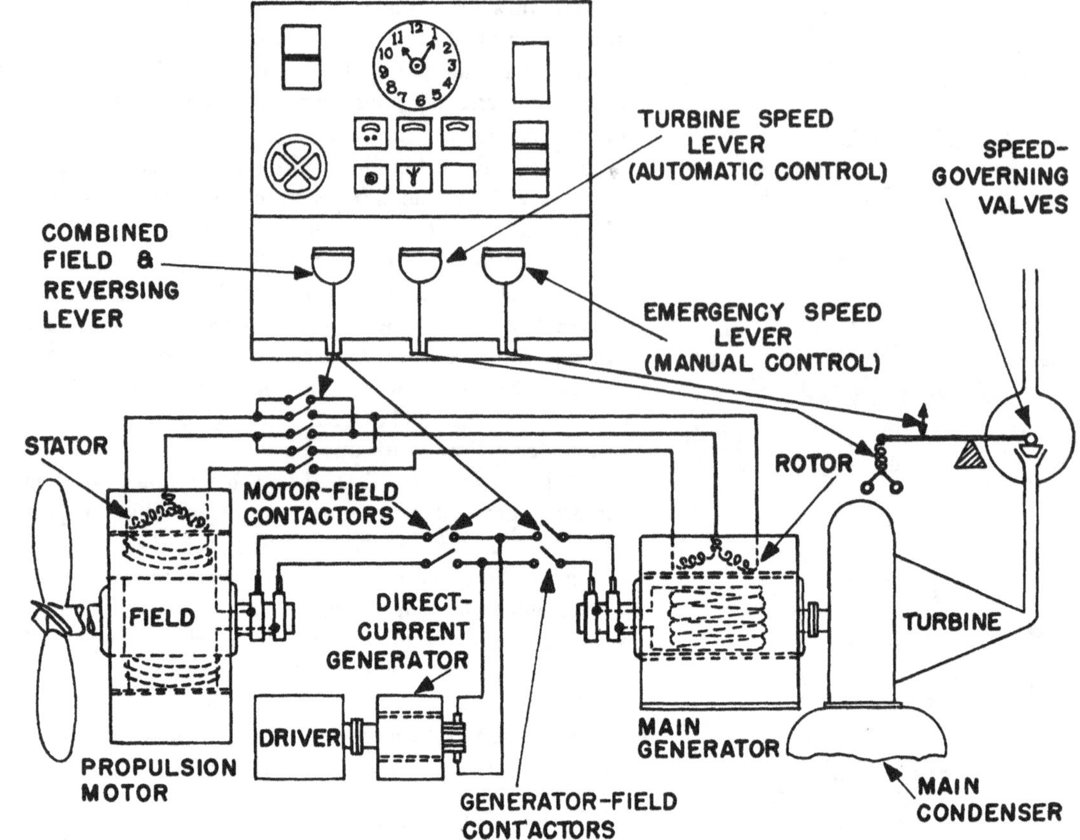

Figure 5.8 Turbolelectric drive arrangement

The three quantities on the right-hand side are recognized as change of enthalpy in the working fluid, kinetic energy of the fluid leaving the turbine, and mechanical losses. These quantities will now be examined individually.

Enthalpy Change

If the turbine were thermodynamically perfect, the enthalpy change could be represented as an isentropic line on an *h-s,* or Mollier, diagram, where the upper point is fixed by inlet steam pressure and temperature and the lower point by the exit pressure. This is shown by the line connecting h_a and h_e' in Figure 5.9. However, thermodynamic losses, caused by the throttling action of steam leakage, nozzle and blade friction, and other factors previously discussed, result in a displacement of h_e' to the right as shown by the actual point, h_e, in the figure. The solid line connecting h_a and h_e is called the "state-line" or "condition-curve." Each point on it represents the actual state of the steam at the corresponding point in the turbine. The line is nearly a straight line, and little error results by ignoring its curvature in turbine analysis. Thus it is very closely determined by the location of its lower end, called the "state line end point," abbreviated "S.L.E.P."

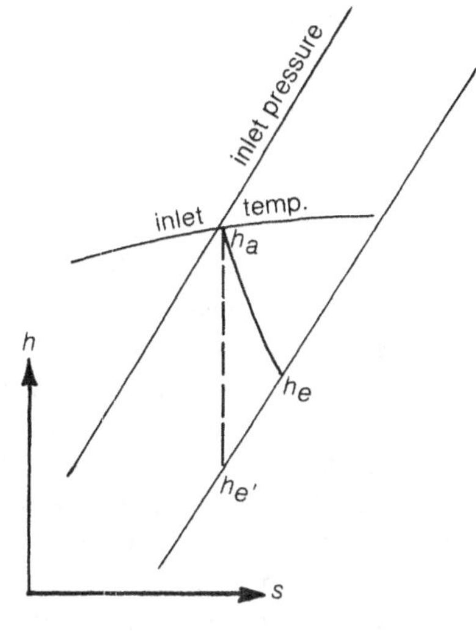

Figure 5.9 An *h-s* (Mollier) diagram for a turbine

The position of the S.L.E.P. will vary with different loads, representing decreased efficiencies at loads away from the point of maximum turbine efficiency. As shown in Figure 5.10, one-half power S.L.E.P., or $h_{e1/2}$, is not at the same pressure as h_e. At full power, the water velocity through the condenser tubes and the steam flow are both greater than at cruise speed. On the Mollier diagram, if they are only 1″ Hg apart, efficiency is affected. In turbines having a velocity compounded inlet stage, there is characteristically a break in the line after the first stage, indicating the lower efficiency of this stage. Also, there is a rise in enthalpy in the crossover pipe between the cruising and high pressure units, and between the high and low pressure units. This is due to reconversion to thermal energy of a part of the kinetic energy present in the steam when it leaves the blades of the stage preceding the crossover pipes. This is discussed more fully under "Exit Losses," below. Some throttling is usually present between the inlet valve and the steam chest, giving a short, horizontal portion of the top of the curve. These features are indicated in Figure 5.10. Mollier charts, containing the state lines, will be found in most turbine instruction books on board ship.

For analysis of overall performance, sufficient information for our purposes is given by the location of the inlet condition point and the S.L.E.P.

Assuming the S.L.E.P. to be known at full power (marked h_e in Figure 5.10), the internal efficiency of the turbine may

Figure 5.10 Mollier diagram for an actual turbine

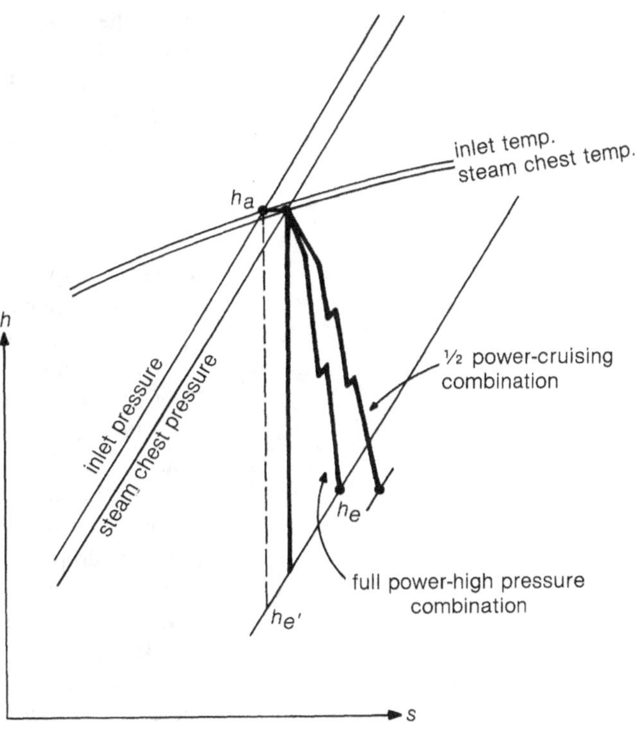

be found. Considering the throttle as part of the turbine, the energy available to the turbine is $h_a - h_e'$. The energy delivered to the wheels is $h_a - h_e$. Therefore the internal efficiency is:

$$\eta_i = \frac{h_a - h_e}{h_a - h_e'}$$

Exit Losses

Exit losses for individual stages, represented by the kinetic energy of the steam leaving the stages, have already been discussed. These individual stage losses are not necessarily lost to the turbine as a whole. Normally, about 75% of the kinetic energy is effective energy "carry-over" to the next stage. An exception to this is the velocity compounded stage, which usually has a larger diameter than the subsequent stage, and a generous dead space between its exit and the next stage inlet (see Figure 5.1). Even in this space the kinetic energy is converted back to thermal energy, so that the major loss is that associated with an entropy increase which restricts the subsequent availability of energy from the steam. Entropy increase and energy conversions are the controlling factors in considering the exit loss of the last stage of the high pressure turbine, though here the effect of entropy increase is even more marked. The major loss occurs at the exit of the last stage of the low pressure turbine, for here the kinetic energy is lost to the turbine. If the exit velocity, V_e, is known or computed from a velocity diagram, the loss is readily calculated. If the diagram is not known, the velocity can be determined approximately by consideration of the last stage steam flow area, the quantity of steam flowing, and its specific volume, from the continuity relation, as follows:

$$V_e = \frac{\dot{m}v}{A}$$

where V_e = exhaust velocity (normal to blade row), fps
     $\dot{m}$ = quantity of steam flow, lb/sec
     v = specific volume of steam at exit, ft³/lb
     A = annular area of last stage, ft³

This calculation is approximate because it gives only the axial flow component. However, the exhaust component is nearly axial at high powers when $\dot{m}$ is large, so the equation is useful in this simplified form.

Another exit loss component is the "hood" loss due to pressure drop between the last stage blades and the exit flange to the main condenser.

Example 5.1: The annular exhaust area of an LP turbine is 4.16 square feet. The S.L.E.P. at full power is 5″ Hg and 4.7% moisture. Steam flow is 144,600 lb/hour. Estimate the exit loss in BTU/lb.

Solution: Using steam tables:

$$p = 5'' \text{ Hg}$$
$$m = 4.7\% \text{ (moisture)}$$
$$v = 136.5 \text{ ft}^3/\text{lb}$$

Also, $\dot{m} = 144{,}600/3600 = 40.2$ lb/sec

Using the equation for V_e:

$$V_e = \frac{40.2 \times 136.5}{4.16} = 1319 \text{ ft/sec}$$

$$\text{Exit loss} = \frac{V_e^2}{2g_cJ} = \frac{(1319)^2}{64 \times 778} = 34.9 \text{ BTU/lb}$$

Mechanical Losses

The mechanical losses in the turbine are those associated with friction in the turbine bearings and shaft packing. Heat thus produced is carried away by circulating lube oil and is not available for conversion to work. This loss is small. Its exact determination can be made only by calculations based on test measurements, but a fair approximation is given by the following relation, adapted from Brown and Drewry:*

$$L_m = \frac{1.46(h_a - h_e)}{\sqrt{HP(\text{rated})}} \text{ BTU/lb}$$

where: $h_a - h_e$ = actual enthalpy drop at rated power.

Example 5.2: A 40,000 HP turbine at full power receives steam at 1050 psig and 940°F and discharges to a condenser at 5'' Hg pressure. The quality of steam at exit is 92.5%. Estimate the mechanical losses.

Solution:

$h_a = 1469$ (approx.; from steam tables for inlet condition)

$h_e = 1043$ (approx.; from steam tables for S.L.E.P.)

$$L_m = \frac{1.46}{\sqrt{40{,}000}} (1469 - 1043) = 3.11 \text{ BTU/lb}$$

This loss may be considered constant (on a per pound of steam flow basis) at all powers, but note that the formula is valid only when rated power values of enthalpy change and horsepower are used.

Summary

In summary, the *mechanical efficiency* of the turbine may be defined as the energy at the turbine shaft divided by the energy available to the turbine:

$$\eta_c = \frac{(h_a - h_e) - \dfrac{V_e^2}{2g_cJ} - L_m}{h_a - h_e'}$$

*Brown, E. H., and Drewry, W. K., "Economy Characteristics of Stage Feed-water Heating by Extraction," *Trans. A.S.M.E., 1923.*

where the symbols have the meaning previously given. This is often called *engine efficiency.*

Example 5.3: The turbine above (for exit loss calculation) delivers 20,000 SHP at full power. Its steam inlet conditions are 540 psig and 840°F. Find internal efficiency and engine efficiency.

Solution:

h_a = 1432 BTU/lb (from steam tables, for inlet conditions)

h_e = 1071 BTU/lb (from steam tables, for S.L.E.P.)

h'_e = 977 BTU/lb (from steam tables, for isentropic drop to exh. pressure from inlet conditions)

$$\eta_i = \frac{h_a - h_e}{h_a - h'_e} = \frac{1432 - 1071}{1432 - 977} = 79.3\%$$

$$L_m = \frac{1.46(1432 - 1071)}{\sqrt{20,000}} = 3.7 \text{ BTU/lb}$$

$$\eta_e = \frac{(1432 - 1071) - 34.9 - 3.7}{(1432 - 977)} = \frac{323}{455} = 70.9\%$$

5.3 Turbine Control

Control of the quantity of steam permitted to flow through the turbine is the obvious means of controlling speed and power. The most apparent way to do this is by use of a simple throttle valve to restrict the flow. However, such a valve also decreases the pressure on its downstream side, and reduces the amount of energy per pound of steam available for conversion to work. One may recall that the throttling process is one of constant enthalpy. It is represented in Figure 5.11 by the horizontal process line connecting h_a, representing throttle inlet conditions (at steam supply pressure and temperature), with h'_a. Note that the energy available to the turbine per pound of steam is decreased by the amount $h'_e - h_e$. Therefore, pure throttling is unsatisfactory as a means of control.

If steam can be supplied to the first stage nozzles at line pressure, and the nozzle area can be varied to control flow, the above loss is practically eliminated. This can be accomplished by arranging several valves to open nozzles, or groups of nozzles, in the first stage in sequence. Schematically, such an arrangement is shown in Figure 5.12. The forward end of the turbine is shown with a portion of the steam chest removed to reveal three separate chambers containing nozzle blocks. The three nozzle control valves open sequentially to admit steam to the nozzles. One valve is opened fully before the next commences to open. The valves are all controlled by a single handwheel and operated by a cam and spring arrangement. In some modern installations, the control handwheel operates a power unit, which in turn controls the valves. When any of the valves

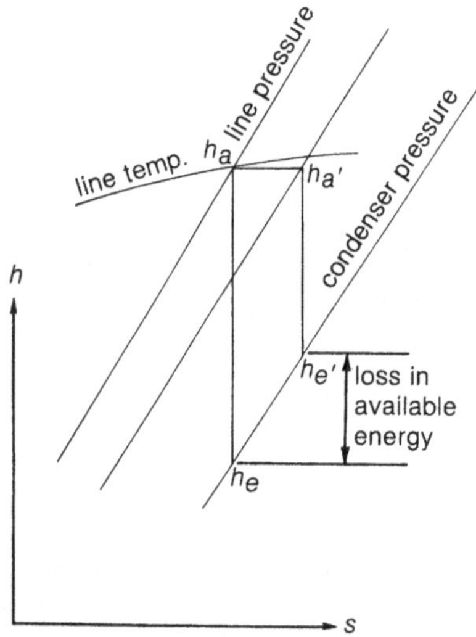

Figure 5.11 Mollier diagram indicating loss in available energy due to throttling

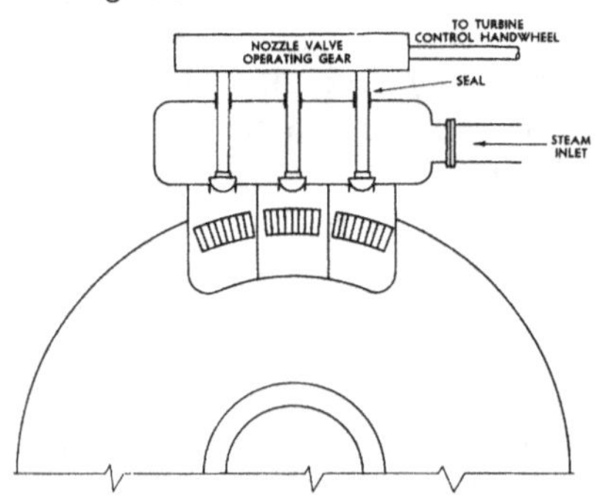

Figure 5.12 Nozzle valve control arrangement

are partially open, some throttling still occurs. From two to five nozzle control valves may be found on modern installations.

An alternative arrangement is shown in Figure 5.13. In this arrangement the nozzle control valves are individually operated by hand, and are either wide open or closed. Speed combinations requiring 4, 6, 7, 9, 11, 12, or 14 nozzles can be set with the three nozzle control valves. Speeds intermediate between these combinations are obtained by necessary adjustment of the main throttle valve.

Another nozzle control arrangement used in newer ships is shown in Figure 5.14. There is no throttle valve; the steam enters the turbine through nozzle control valves, which are operated by a lifting bar mechanism. The valve stems extend through the holes in the lifting bar and are fitted with the shoulders at the top ends. The valve stems are of varying lengths; as the lifting bar is raised, the valves open in succession, the shorter ones first, then the longer ones. When the lifting bar is lowered the valves come to rest upon their seats and are held down by the bar and steam pressure in the steam chest.

The valve seats are sealed against steam leakage by a sequence of piston rings in the bottom part of the seat block. Also shown are the passages in the lifting rod bushings and chest cover for leak-off steam, which is usually taken to the gland steam line.

The problem of efficient turbine control over a wide speed range is not completely solved by the above scheme. Variation in blade speed–steam speed ratios occurs at different wheel speeds. Recall that the enthalpy drop across the turbine depends on the initial pressure and temperature

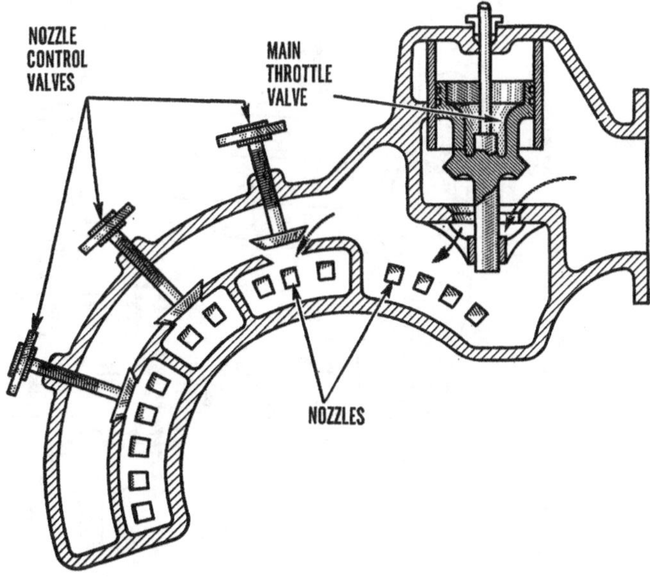

Figure 5.13 Nozzle valve manual control arrangement

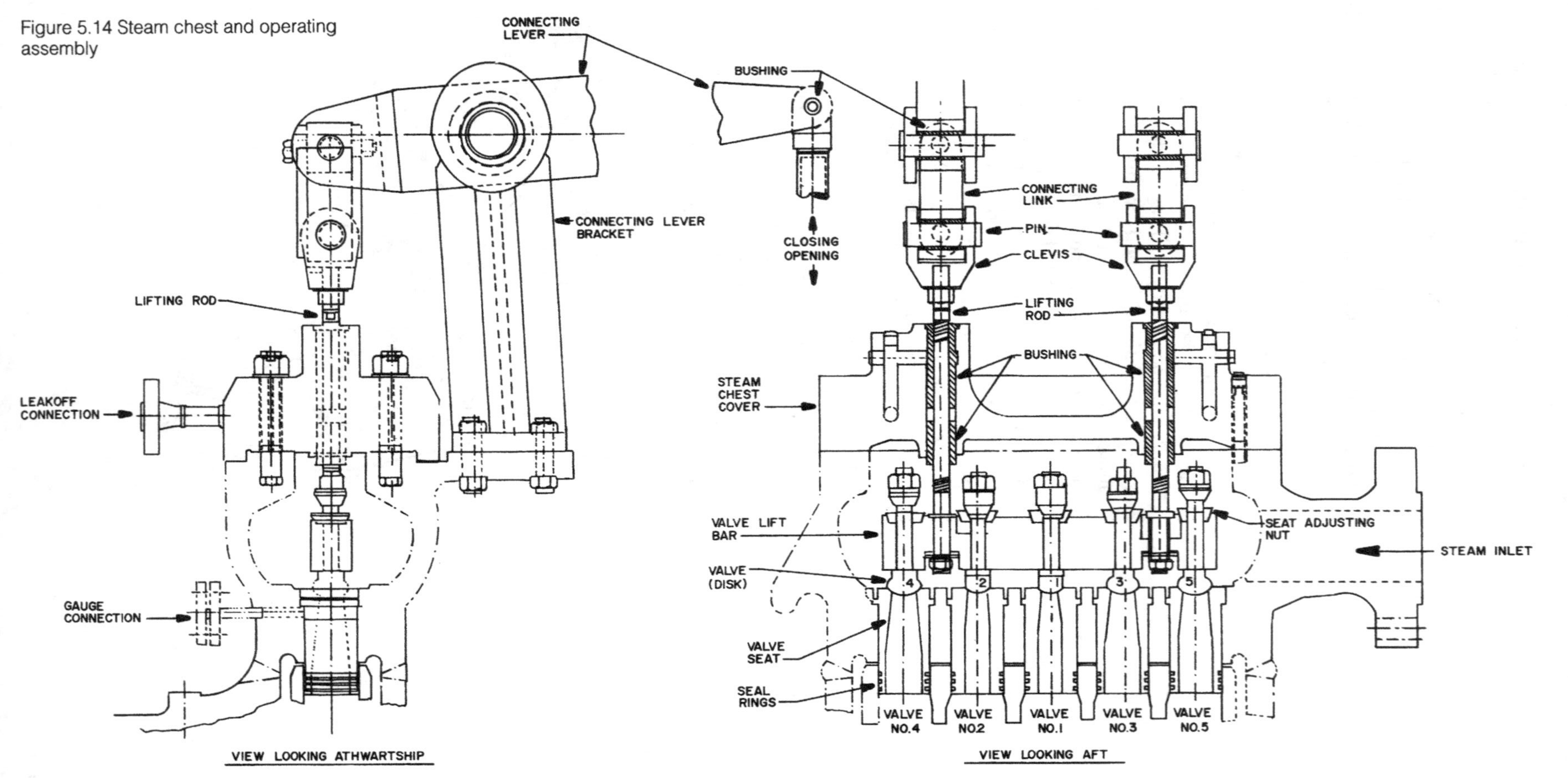

Figure 5.14 Steam chest and operating assembly

of the entering steam and the final pressure of the exiting steam. For an ideal turbine with nozzle valve control, this total drop would be constant at all loads, except for minor variations in initial conditions caused by some throttling, and in the exit conditions caused by slight changes in the condenser pressure with load. The portion of the enthalpy drop taken across the first stage varies considerably with the load, but the remaining drop is divided about equally among the remaining stages, and each stage enthalpy drop varies only slightly with the load. Since the nozzle velocity in each stage is proportional to the square root of the enthalpy drop, the nozzle velocity in each stage varies only slightly at varying loads. On the other hand, the blade speed varies over a wide range. For a turbine designed with the ideal velocity ratio at high powers, it follows that smaller stage enthalpy drops and consequent lower nozzle velocities would be desirable at low powers. With a given total enthalpy drop, the only way to obtain smaller stage drops is to add more stages.

Use of the cruising turbine permits variation in the number of stages. The cruising turbine (see Figure 5.5), placed ahead of the high pressure turbine, is added to the steam flow path at lower power operation and bypassed at higher powers. In addition, in many naval turbines operating at very high powers, the first few stages of the high pressure turbine are bypassed. The combination of these two schemes permits maximum conversion of thermal energy to work in each used stage over a wide speed range. A typical bypassing arrangement is shown in Figure 5.15. Efficiency considerations aside, bypassing also permits steam to enter the turbine at a point where the total nozzle area is greater than in the first stage, and so permits greater flow rates and consequently higher power from a given turbine. The bypassed stages do no effective work. The valve areas are so arranged as to cause a small pressure drop across these stages, so as to provide a continuous flow of steam to prevent a dangerous build-up of temperature due to windage in "dead," or nonflowing, steam.

One disadvantage of bypassing in the high pressure turbine is the exposure of the lower pressure turbine casing to essentially line pressure and temperature, and the turbine blading there to the high temperatures. Avoidance of this condition dictated from the inception of this bypassing mode of operation use of a velocity compounded first stage wheel, as previously discussed. Even with the lower pressure and temperature of the old 600-psig systems that were in common use for many years (600 psig, 825°F), the blading could not be continuously exposed to the temperature, but exposure for the relatively small percentage of total turbine life spent at high powers was acceptable. With the

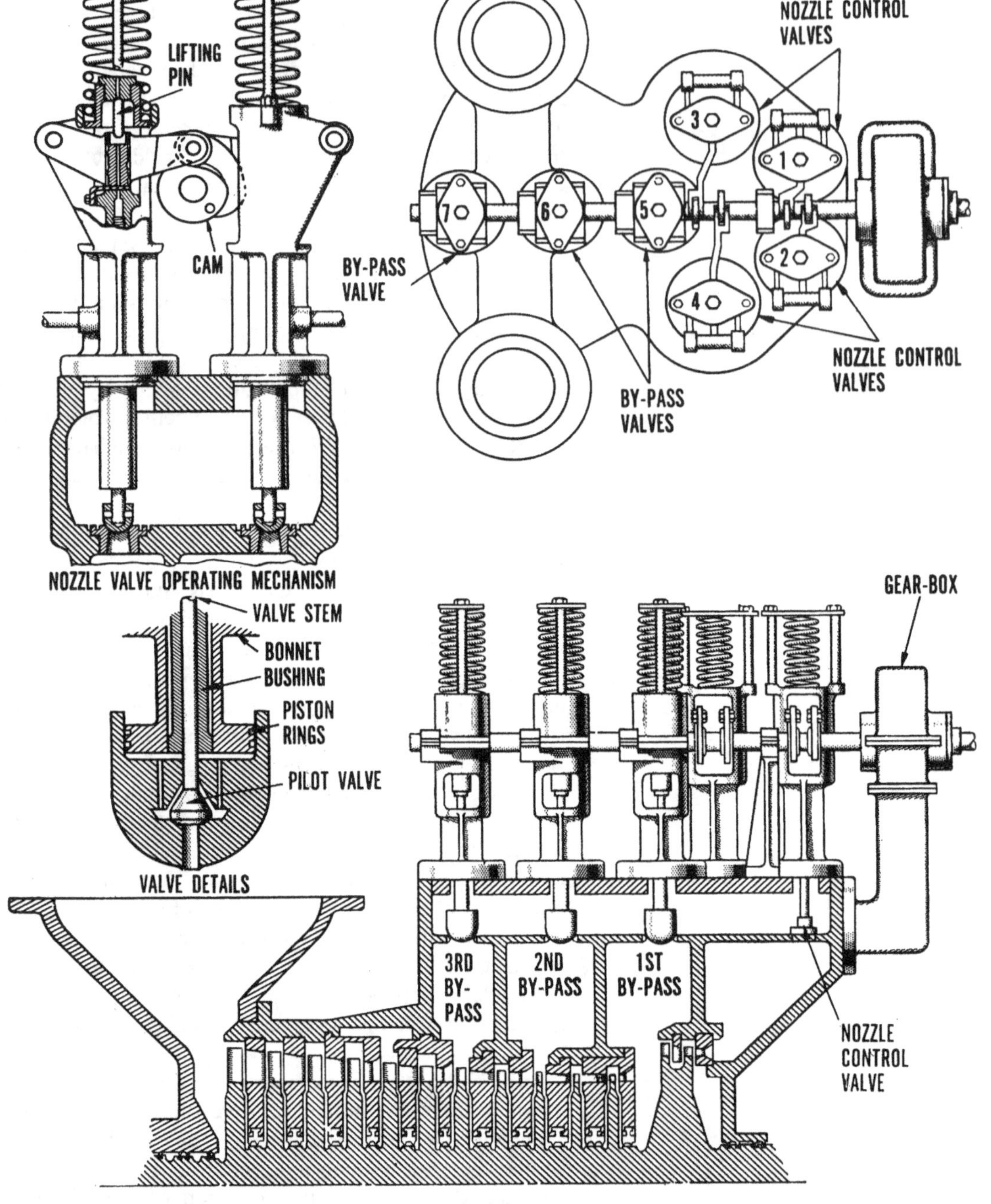

Figure 5.15 Nozzle control and bypass valve arrangement

more recent 1200-psig systems (1200 psig, 950°F), even short periods of exposure are unacceptable, and the velocity compounded wheel is not bypassed.

Another scheme for varying the number of stages is embodied in the CV-60 turbine, shown in Figure 5.16. In this turbine, line steam is admitted in the center of the high pressure–intermediate pressure unit. At low powers, only the nozzle control valves to the HP section are open. Steam flows through the velocity compounded wheel and the five remaining stages of the HP section, then through a transfer line to the seven stages of the IP turbine, from which it passes to a conventional low pressure turbine. At high powers, the nozzle control valves to both sections are opened. The entering steam divides, and a portion passes through the HP and IP sections simultaneously. Valves in the transfer line are arranged to join the exhausts from these two turbines to pass together into the low pressure unit. A large enthalpy drop is taken in the first stage IP (note the larger wheel diameter, giving greater blade speed), although it is not velocity compounded. This turbine also has an extraction opening to provide steam to the DFT for feed heating, a feature seldom found on naval turbines, although common in commercial units.

Good turbine control, then, embodies a steam admission arrangement to minimize throttling losses, and a scheme for varying the number of stages. Such complex controls are seldom found on small auxiliary turbines, which are usually controlled by a simple throttle valve, the higher percentage losses being acceptable because of their relatively low power. Ship service generator turbines, being larger, usually have nozzle valve control, but no bypass arrangement.

Figure 5.16 CV-60 HP-IP turbine

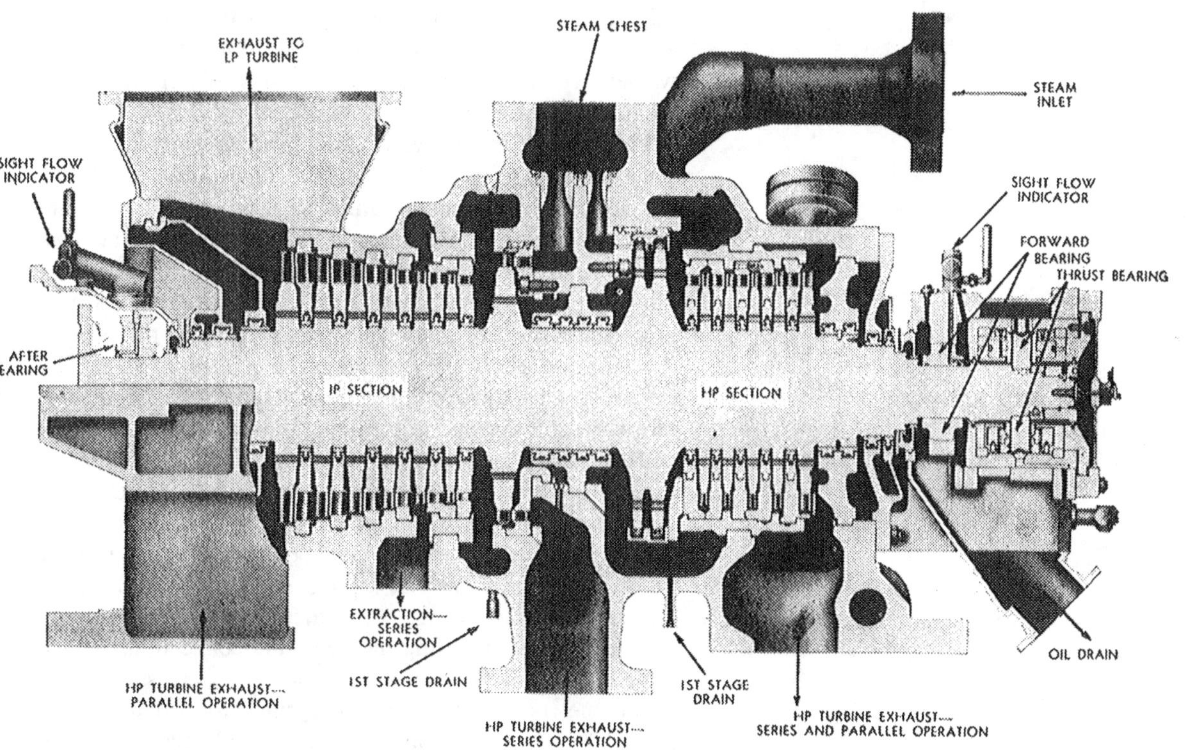

5.4 Turbine Components and Mechanical Features

The principal components and mechanical features of steam turbines to be discussed are: (a) the casing; (b) the rotor, its support and positioning features; (c) the seals or glands used to keep the steam in the turbine and the air out; (d) the blading (although the principles were treated above, the mechanical features such as attachment methods are discussed below); and (e) elements of lubrication.

There are many other features important to turbine operation and maintenance, such as valves, pumps, valve gear operation, removal and replacement of parts, rotor balancing, and so on; but these topics are beyond the scope of this book. Moreover, detailed descriptions and instructions for maintenance and overhaul are available in equipment-specific reference manuals on board each ship.

Casing

The turbine casing is a containment vessel supporting the rotor bearings and providing the structure needed to guide the steam flow through the nozzles and buckets. Casings are usually split horizontally to permit easier access for inspection and repair. The flanges are accurately machined to ensure a steam-tight metal-to-metal fit, and the flanges are strongly bolted together. Flange seals, except for corrective maintenance measures, are not normally used. In some (high temperature) units, both top and bottom casing halves are made of two vertical castings bolted together and seal-welded. The inlet end is made of alloy steel (Cr-Ni), while the exhaust end is made of carbon steel.

Each casing has a steam chest to receive the incoming steam and to deliver it to the first stage nozzles. In very large and high temperature units the steam chest is separate from the main turbine structure. In marine turbines the steam chest is mounted directly on the casing. Since a reaction force acts on stationary nozzles and blading, the turbine casing must be securely fixed to a foundation to resist these forces and to prevent the turbine from moving. The casing, and also the rotor, move and expand due to temperature variation during start-up and shutdown. To allow for both thermal expansion and resistance to reaction forces, usually the low pressure end is firmly secured to ship's structural steel, and the forward end is allowed to move axially, either by use of elongated bolt holes, sliding feet, or by securing the front end to a deep flexible I-beam, arranged below the casing, perpendicular to the turbine axis.

Openings in the casing include drain connections, steam bypass connections, openings for rotor ends, pressure gages, thermometers, relief valves, and holes for balancing the rotor. Casings and steam chests are usually cast steel with molybdenum as the primary alloying agent for 600 psi

(825°F) applications. A 1200 psi unit may be a 2½% Cr, 1% Ni alloy with a percent of Mo and/or Cb each. (The chemical symbol Cb represents the older name colombium for element 41, niobium, in the periodic table. According to international agreement, the periodic table carries the designation Nb. The metallurgical trade, however, where this element is used for alloys for its high-melting-point properties, continues using Cb.)

Rotor

The turbine rotor carries the moving blades and transmits the work (torque) from the turbine. Older disc and diaphragm turbines had rotor wheels and shafts forged separately, and then the wheels were shrunk and keyed onto the shaft. The newer, drum type, units are made of one forging and then milled to receive the blade roots.

The turbine shaft pierces the turbine casing at both ends. Each end passes through a sealing arrangement as it leaves the casing, to be discussed later, and then through bearings, which carry the weight of the rotor. The portion of the shaft passing through the bearing is called the journal, and must be finished with an exceedingly smooth surface. The journal rests in a sleeve type bearing (as opposed to a ball or roller type). Two important types are shown in Figures 5.17 and 5.18. Note that both are provided with a babbitt lining, which is cast into a steel shell. The shell is, in turn, mounted in a housing rigidly attached to the turbine casing and supported on the turbine foundation. The housing is flanged so that its upper portion, or cap, can be removed to take out the bearing. Bearings are made in two halves; the upper half can be lifted off when the cap is removed, and by jacking the rotor weight off the lower half, the latter can be rolled out around the shaft. The upper half is doweled to the cap to prevent rotation of the bearing with the shaft. The spherically seated type in Figure 5.18 is most commonly found on modern installations. The spherical seating feature allows the bearing automatically to seek exact alignment with the shaft. Note the provision for positive lubrication, and the oil deflector rings, which prevent oil loss from the lubrication system and its contamination by steam present on the turbine casing side of the bearing.

Figures 5.19 and 5.20 show the end view of the bearing housing assembly, and the housing assembly with the thrust bearing arrangement, respectively.

In addition to carrying the weight of the rotor, the bearing provides a means of adjustment for its exact radial position. This is important since radial clearances are critical in the turbine labyrinth packing and, in many cases, between the blade ends and the casing. Initial radial adjustment and subsequent compensation for wear can be made by the selec-

Figure 5.17 Cylindrical journal bearing

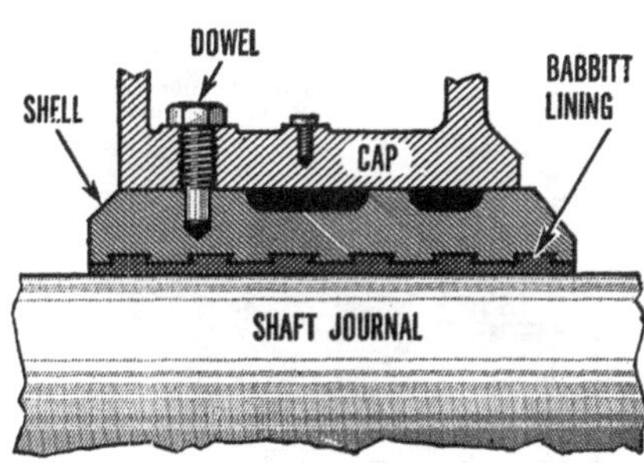

Figure 5.18 Spherically seated journal bearing

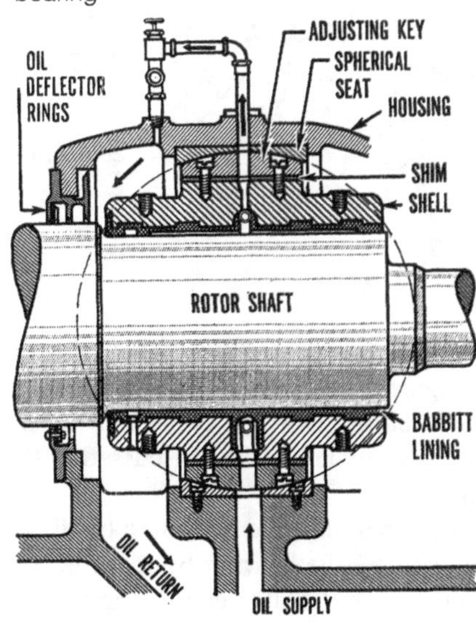

Figure 5.19 HP turbine bearing, end view

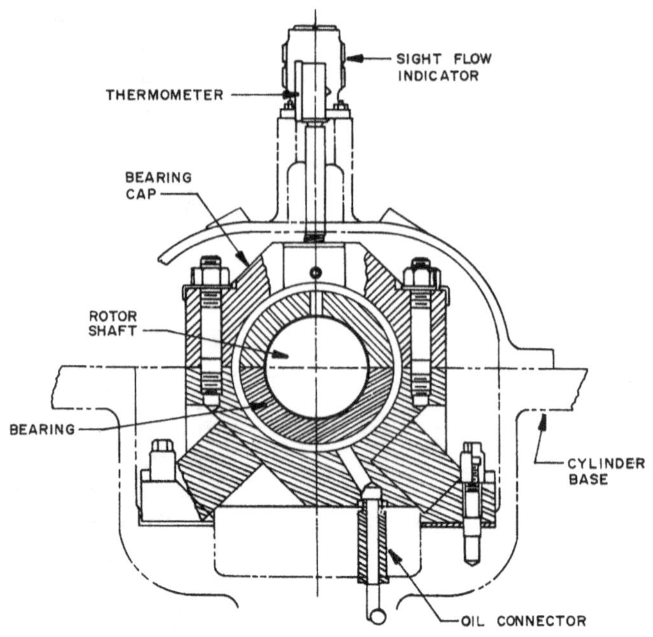

Figure 5.20 Bearing housing assembly, LP turbine

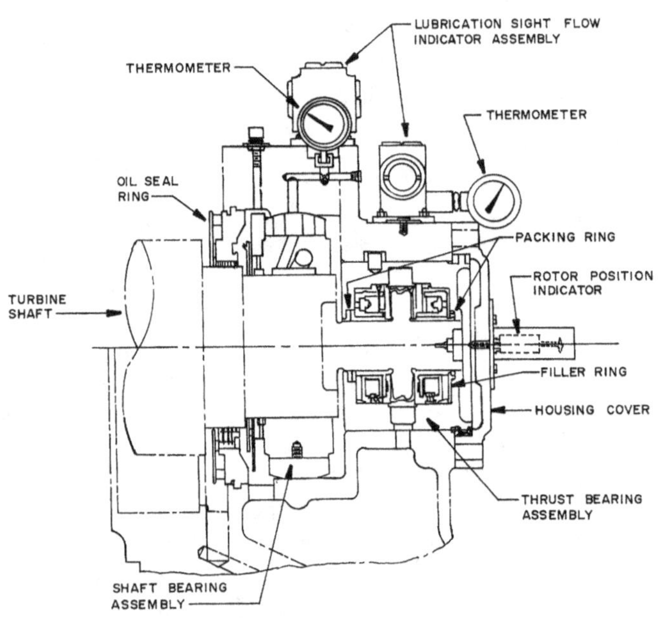

tion of proper shim thicknesses between the shell and the housing, or seat (see Figure 5.18).

Another type journal bearing that has come into recent use is the pivoted shoe type, which will be discussed in the section on lubrication.

Bearing loads of journal bearings are measured by dividing the total weight carried by the bearing by its projected bearing area, the length times the diameter. Bearings are made large enough to limit these unit pressures to about 200 psi. Except for a short period during start-up when a brief metal-to-metal contact exists between the journal and the lining, these surfaces are separated by a thin film of oil, which also transmits the load from the shaft to the bearing.

As already mentioned above, some axial force is always present on the turbine rotor, even in impulse type turbines. It is necessary to absorb this thrust to maintain the close axial clearances required between the moving blades and fixed blades, the shrouds, seals, and nozzles. The turbine thrust bearing performs this function. It usually is located in the forward end of the turbine shaft. Kingsbury or pivoted-shoe thrust bearings are usually used in propulsion turbines (see Figures 5.2, 5.3, and 5.20). The Kingsbury bearing is discussed fully in the section on lubrication below.

In single flow reaction turbines, the axial thrust is partially counterbalanced by the use of a dummy piston and cylinder arrangement and by a set of dummy seals (see Figures 5.2 and 5.21). During normal operation the net force on the blading tends to push the rotor toward the low pressure end. To provide for a balancing force back toward the high pressure end, an equalizing pipe is installed between the turbine exhaust and space A, Figure 5.21. As the shoulder of the rotor is under full inlet steam pressure, the pressure difference between the shoulder and space A provides a balancing force toward space A.

In double flow turbines, where the axial thrust is balanced by equal and opposite flow through the turbine (see Figure 5.3), dummy pistons and cylinders are not required.

In order to protect the turbine rotor and hydrodynamic bearing material at times when the turbine is stopped while under steam, all turbines are equipped with a motor-driven turning or jacking gear (see Figure 7.10, below). On geared turbines the jacking gear motor is mounted on the reduction gear casing. It is also used to turn the rotor, and hence the reduction gear, during gear teeth inspection. The jacking gear is normally used during warm-up and securing periods while the turbine is under partial steam pressure, to permit the rotor to warm and heat evenly. The heavy rotor of a hot turbine, if left stationary, will sag and bow in a few minutes, causing heavy rubs and damage to blading and seals. The

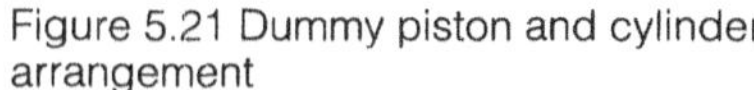

Figure 5.21 Dummy piston and cylinder arrangement

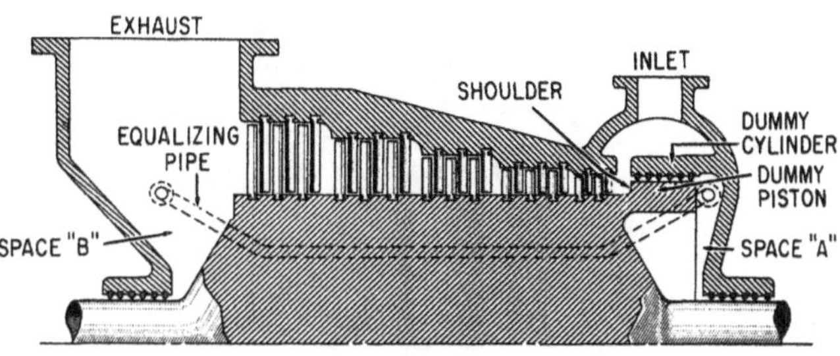

jacking gear motor is equipped with an engage-disengage clutch and a locking device for locking the propeller shaft. The normal operating speed is one revolution of the shaft every few minutes.

Rotor position indicators are provided at the end of both HP and IP turbine shafts (see Figures 5.2 and 5.3). The rotor position indicates whether the rotor is excessively displaced in the cylinder, and also the clearances on the thrust bearing. The LP turbine, which contains the astern elements, must be monitored often during change-over of speeds from ahead and astern, as maximum differential expansion will be encountered then due to changing steam and temperature conditions in the turbine. The dial position indicator is mounted to the bearing housing base (Figure 5.20), and the readings are taken by pressing an extension rod into the end of the turbine shaft. Otherwise the extension rod is retracted to reduce the wear to a minimum.

Seals, Glands

The necessity for sealing between stages has been previously discussed. The leakage path in the impulse stage

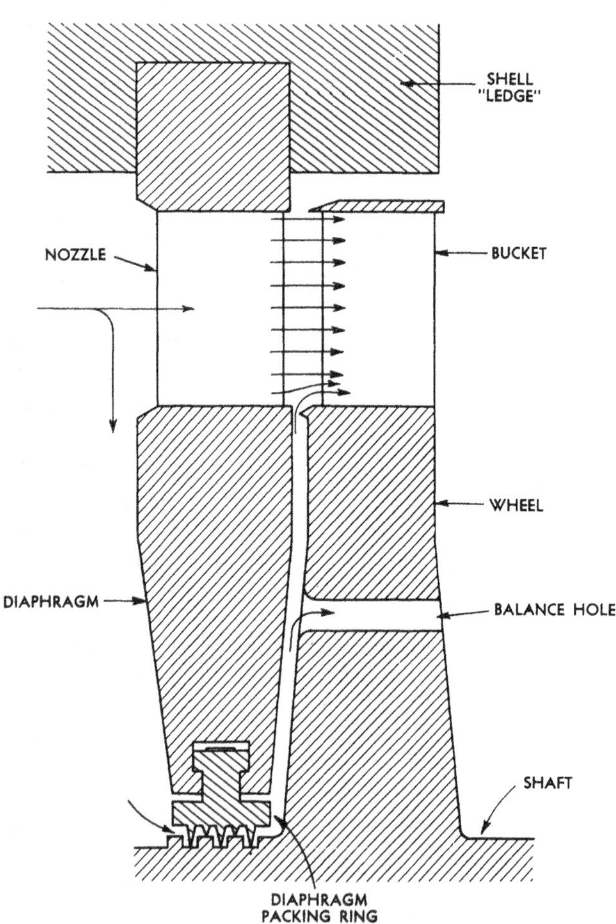

Figure 5.22 Steam leakage in impulse stage

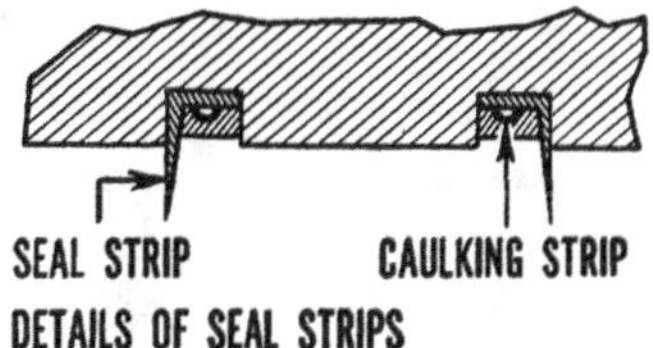

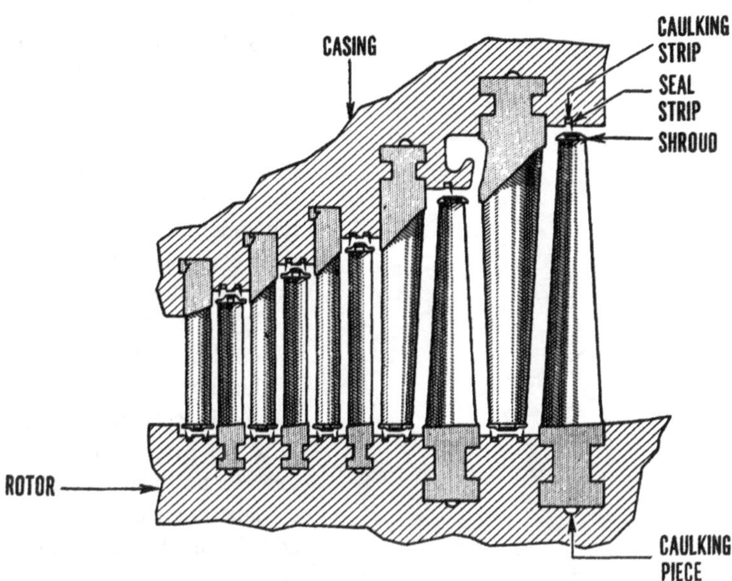

is between the rotating shaft and the fixed diaphragm, as indicated in Figure 5.22. In the reaction stage, the leakage path is around the ends of the fixed end moving blades, as shown in Figure 5.23. In both cases the small annular passages between the sharp projections and the shaft, or casing, act as flow-restricting orifices. These are effective to the extent of reducing to about 4–5% the amount of steam that would pass through a minimum friction passage (or nozzle) with the same cross-sectional flow area and the same pressure drop. The effectiveness increases with the number of "teeth" in the packing, and longer labyrinths are placed between stages where the designed pressure drop is higher. The cross-sectional area is kept small by limiting the clearance between the packing projections and the shaft to about 0.010″.

Another vital sealing point is the juncture of the shaft and turbine casing. The seal here serves two functions: (a) it minimizes steam leakage from the casing when the casing (shell) pressure is higher than atmospheric; (b) it prevents the leakage of atmospheric air into the turbine (and thus into the condenser) when the shell pressure is less than atmospheric. Both possibilities may exist at one time or another at all seals, except at the low pressure exhaust end, where the shell pressure is always lower than atmospheric.

It is very important to prevent all air from leaking through the shaft end seals, called glands, as the air leaking into the

LP turbine will find its way into the condenser and destroy the vacuum. In addition, it will contribute to corrosion of high carbon steel parts. Shaft sealing is accomplished by means of gland sealing steam, which is supplied to the glands from the gland seal system.

Two types of gland seals, labyrinth and carbon packing, are used. Typical labyrinth seals are shown in Figure 5.24. Labyrinth type seals, called labyrinth packing, are used alone or in conjunction with carbon packing rings at points where high temperatures and pressures may be present. The carbon packing, when used with labyrinth packing, is always on the outboard end of the latter, where pressures and temperatures are lower. Figure 5.25 illustrates such an arrangement. At other points such as the LP turbine exhaust end, carbon packing alone may be used on some installations, although often labyrinth packing is used here also.

Figure 5.25 shows the location of the leak-off points. When the shell pressure is above atmospheric, they bleed off leakage steam to prevent its loss. Steam throttled from the auxiliary exhaust line is supplied under low pressure, ¾ to 1½ psig, to the gland seal connection. This steam divides, a part leaking back along the labyrinth packing to the exhaust connection, a part through the carbon packing (or additional labyrinth packing in many cases) to the shaft end.

Figure 5.24 Labyrinth seals. A) High pressue stepped labyrinth B) Low pressure straight-through labyrinth

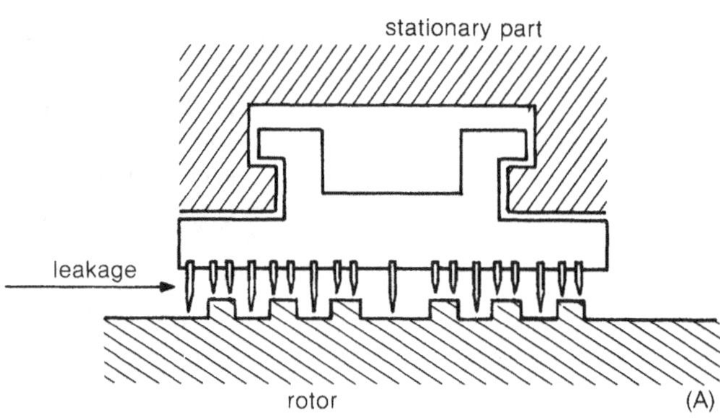

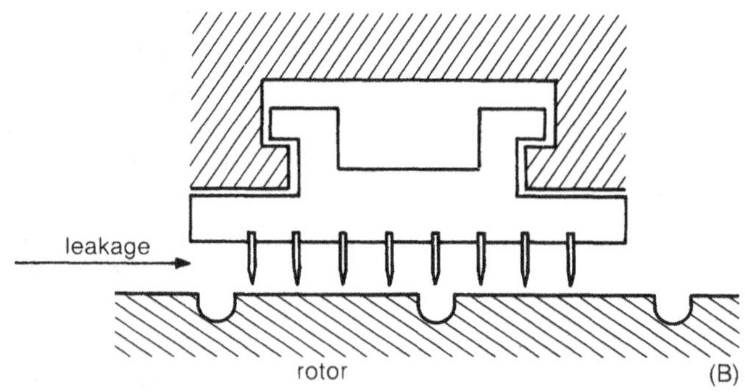

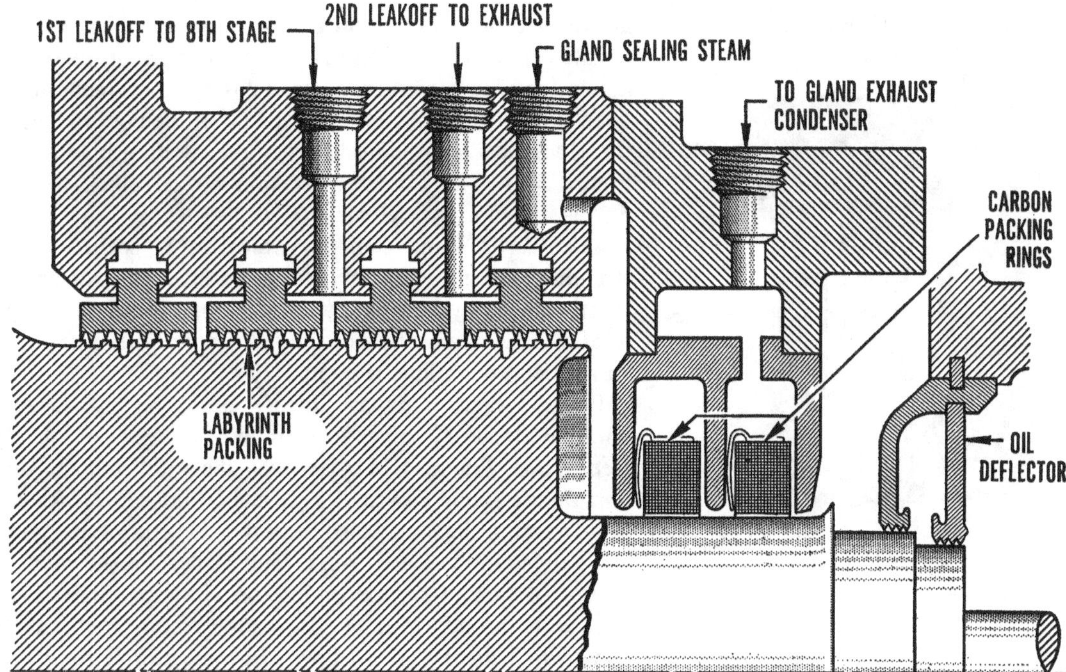

Figure 5.25 Typical turbine gland

This is a small quantity, which condenses here and drips off, the remainder passing into the atmosphere. This steam is supplied from a system so valved that it supplies steam only when the shell pressure is lower than atmospheric, its flow effectively preventing the leaking of air through the packing. The detailed arrangement of the gland seal system differs considerably on various installations. The carbon packing rings are made up in segments, held tightly around the shaft and against vertical sealing faces by springs.

Low temperature rotors may be made of hot rolled carbon steel bar stock. High pressure and temperature rotors and newer LP rotors are alloy steel forgings containing less than 0.5% carbon, between 1 and 2% manganese, chromium, molybdenum, nickel, silicon, sulfur, and phosphorus. The seals and glands are made, in addition to carbon packing, of stainless steel, nickel brass, and corrosion-resistant chrome molybdenum alloys. The seal springs are inconel, monel, and stainless steel. Casing bolts in high pressure turbines are subject to creep and must be tested periodically. To help resist elongation under stress and thermal cycling, bolts are made of varying alloys containing chrome, molybdenum, vanadium, and tungsten.

Blading

Turbine blading design and types have been discussed above in several sections. Still to be covered are the methods of blading attachment to the casing and the rotor.

The HP turbine nozzle blocks (see above, Figure 4.10)

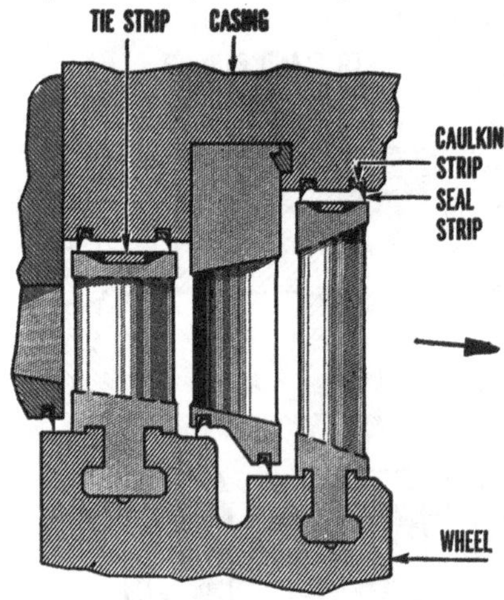

Figure 5.26 Impulse blading

Figure 5.27 Methods of fastening blades. A) Fir- or pine-tree B) Dovetail C) Bulb and shank D) T-slot E) Straddle-T F) Straddle fastening G) Outside straddle H) Inverted circumferential dovetail I) Pine tree dovetail

are formed of two concentric rings secured by bolts to the nozzle chamber. Since all marine turbines are of the partial admission type (steam enters only through a segment of the nozzle block), much of the top and all of the bottom half of the block acts as a baffle between the rotating blades (see Figure 5.2). The LP astern section nozzle block (Figure 5.3) is bolted to the cylinder.

The Curtis stage stationary blading is secured in a straight-sided groove machined in the blade ring or casing (Figure 5.26) by keys or caulking strips. The Curtis stage rotating blades are often of the side entry type inserted in various configurations (Figure 5.27).

Rateau stage stationary blades are arranged as diaphragms, made in halves to surround the rotor completely (see above, Figure 4.11). Between the rotor and the diaphragm are found labyrinth type seals (Figure 5.2). Rateau stage rotating rings may also be of the side entry type, similar to those in the Curtis stage. The blades are shrouded in groups of four to six, depending on vibration characteristics, with shroud strips fitted over the tenons at the blade ends, which are then riveted over. In some turbines only two shroud bands (each a semicircle) have been used. The rotor discs for Rateau stages are machined with axial holes to equalize pressures on disc sides (Figures 5.2 and 5.7). Interstage steam leaking is minimized by the stationary blading labyrinth seals.

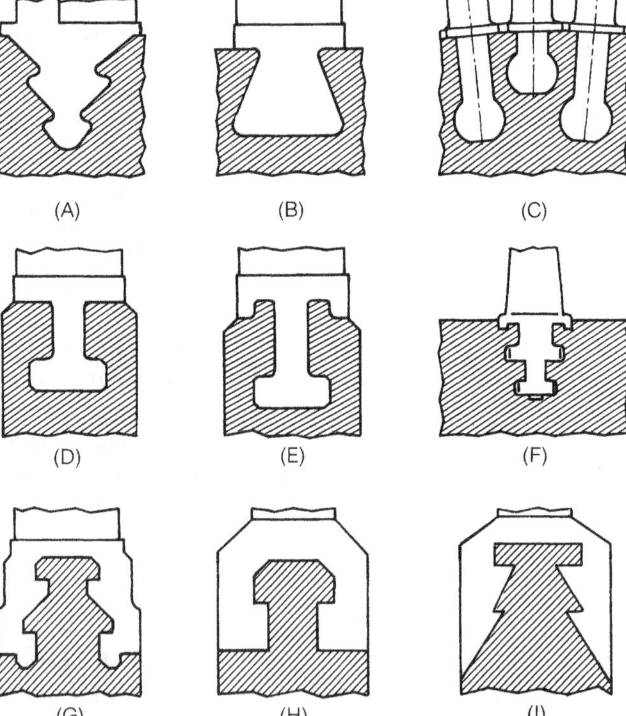

(A)   (B)   (C)

(D)   (E)   (F)

(G)   (H)   (I)

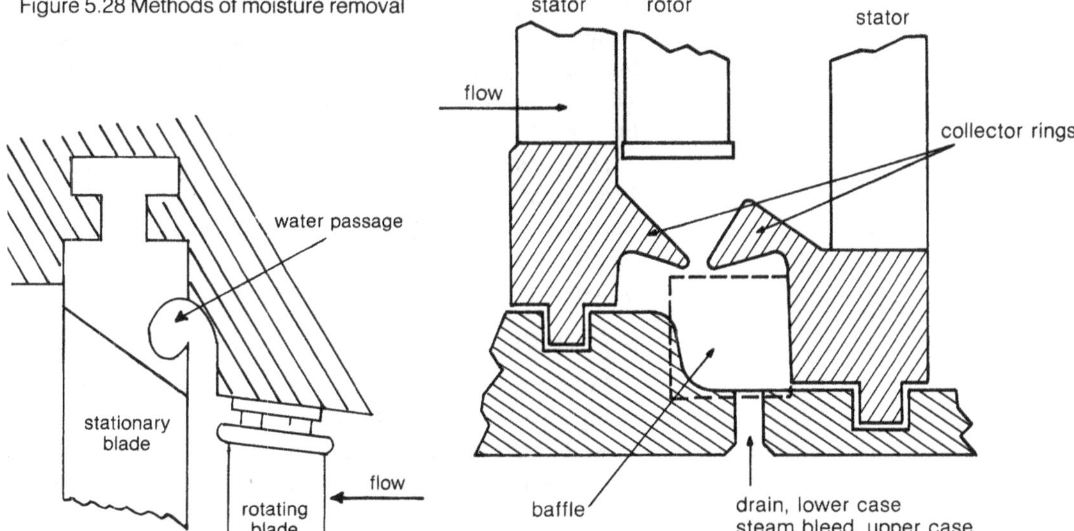

The reaction stage (LP) stationary blades are made up similar to diaphragms with a 180° outer ring, blades, and inner shroud strips against the rotor. They are secured in the cylinder with caulking strips. The rotating blades are attached to rotor machined grooves, usually with T-type blade roots (Figure 5.27). The blades are again fitted with shrouds over tenons.

A serious problem in all steam turbines is blade erosion due to condensed moisture in the last stages. When the moisture content in the steam reaches about 10%, enough moisture condenses out in sufficiently large droplet sizes for it to be deposited on the leading edges of the stator blades. Since the initial droplets are very fine, little or no damage is caused to the stator blades. However, the passing flow and continuous accumulation will drag the moisture to the stator trailing edge, where much larger droplets will be torn off, which then impinge upon the rotor blades. Because of the large relative velocity between the droplets and the rotating blades, serious erosion damage is caused near the leading edge of the tip portion of the rotating blades (see previous discussions in Sections 4.7 and 4.9). To minimize erosion damage, stellite shields are installed over the outer span of the rotating blades (see Figure 4.28, page 83). The last two stages of the LP turbine are often provided with circumferential grooves at the inlet side of the diaphragms to allow condensed moisture, gathered on rotating blades, to be thrown out through drainage holes into exhaust passages. Drain holes are also provided from the inlet casing and the astern cylinder into the exhaust passage.

Figure 5.28 shows two methods that are used to remove moisture from blade rows. In nonnuclear cycle turbines, where the steam is initially admitted to the turbine in a superheated state, only the last few LP stages need protection from erosion-causing water droplets. In nuclear cycles, the steam is initially moist (1–2%), and most of the turbine blading operates in steam that contains 5–12% moisture. This condition requires moisture-extraction devices and stellite shields in most rows of the turbine.

High temperature nozzles and diaphragms are rolled or forged chromium steel with vanes machined and welded into diaphragms. Rotating blades are usually high chrome or Ni-Cr steel. High temperature buckets are often hot rolled. Shroud bands are stainless steel.

5.5 Lubrication

Lubrication plays an important role in efficient operation of the main steam propulsion turbine and auxiliary turbines, as well as gas turbines. One of the reasons for including it in the chapter on steam turbines is that the steam turbine thrust bearing operation is based on proper oil film formation on the thrust collar, and lubrication provides a vital cooling function in the relatively large steam turbine journal bearings. Gas turbine bearings, including the thrust bearings, are frequently of the ball bearing type. Nevertheless, the basic concepts still apply.

Friction

When two solids are in contact under even a small amount of pressure, forces that tend to prevent their relative motion may arise from three sources: adhesion, shearing, and plowing. In Figure 5.29a, the adhesive action is shown. The two pieces of metal in all three parts of the figure represent finely machined surfaces, with their surface profiles greatly exaggerated in the picture. The force necessary to move one piece relative to the other must be sufficient actually to shear the metal at the cold-welded points. Electrical conductivity tests have shown that the area of the welded surfaces is proportional to the load normal to the interface, so that the contact stress (normal load divided by the contact area) remains practically constant *regardless of the total area of the rubbing surfaces.* This explains why total contact area has negligible effect on the friction force. (It may be recalled that in the standard inclined plane friction experiment, a rectangular block will commence sliding down the incline at practically the same angle, irrespective of whether it rests on either of its sides or its end.) Furthermore, since the shearing force is proportional to the total welded area, and since the total welded area is proportional to the normal load, the shearing force is also proportional to the normal load. This proportionality constant is, of course, the coefficient of friction. Carefully controlled experiments and microscopic examination have shown that practically perfect welds actually occur, even between dissimilar mate-

Figure 5.29 Three types of friction

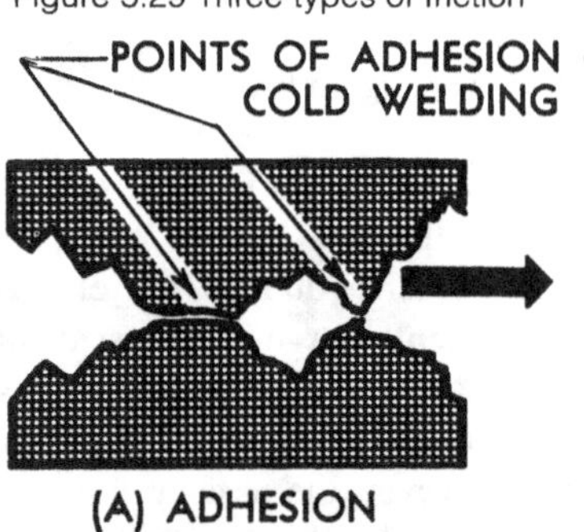

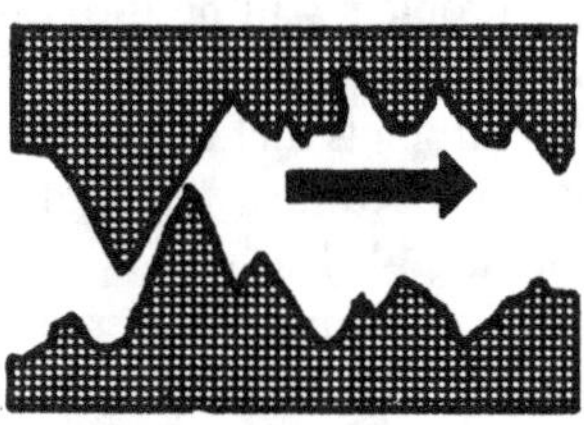

rials. This process is believed to continue even when relative motion exists between the contact faces. Adhesion is thought to be the primary cause of friction between well-machined surfaces.

Shearing, shown in Figure 5.29b, occurs when mutually interfering projections on the surface are broken off. This action is believed to occur particularly with rough surfaces, and especially in "new" contact areas, being significantly present in new bearings and disappearing somewhat after the bearings are well "run in." The action doubtless is accompanied by adhesion, since the fracture energies of the projections cause high localized temperatures.

The third major friction type is plowing, and is present when the two surfaces have different hardnesses (see Figure 5.29c). Plowing may plastically deform the softer material without fracture.

The presence of foreign solid material between the two areas can cause any or all of the above actions.

In rolling friction, the latter two types of friction are probably largely eliminated, but adhesion forces still exist.

Mechanisms of Lubrication

It is the function of lubrication to separate the two metallic surfaces so that they do not touch, and thus to substitute smaller forces of fluid shear for the types of force described above. Three mechanisms of lubrication are well understood: *hydrostatic, hydrodynamic,* and *boundary lubrication.*

Hydrostatic lubrication depends on oil pressure supplied from an outside pump to keep the surfaces apart. If the surfaces shown in Figure 5.30 are rectangular, and high pressure oil is forced into the supply orifice, the surfaces will separate so that the peripheral area around the step in the lower plate will be large enough to accommodate the supply flow. Tremendous loads can be accommodated in this manner. The moving part of the Hale telescope, weighing approximately 1,000,000 lb, is supported on three such bearings or pads, with a combined recess area of 84 square inches, and it is operated by a 1/12 HP clock motor! When several such bearings support a weight, it is a virtual necessity that a separate pump be supplied for each bearing, so that maintenance of the flow to each bearing can be ensured. Tilt of the load to one bearing would allow all the flow of a single pump to pass through the others. Therefore, this system has not been adopted for naval machinery use, as the additional necessary pumps would greatly complicate the reliability requirement.

Hydrodynamic lubrication also depends on oil pressure separation of the bearing surfaces, but here the pressure is developed by the motion, or pumping action, of the bearing

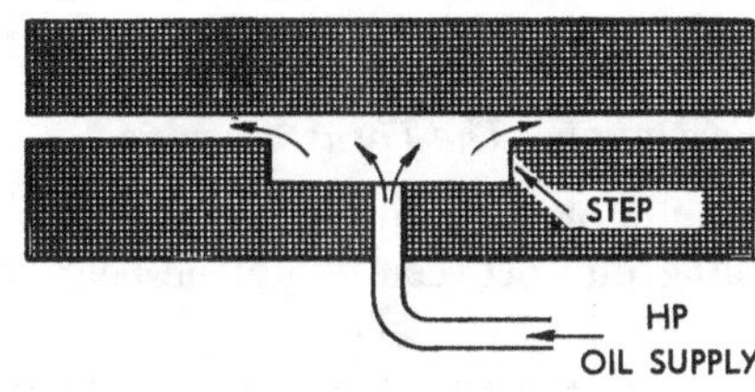

Figure 5.30 Hydrostatic lubrication

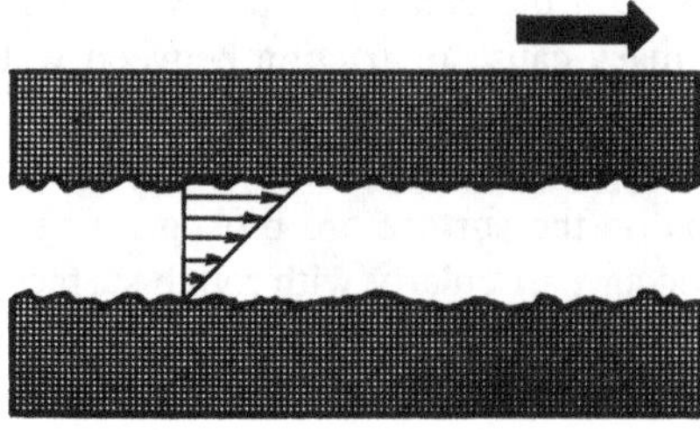

Figure 5.31 Oil film velocity distribution

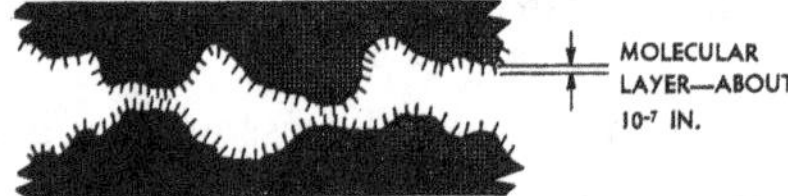

Figure 5.32 Boundary lubrication

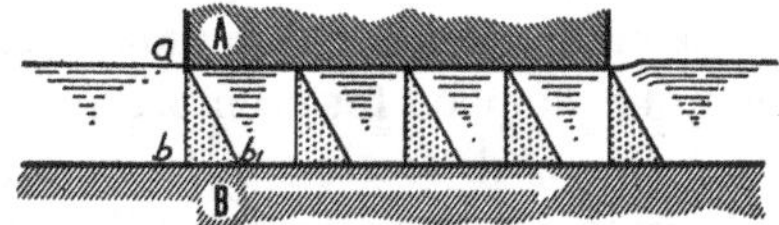

Figure 5.33 Velocity distribution due to motion only

itself. When two surfaces move past each other, as in Figure 5.31, oil clings to both surfaces. The oil at the upper surface has the velocity of that surface; the oil at the lower surface is at rest. The velocity of the molecules of oil between the surfaces varies from one surface to the other. It will simplify the subsequent discussion to assume a linear velocity distribution as shown in the figure. Since flow is set up, there is a pressure build-up. This is the type of lubrication found in nearly all large shipboard bearings.

Boundary, or thin-film, lubrication results from an extremely thin layer of oil, on the order of one ten-millionth to a maximum of one forty-thousandth of an inch in thickness, which is "bonded" to the metal by adsorption or, in some cases, by chemical reaction. Furthermore, the molecules of oil in contact with the surfaces are capable of resisting displacement to a great degree. The molecules (long hydrocarbon chains) are believed to be bonded at one end to the surface and to orient themselves approximately perpendicularly. "Oiliness" is the term that describes the ability of a lubricant to form such a boundary film. It is found to a high degree in animal and vegetable oils and the chemical family of fatty acids, but to an extremely limited degree in natural mineral oils. However, small amounts of additives to mineral oils are sufficient to impart the property of oiliness. Boundary lubrication is schematically illustrated in Figure 5.32. The film gradually squeezes out under load and with time. This film is the only lubrication available to the friction surfaces of shaft journals during start-up. For this reason, it is important to keep it established by periodic "turning over" or operation of machinery. This type of lubrication is important also in ball and roller bearings, and is the principal type of lubrication in gear meshes.

Hydrodynamic Film Formation—The Thrust Bearing

It is useful to examine in more detail the formation of a hydrodynamic lubricating film between two plane bearing surfaces.

Imagine that Figure 5.33 represents the cross section of two plane surfaces A and B, both of which are infinitely long in a direction perpendicular to the plane of the paper. Surface A is fixed, while B is moving at constant speed. Assume for the moment that the two surfaces are separated by a continuous film of oil, and that there is no force tending to cause one surface to move nearer the other. Now, the oil will cling to both surfaces, and hence the particles of oil adhering to A will be stationary, while those adhering to B will have the same velocity as B. Particles lying between A and B will have a velocity proportional to their distance from A. Hence if bb_1 represents the velocity of B, the triangles abb_1 are velocity triangles, and the dotted area represents the vol-

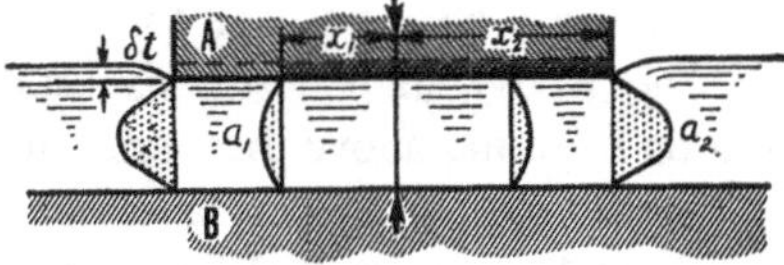
Figure 5.34 Velocity distribution due to pressure only

Figure 5.35 Velocity distribution due to velocity and pressure—surfaces parallel

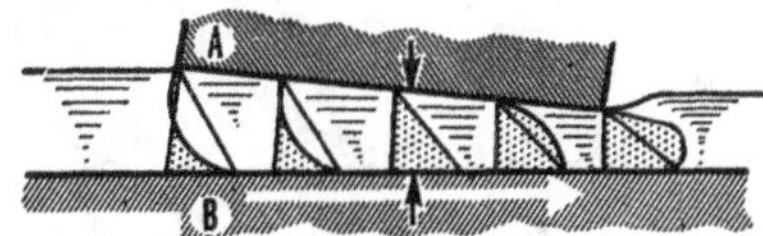
Figure 5.36 Velocity distribution due to velocity and pressure—equilibrium condition

ume of oil passing any section per unit depth perpendicular to the plane of the paper. Hence the same volume of oil flows past each section, and the arrangement simply constitutes a pump.

Assume, now, that both surfaces A and B are at rest, but that a force is applied which will gradually cause the two surfaces to approach each other and to squeeze out the oil film. If the dashed line represents the initial conditions in Figure 5.34 and the full lines the conditions at some other time, it is clear the areas a_1 and a_2 of the bulges will be equal respectively to $x_1 \Delta t$ and $x_2 \Delta t$. It will be understood that the curved bulging lines represent successions of oil molecules which were originally represented by the straight lines. The existence of the bulging lines is evidence of an internal pressure in the oil, which is momentarily balancing the external forces applied to A and B.

Now consider that the conditions assumed in the foregoing considerations exist simultaneously, i.e., that B is moving and that a force is applied to A and B tending to bring them together as in Figure 5.35. It is to be further assumed that the surfaces are so constrained as to remain parallel. From what has been said before, it will be apparent at once that the dotted areas in Figure 5.35 represent the volume of oil passing any section of unit depth in unit time. But these areas are not equal, and it would seem that more oil is passing out from between A and B than is entering, and therefore that in a short time no oil will separate the surfaces A and B, and they will be in metallic contact.

The conditions assumed are those which exist in that type of thrust block where one or more collars, turned on the shaft, abut fixed shoes. The shoes are lined with white metal, and the oil is fed into the space between the shoes and collars at various points. In actual practice the performance of this type of thrust bearing is notoriously bad at high speeds, and the maximum allowable pressure is only about 40 to 50 lb per square inch of effective area.

Imagine, now, that block A is freed from constraint and allowed to tilt at any angle to surface B, as shown in Figure 5.36. Then if the angle is such that the dotted areas are equal at all points, a constant quantity of oil will flow through the bearing; and since the oil is dragged in because of its adherence to surface B and its internal viscosity, surfaces A and B will always be separated by a film of oil.

The principle of the pivoted-shoe thrust bearing is that of the wedge-shaped film described above (Figure 5.36). When the film is supplied constantly with fresh oil, there is a complete separation of the surfaces, and hence no wear.

The pivoted-shoe thrust bearing (Kingsbury thrust bearing) makes possible the automatic formation of wedge-shaped oil films under a thrust load. This result is achieved

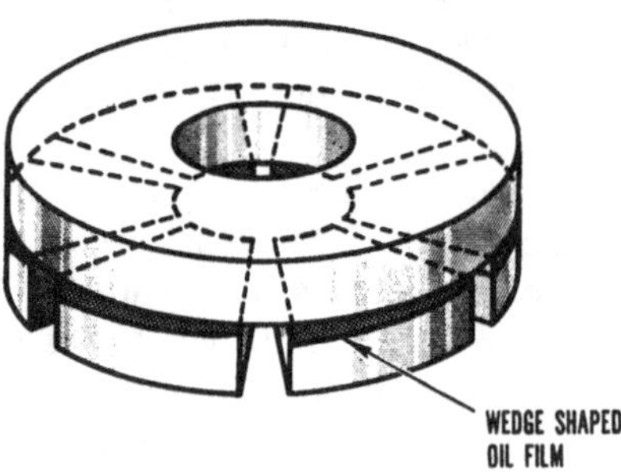

Figure 5.37 Thrust collar and shoes

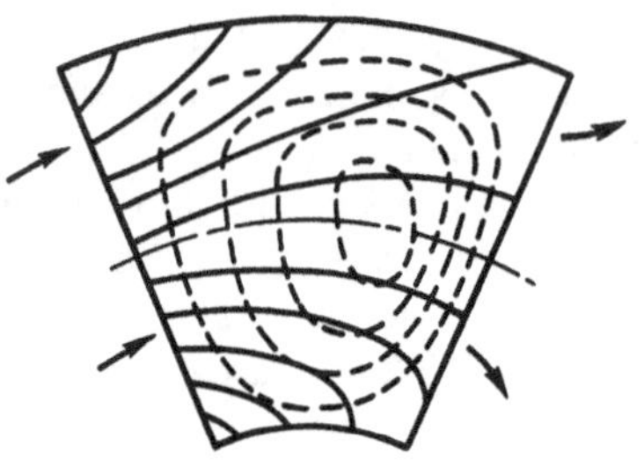

Figure 5.38 Flow lines and pressure distribution on shoes

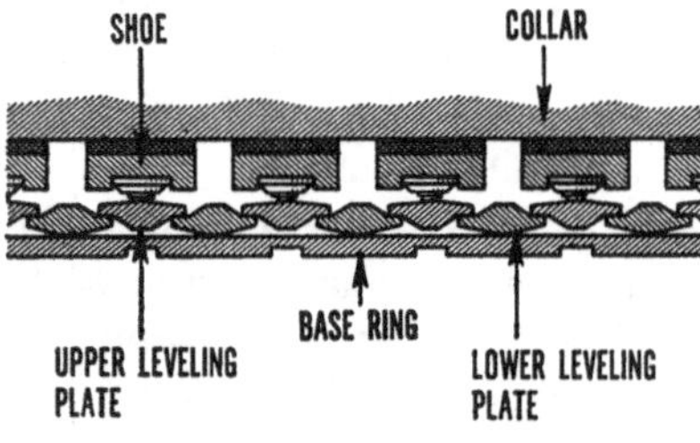

Figure 5.39 Schematic cross section of leveling plates

by dividing the stationary bearing element into segments (called shoes), usually three or six in number (Figure 5.37). These segments are so supported and pivoted that they are free to tilt slightly. The rotating bearing element (or collar) is attached to, and turns with, the turbine or propeller shaft. The entire assembly is submerged in oil. When the shaft rotates, some oil is dragged in (between the collar and each shoe) at the leading edge of the shoe. The thrust on the shaft or collar naturally has a tendency to squeeze the oil out again. It can leave at three sides of the shoe. Hence the trailing edge need present a much narrower passage for the oil than the leading edge. The wedge-shaped film is established. It should be obvious from the above discussion that the oil film automatically assumes whatever taper is required by the speed, load, and oil viscosity. The bearing pressure of a typical pivoted shoe bearing is about 250 lb per square inch of effective area. The flow is as indicated in Figure 5.38, which also shows the lines of equal pressure, and indicates the region of peak pressure.

The essential elements of a Kingsbury thrust bearing are: (1) the thrust collar, made of steel and keyed to the turbine or propeller shaft, which transmits the thrust from the shaft to the oil film and the shoes; (2) the tilting segments, called shoes, which maintain the proper oil wedge between the collar and the shoes and transmit the thrust to the leveling plates; (3) the upper leveling plates, upon which the shoes rest, and the lower leveling plates, upon which the upper leveling plates rest—the function of these leveling plates is to equalize the thrust load among the shoes; and (4) the base ring, upon which the lower leveling plates rest. The base ring is the support which holds the leveling plates in place and transmits the thrust on them to the ship's structure.

Figures 5.39 and 5.40 are developed and exploded views, respectively, of a Kingsbury thrust bearing.

Figure 5.40 Exploded view of pivoted-shoe thrust bearing

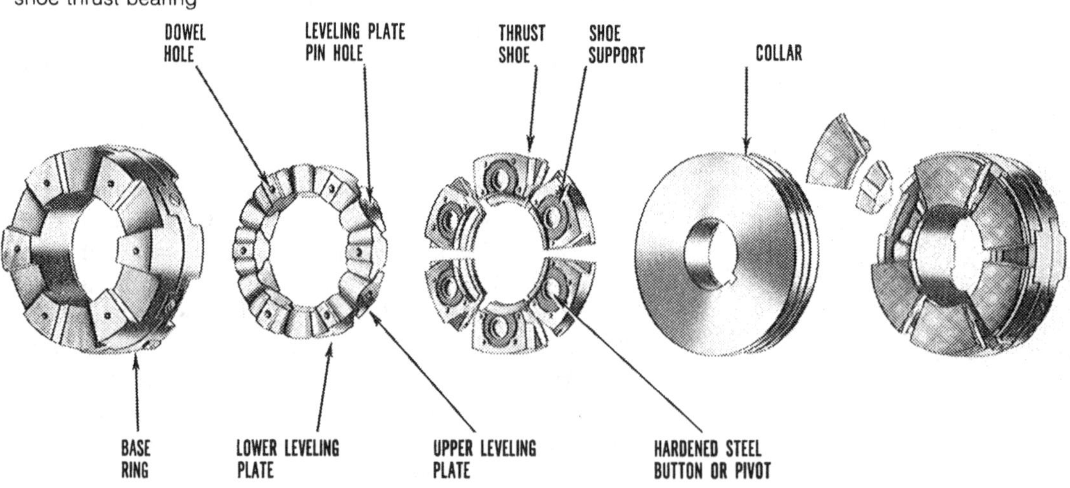

Hydrodynamic Film Formation—The Journal Bearing

The formation of the wedge-shaped film in a journal bearing is very similar to that in the thrust bearing. Figure 5.41 shows the progressive formation as the journal accelerates from zero to normal speed. In (a) the journal is at rest. In (b) it is rotating slowly, and owing to high friction is "climbing up" the left side of the bearing, stopping at a position where the upward component of friction force balances the weight. (The position in the figure is exaggerated for clarity.) Meanwhile the journal is dragging or pumping oil into region A; increased pressure here tends to move the journal back to the right, resulting in the position shown in (c). But the journal continues to pump oil into region A, and the shaft seeks its final position at running speed to the right of the center of the bearing. Figure 5.41d shows this position together with the pressure distribution in region A, which yields a resultant force R. This force holds the shaft in equilibrium up the "incline" of the right side of the bearing.

Hunsaker and Rightmire (see Bibliography) have shown that the resultant force due to the film pressure is always at right angles to the line of centers. The journal, therefore, always positions itself in the bearing so that the line between the center of the journal and the center of the bearing is at right angles to the direction of the load. The magnitude of the load determines the degree of eccentricity of the journal center from the bearing center, larger loads causing a larger eccentricity. The above statements hold true for any shaft carrying a load that is constant in magnitude and direction. They do not hold true in cases where the magnitude varies, as in the crankshaft of an internal combustion engine.

In recent years, the pivoted-shoed bearing has been adapted to journal bearings. If the shoes are properly pivoted, they have a freedom of motion that permits them automatically to align themselves to the shaft, and yet hold the shaft very close to its proper axial position. The motion of the center of the journal in Figure 5.41 may amount to several thousandths of an inch. A pivoted-shoe bearing, depicted schematically in Figure 5.42, is capable of restricting this motion to a much smaller amount.

Although theoretical considerations indicate a slightly eccentric arrangement, as shown in Figure 5.42a, for proper film formation, practical considerations require that the pad be pivoted at its centerline, as in Figure 5.42b. Changing viscosity tends to compensate adequately in proper film formation. Moreover, one of the pads should be mounted at the bottom of the housing (as shown in Figure 5.42b), to be located under the direct vertical load of the shaft while at rest.

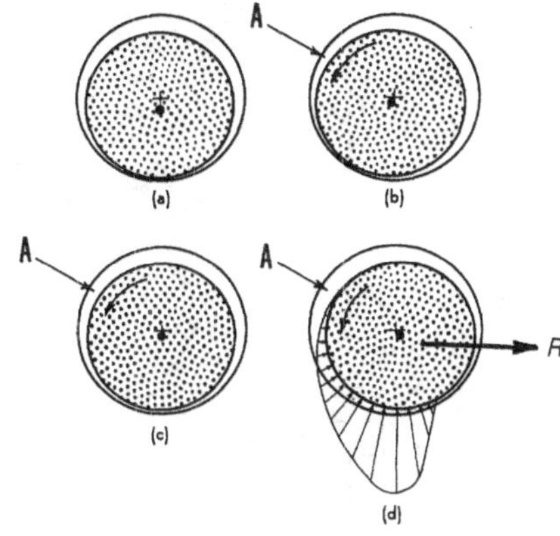

Figure 5.41 Formation of film in journal bearing

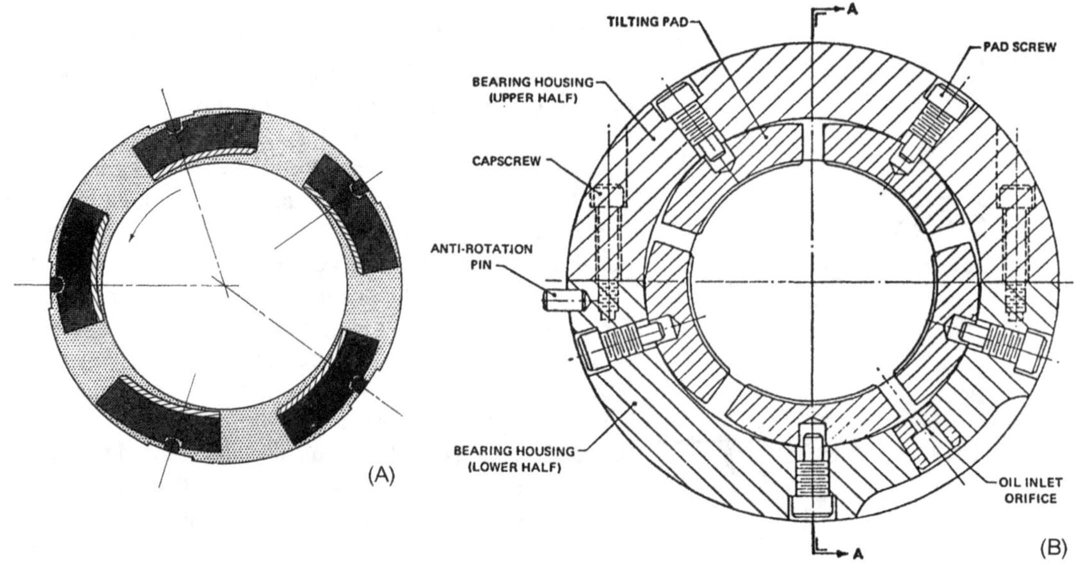

Figure 5.42 Pivoted-shoe (tilted pad) bearing. (A) Schematic of bearing. (B) Bearing arrangement

Experience with eccentrically placed pads shows that bearing seal wear will take place, since the seals, not the pads, will support the shaft at rest.

A phenomenon occurring in high speed, lightly loaded journal bearings is "oil whirl," or "oil whip." It results from inability of the forces present (see Figure 5.41) to reach equilibrium, with the consequence that the shaft whirls rapidly around in the bore of the bearing, setting up intense vibration. The shaft journal and bearing may be destroyed under these conditions. The pivoted-shoe journal bearing, when properly applied, has been a successful remedy for this condition.

Lubricating Oil Properties

It is evident that certain fluid properties are associated with proper lubrication. "Oiliness" has been discussed. Viscosity is another highly important property of a lubricant. Viscosity may be defined as the resistance of a fluid to shearing forces. High viscosity fluids will support heavier loads than low viscosity ones, but have a greater resistance to shear, or a greater internal friction. High viscosity fluids require larger bearing clearances. On the other hand, low viscosity fluids require less bearing clearance and have a smaller degree of internal friction. The selection, therefore, is a compromise, considering load, clearance, and operating temperature. The temperature factor is important because viscosity decreases with an increase in temperature. Too high an operating temperature may result in an unacceptably low load-carrying ability. Lubricants are now treated with additives that tend to suppress the change of viscosity with temperature.

As a practical matter, a good lubricating oil for turbines and reduction gears must have certain other characteristics. For one thing, it must resist oxidation, which produces organic acids that may react chemically with the bearing materials to corrode the bearing surfaces. Oxidation is promoted by high temperatures and the presence of metal oxides or other impurities that may catalyze the reaction. The presence of water in the lubricating oil promotes rust, or oxidation of metal surfaces, and thus indirectly promotes oxidation of the oil itself. Additives acting as antioxidants and rust inhibitors aid in this problem. Since it is difficult to prevent the entry of any water into the lubrication system, emulsification of water and oil may occur, the resulting fluid having poor lubrication qualities, and promoting corrosion. Thus another quality needed in a good oil is high demulsibility, or ability to separate rapidly from an emulsion.

In summary, the requirements of a good turbine lubricant are:

1. That it will readily form and maintain a boundary film (adequate oiliness).

2. That it will form a hydrodynamic film with low friction losses (proper viscosity).

3. That it will be resistant to chemical and physical changes (slow to oxidize and has good demulsibility).

The foregoing discussion has been concerned with the use of lubrication to reduce friction and accompanying power losses. Though friction cannot be completely eliminated, it has been shown that fluid friction replaces dry friction to considerable advantage. But power is still lost through the generation of heat by lubricant friction in the bearings. Lightly loaded or slow-moving bearings can dissipate the heat generated by radiation and conduction, but heat energy equivalent to a few hundred horsepower may accumulate in the bearings of large turbines at high powers. Thus, an additional important function of lubricating oil is to serve as a coolant. This is accomplished by circulation and cooling in a heat exchanger. The circulation has a twofold effect—it disposes of the heat accumulated from all sources in the bearings, while simultaneously preventing the oil from reaching a temperature at which a change in its physical or chemical properties would endanger its lubricating properties.

In lubrication, every oil system builds up an internal vapor pressure. In addition, the rotating elements shed and stir a very fine circulation of mist (small droplets). These vapors are normally exhausted into the engine room rather than into the turbine. Thus, the engine room decks, handrails, and other surfaces accumulate an oily film. Early attempts to remove this film with mist eliminators of the inertial or mesh type were unsuccessful; but in recent years, electrostatic precipitators have been applied in a highly successful fashion, enhancing cleanliness in the engine room, eliminating slippery surfaces, and reducing fire hazard.

5.6 Auxiliary Turbines

The discussion in this chapter so far has been confined to main propulsion turbines. However, there are a number of auxiliary turbines used on board ships to drive a variety of auxiliary machinery. They include ship's service generators, forced draft blowers, and a number of pumps, such as main feed pumps, condensate pumps, condenser circulatory pumps, booster pumps, fuel oil service pumps, and lube oil service pumps. Although these turbine drives could be considered smaller cousins of the main propulsion turbines studied above, a number of significant variations not yet considered are worthy of special description here. Some special features also arise owing to use of auxiliary steam.

The auxiliary steam system was briefly discussed in Section 2.3. It will be recalled that this system furnishes steam to drive the propulsion auxiliaries, and for other functions necessary to the operation of the plant. Ship's service turbo-generators are very often driven by superheated steam. The pressure and steam conditions of the auxiliary system vary, depending on the installation. Until recent years, the auxiliary system carried 600 psig saturated steam. With present-day higher boiler pressures, many of the major auxiliaries are driven by superheated steam at the higher boiler pressure. In some other instances, a portion of the main steam is desuperheated and reduced in pressure to 600 psig to provide steam for driving certain auxiliary units.

As an example, the CV-60 nmachinery plant provides boiler pressure auxiliary steam at 1200 psig to the main feed pumps and the turbo-generators (ship's service and aircraft-starting). In addition, a quantity of 1200 psig steam emerging from the superheater is led back through desuperheater coils in the boiler steam drum, thence to reducing valves, where it is reduced to 600 psig auxilary steam. This then supplies the condensate pumps, main feed booster pumps, fuel oil service pumps, forced draft blowers, main circulator, lubricating oil service pumps, and fire pumps, as well as reducing stations for lower-pressure systems.

In nuclear ships all turbines (propulsion and auxiliary) operate on wet steam, about 1% inlet moisture. The auxiliary turbines on these ships operate over a wide range of pressures. On nuclear submarines the range is from 300 to 750 psi; on the carrier *Enterprise* the range is 600 to 1050 psi. The generator turbines, with the exception of a number of the submarines, exhaust to their own separate condensers. Other auxiliary turbines exhaust to the main condenser. As pointed out in Section 5.1, many auxiliary turbines operate noncondensing, at or above atmospheric pressure, with the exhaust steam being used for feed water heating. This is by far the most common practice in commercial shipping.

It should be pointed out that the auxiliary machinery could potentially be powered by electric motors, and a number of smaller pumps indeed use electric motors as prime movers.

The selection of the type of prime mover is subject to many considerations. The generation of electric power in an efficient multistage turbo-generator, which operates condensing, and the use of electric power to drive pumps with well-designed electric motors, is generally more efficient than driving the same pumps with small few-stage turbines which operate noncondensing. However, there are several considerations that make the choice of steam turbine drive preferable. First, above about 50 HP, steam turbines begin to have a decided advantage over electric motors in weight and space per horsepower. Second, even for smaller sizes, the necessity for plant reliability makes the use of steam power advantageous to prevent stoppage of the entire plant in the event of an electrical casualty of even short duration. For instance, if the fuel oil servide pumps were electrically driven, the ship might have to stop, or at least slow down, for such a minor casualty as a circuit breaker trip caused by a temporary electrical overload. Because of this, all auxiliaries vital to propulsion either are normally steam-driven, or are installed in duplicate or triplicate, with at least one independent of electric power. Third, some auxiliary units are required to operate at speeds above those feasible for electric motors, or at varying speeds, which may be difficult to achieve with electric motors. For example, main feed pump speeds may be as high as 10,000 rpm, and forced draft blower speeds of over 5000 rpm may be required. Finally, steam at low pressure for DFT heating and deaerating and other propulsion plant needs is required. In these instances it is thermodynamically more efficient to use steam that has had its pressure decreased by the accomplishment of work than to use steam that has had its pressure decreased by a straight throttling process.

Auxiliary turbines are constructed to be rugged and reliable. These two considerations have led to a simplicity in design at the expense of efficiency. At one time, practically all auxiliary turbines except the ship's service generator turbine were single stage units. More recent plants incorporate one-, two-, or three-stage turbines, all of which are of the impulse type. They are characterized by a large pressure drop across the stages, high rotor velocities to keep the blade speed–steam speed ratio in a reasonable range, and high exit losses.

Older auxiliary turbines are fitted with a casing relief valve to prevent the casing from being subjected to full line pressure in the event that the throttle valve is inadvertently opened before the exhaust valve. This casing relief valve usually relieves directly to the machinery space atmosphere and is usually set to open at 25 to 30 psig. It is current practice to provide a spring-loaded exhaust valve which cannot be completely closed by hand. The valve is designed to open automatically when the steam pressure in the turbine casing exceeds the exhaust line pressure by more than about 2 psi. Whatever the arrangement, all auxiliary turbines are equipped to prevent the turbine casings from being subjected to line pressure. This permits much lighter construction of the turbine casing than would otherwise be possible and prevents damage to the shaft seals.

Auxiliary turbines are usually of the radial, helical, or axial flow type. They may be one-, two-, or three-stage turbines, except for the ship's service generator turbine, which will be discussed separately. Often the first stage will be velocity compounded to permit a large pressure drop across the first stage nozzles without an inordinately high wheel speed. The particular type turbine used in any given application will vary according to the choice of the supplier of the unit. Normally, only operating characteristics are speci-

Figure 5.43 Radial flow, single stage, re-entry turbine

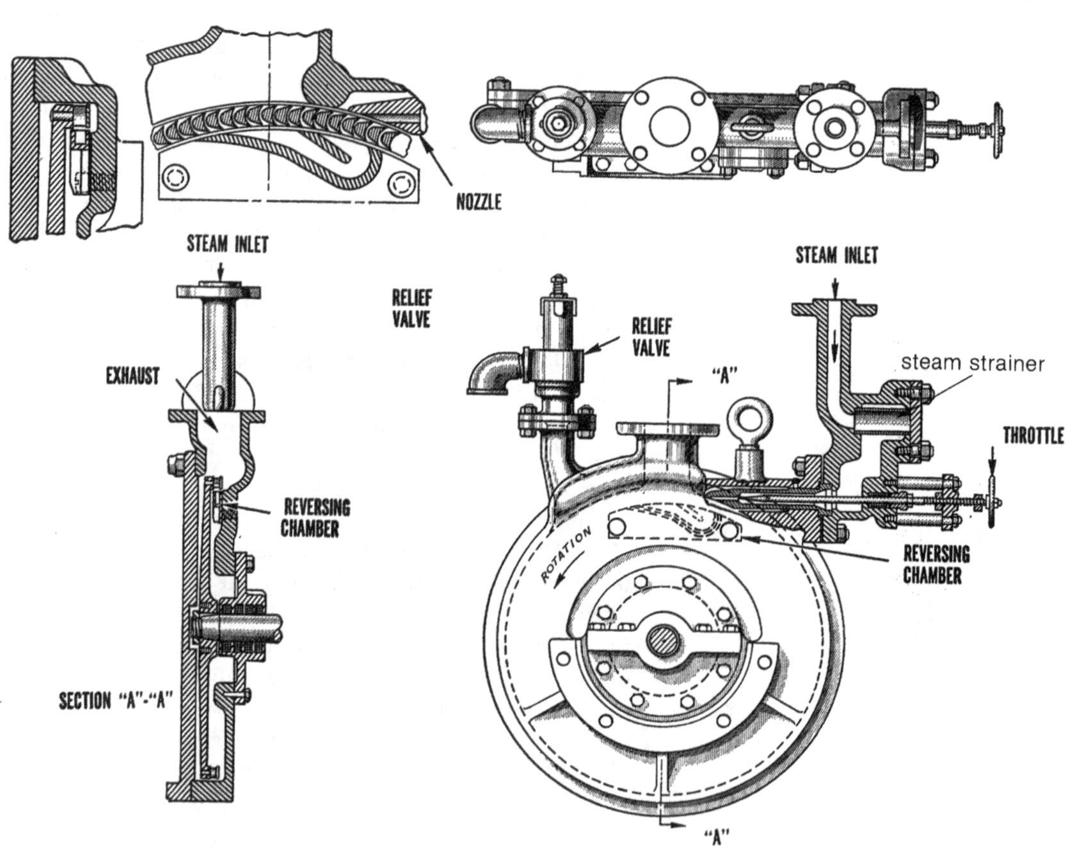

fied; within these limits the contractor has a wide choice of types. It may be found that the same auxiliary components on sister ships will be driven with different type turbines.

Figure 5.43 shows a radial flow, single stage, re-entry turbine. The re-entry feature effects a velocity compounding. Note the needle valve control, a feature found on many auxiliary turbines. The needle valve arrangement approaches the efficiency of a nozzle in the conversion of thermal to kinetic energy, and avoids much of the thermodynamic loss associated with pure throttling. The needle serves, in effect, to vary the flow through the nozzle by changing the throat area, thus varying the flow without decrease of pressure at the nozzle inlet, as would be caused by an upstream throttle valve. The single penetration of the casing by the shaft is also typical; this simplifies the shaft sealing problem, solved in this case by four rings of carbon packing. Figure 5.44 is another arrangement of the radial flow turbine. This one has two rows of moving buckets to effect the velocity compounding.

Helical flow, or tangential flow, turbines are often found in various applications. Figure 5.45 is a cut-away view of this type. Note the four reversals, an arrangement which is equivalent to a velocity compounded stage with four sets of fixed blades. Several sets of nozzles may be mounted around the periphery of this type, providing a wide range of power capability for the single turbine wheel. Another view, showing the whole unit, appears in Figure 5.46.

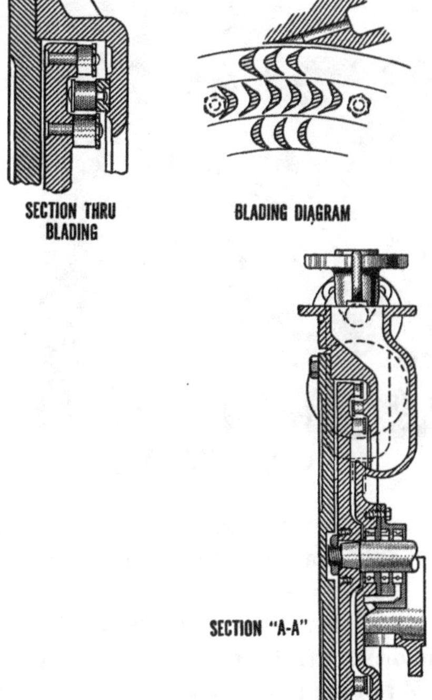

Figure 5.44 Radial flow, velocity compounded turbine

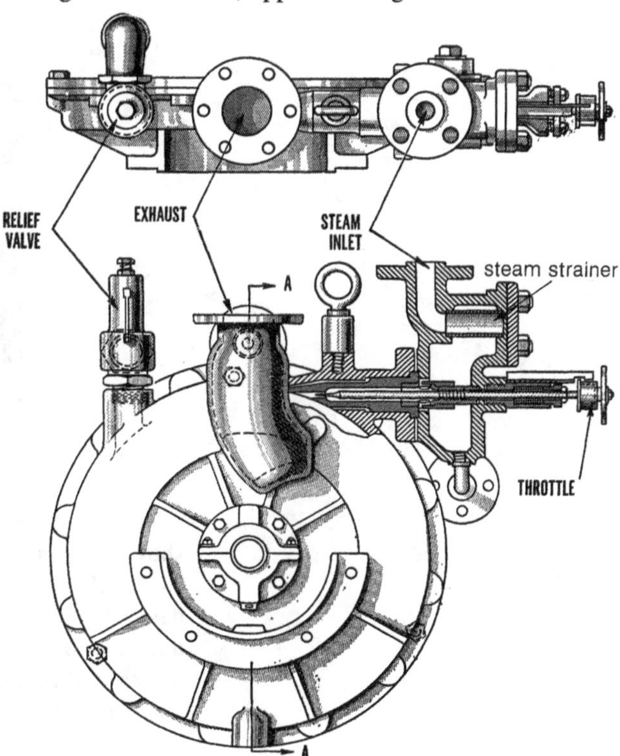

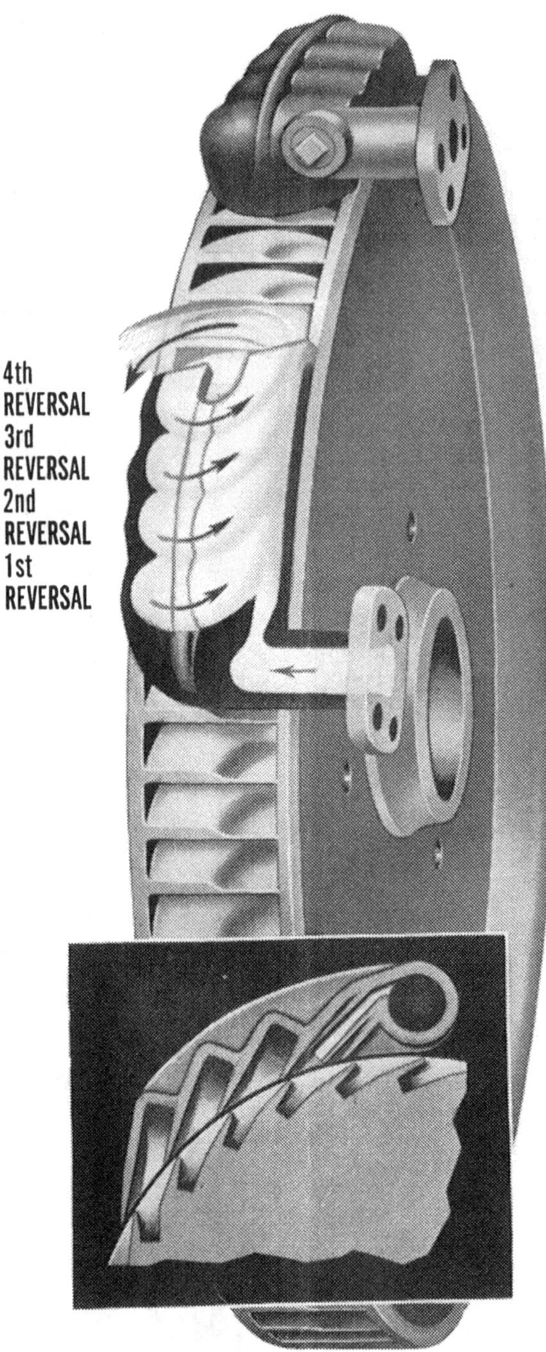

A typical velocity compounded, axial flow turbine is shown in Figure 5.47. This is a vertically mounted turbine (with reference to the shaft), whereas the others discussed have been horizontally mounted. In its application here as a main circulating pump drive, the thrust on the pump impeller, as well as the weight of the turbine and pump rotors, is downward. Consequently a large thrust bearing is provided,

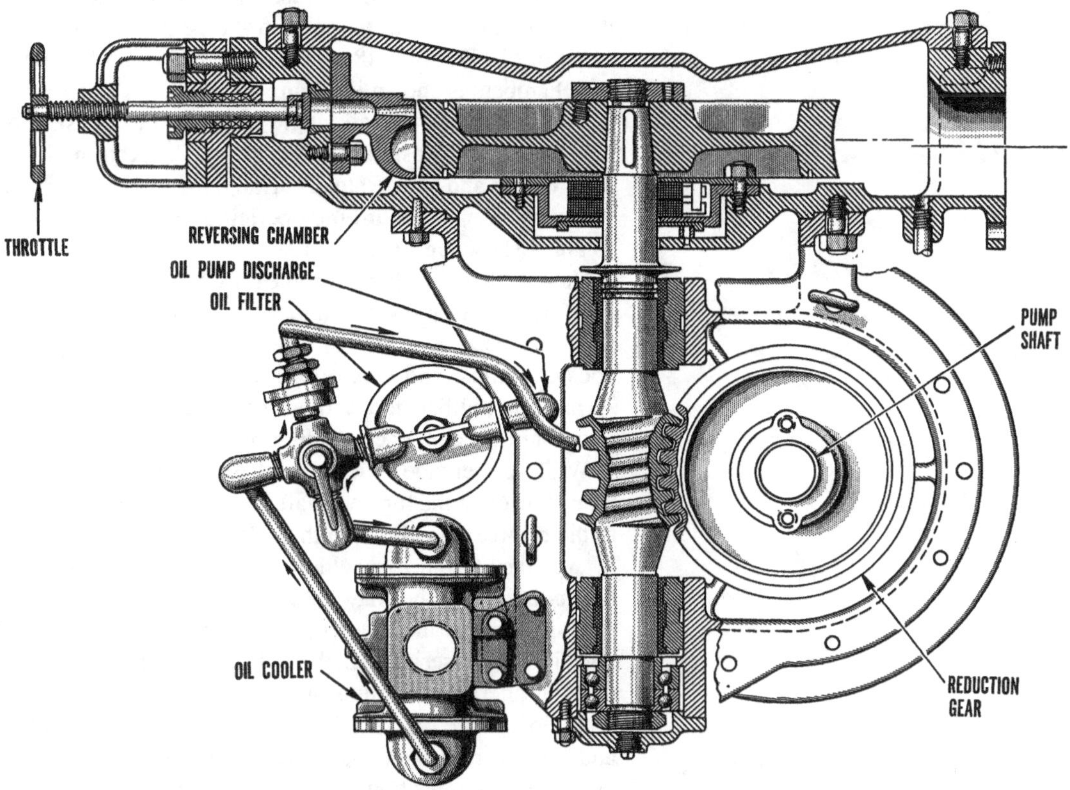

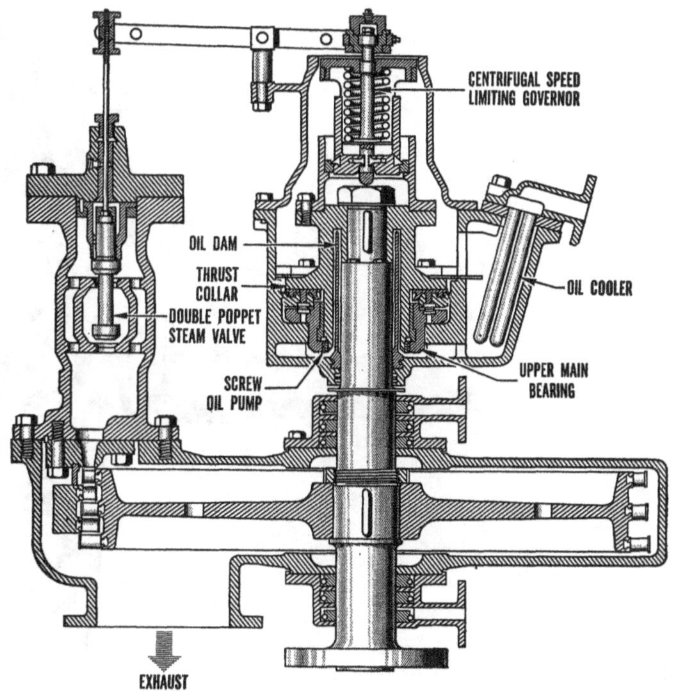

Figure 5.46 Helical flow turbine, sectional view

Figure 5.47 Axial flow velocity compounded turbine

as shown, to support the thrust and weight. Another axial flow arrangement of the re-entry type is depicted schematically in Figure 5.48. An arrangement with two reversing chambers is shown in Figure 5.49.

The preceeding are all single stage turbines designed to operate with 600 psig saturated steam. A recent multistage (one Curtis and four Rateau) arrangement for a ship's service turbo-generator turbine operating with 1050 psig inlet steam is shown in Figure 5.50.

Helical flow auxiliary turbines are used to drive pumps and forced draft blowers. Radial flow turbines, as in Figure 5.44, are used as drives for main condensate pumps, feed booster pumps, and lube oil service pumps on older ships. Axial flow multistage Curtis stages (Figure 5.47) are used with main condenser circulating pumps.

While the engine efficiency of the main propulsion turbine (energy at the turbine shaft divided by energy available to the turbine) may be as high as 75%, the engine efficiency of auxiliary turbines is seldom greater than 60% under the best conditions, and may go as low as 17–20%. However, the failure of the turbines to utilize more of the energy available is not as serious as these figures indicate, since the thermal energy remaining in the exhaust steam is not normally lost to the plant, as in the case of the main propulsion and ship's service generator turbines, but is used for various purposes in connection with the auxiliary exhaust system.

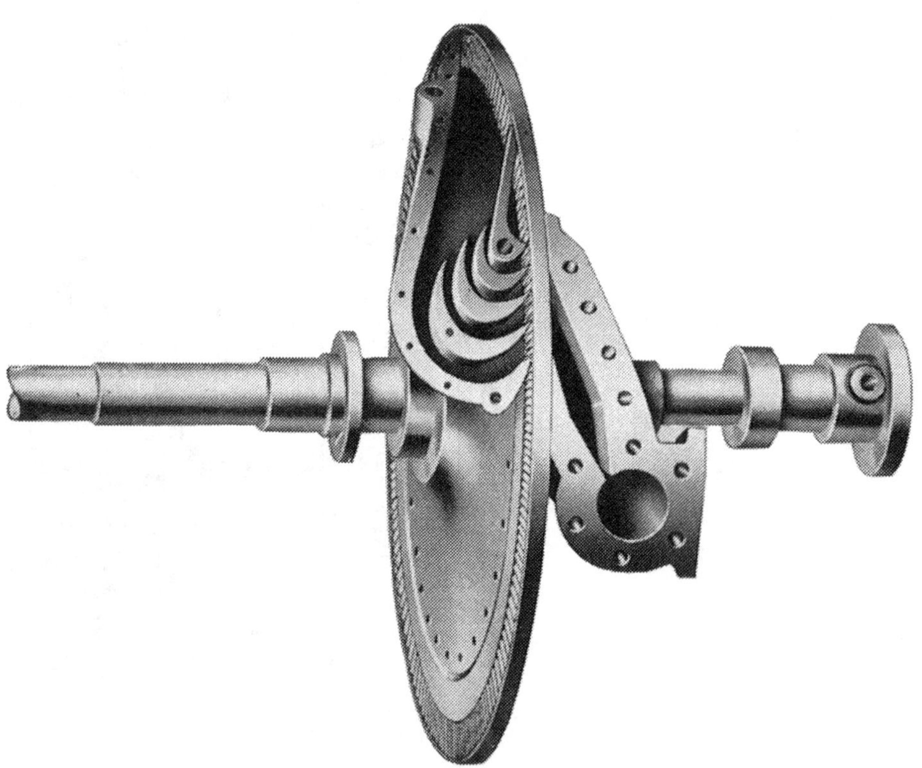

Figure 5.48 Axial flow single re-entry turbine

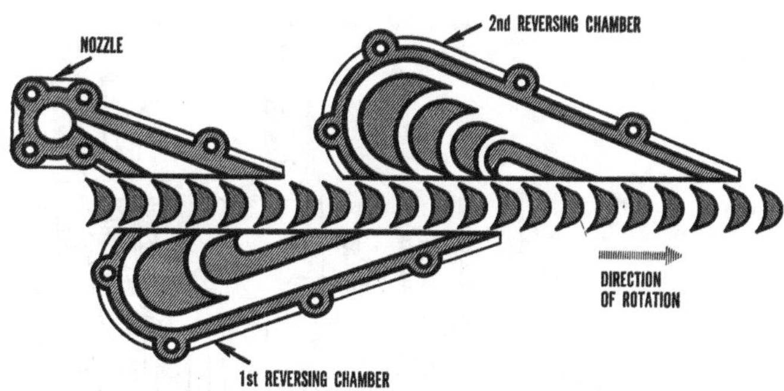

Figure 5.49 Axial flow double re-entry turbine

This completes the general discussion of steam turbines, their essential features, and their most essential components. A detailed treatment of controls, protective devices, instrumentation, and so on, is outside the scope of this book, as their modern versions are worthy of consideration in an extensive separate study.

Some operational features will be further considered in comparison with gas turbine operations in Chapter 7. Table 5.1, (page 135), summarizes the most significant features of steam turbines discussed in this and previous chapters.

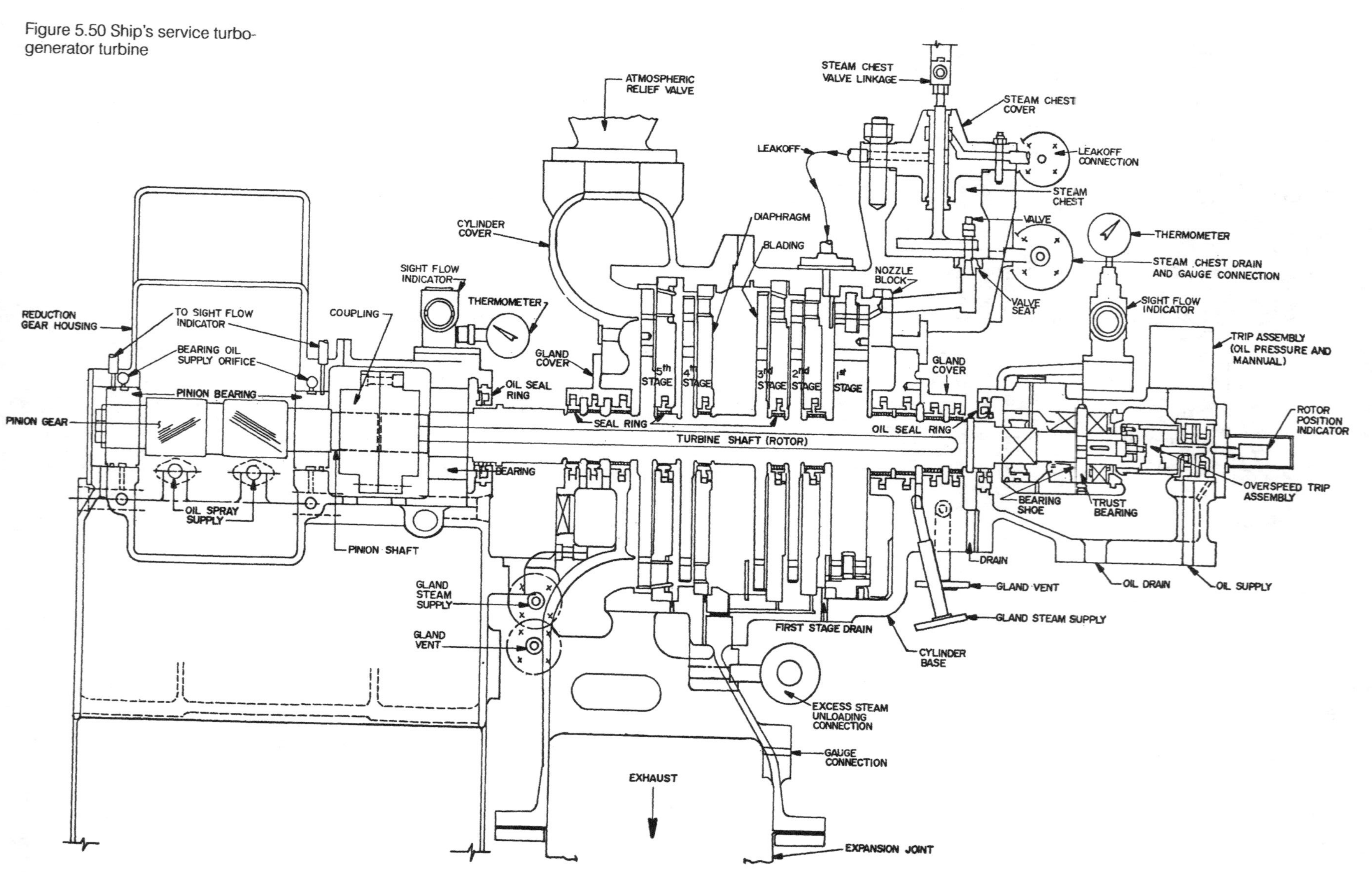

Figure 5.50 Ship's service turbo-generator turbine

Table 5.1 Comparison of Essential Features of Steam Turbines

	Impulse Turbine	*Reaction Turbine*
Method of staging	Simple impulse, velocity-compounded impulse, pressure-compounded impulse, pressure-velocity-compounded impulse, and velocity-pressure-compounded impulse.	Pressure-compounded.
Nozzles	Located in steam chest and diaphragms. Purpose: (1) to expand the steam and thereby give it velocity, (2) to direct this moving steam at the correct angle against the moving blades.	Has no separate nozzles. The space between consecutive blades in a row is in reality a nozzle.
Fixed blades	Attached to the casing and located between the rows of moving blades of a velocity-compounded stage. Purpose: To change the direction of the steam and to direct it at the proper angle against the succeeding row of moving blades.	Attached to the casing and located between the rows of moving blades. Entering steam passes through fixed blades first. Purpose: Same as nozzles for impulse.
Steam pressure	Decreases in every nozzle. Remains constant in moving and fixed blades.	Decreases in every row of fixed and moving blades.
Absolute steam velocity	(1) Increases in nozzles, (2) decreases in moving blades, (3) practically constant in fixed blades.	(1) Increases in fixed blades, (2) decreases in moving blades.
Stage definition	A set of nozzles discharging into one or more rows of moving and fixed blades. Any part of an impulse turbine in which only one pressure drop takes place with generation of kinetic energy. Pressure is the same at all points beyond the nozzle. Velocity changes.	One row of fixed blades and the succeeding row of moving blades. Pressure drops in both fixed and moving blades.
Expansion definition	Similar to a stage. (The term "expansion" not generally used with impulse turbines.)	A number of rows of fixed and moving blades that have the same length and pitch diameter.
Steam admission	Nozzles for first stage extend over only part of the circumference of the first stage wheel and cover a greater part of the circumference of each succeeding wheel. They cover entire circumference of the low pressure stages.	Steam enters around the whole periphery of the rotor.
Method of allowing for increased volume of steam due to expansion	(1) By increasing the number of nozzles in the later stages, (2) by increasing the nozzle cross-sectional area, (3) by increasing blade lengths, (4) by increasing blade interval.	(1) By increasing blade lengths, (2) by gauging the blades, i.e., widening the discharge opening, (3) by increasing diameter, (4) by increasing blade interval.
Diaphragms	Situated between wheels to separate different pressure stages and to carry nozzles	Has none.
Shaft glands	Serve same purpose in both types: (1) To prevent escape of HP steam to the atmosphere where the shaft passes through the casing; (2) to prevent entrance of air to turbine, where the shaft passes through the casing when the pressure within the casing is below atmospheric pressure.	
Speed control	(1) By means of throttle valve, (2) by cutting in or out more nozzles to first stage, (3) by leading HP steam to later stages.	(1) By means of throttle valves, (2) by using "cruising stages" at low speeds, (3) by bypassing them as speed increases.
Efficiency	High at both high and low pressure.	High at low pressure only.
Theoretical efficient speed	For simple impulse, blade speed should equal $V(\cos \alpha/2)$. For velocity-compounded turbine with two rows of moving blades, the speed should equal $V(\cos \alpha/4)$.	For pure reaction, blade speed should equal $V \cos \alpha$.

Chapter Six

Gas Turbines

6.1 Introduction

The gas turbine, as a prime mover, is a relative newcomer in the field of marine propulsion. Its first application was by the Royal Navy in 1947 in the form of a 2500 HP gas turbine unit driving one of the three shafts of the RN motor gun boat MGB-2009. Since the success of MGB-2009, gas turbines have been installed in the ships of a number of navies and have found their way into commercial shipping as well. The first generation of gas turbine units used in ships represented a series of modifications of aircraft jet propulsion units. Since the time to develop a gas turbine unit is long (four to six years) and costly, they numerically represent only a relatively small portion of gas turbine units presently in use in both aircraft and stationary service. These engines are referred to as marinized aircraft derivative units.

The purpose of this chapter is to review the fundamentals and performance of the gas turbine units and to study their individual components and mechanical features, as well as the problems in adapting gas turbine units for marine use, i.e., marinizing. As steam and gas turbine units will coexist for some time to come, the last chapter will attempt to put their respective advantages and disadvantages into proper perspective, pointing out their unique and common features.

Before proceeding to the discussion of the main components that make up a gas turbine unit, it is important to consider the nature of the working fluid that will flow through the machine. In a steam turbine, the working fluid is essentially water which will experience phase change twice: from water to steam in the boiler and from steam to water again in the condenser. The properties necessary for performance calculation are well known and given in tables or in the Mollier diagram. Knowing pressure and temperature permits determination of all other properties.

In a gas turbine the working fluid is essentially air which will remain unchanged in phase throughout the cycle, although it will go through a combustion process before expansion through the turbine. To analyze the processes that would approximate the gas turbine open (Brayton) cycle, the so-called air standard cycle was used in Chapter 3 and the following assumptions were tacitly introduced:

- The working fluid is air, an ideal perfect gas with constant specific heats.
- There are no inlet or exit processes.

- There are no losses in the system.
- The combustion process is replaced by an external heat transfer process.
- The cycle is completed by heat rejection to the surroundings outside the engine.

Thus, knowing inlet values and either temperature or pressure ratios permits determination of all the other properties, as well as cycle performance.

In an actual gas turbine, only the last assumption is satisfied. The results obtained from the air standard cycle will differ greatly from those of the actual engine due both to large temperature variation and fuel being burned in the combustor. The latter will cause the temperature to rise and convert the incoming air into a not very precisely defined mixture of various combustion products. An exact analysis of a gas turbine performance requires highly detailed thermodynamic and aerodynamic calculations throughout all the components, and rather laborious combustion caculations to predict the properties of the flow entering the turbine section.

For practical calculations, it has been found that the following process gives quite satisfactory results:

- In the "cold" section before the combustion chamber constant specific heat and a constant k are used. In high pressure machines where the compressor exit temperatures approach 1000°F, variable or average property values should be used.
- From the combustion section onward a very lean mixture (200–400% theoretical air) is used with c_p and k determined by the local flow temperature.
- The combustion is handled as a thermodynamic bulk process for a given fuel, usually a diesel fuel, which determines the amount of fuel needed to achieve the desired turbine inlet temperature, or the fuel-air-ratio. This will be further discussed in Section 6.3.

Figure 6.1 shows the values of c_p and k for air and a 400% excess air mixture as a function of the temperature. Although these parameters are also functions of pressure, this latter effect is small for temperatures below 3000°F. The effect of assuming c_p and k constant at higher temperatures is evident and will be shown in examples later.

6.2 Compressors

The essential components of a gas turbine unit were discussed in Chapter 2, Section 2.5. The turbine, as a work extraction device, was treated in detail in Chapter 4. The inlet section, the compressor, and the combustion section are the remaining principal components to be studied. The inlet diffuser section in aircraft jet propulsion engines is

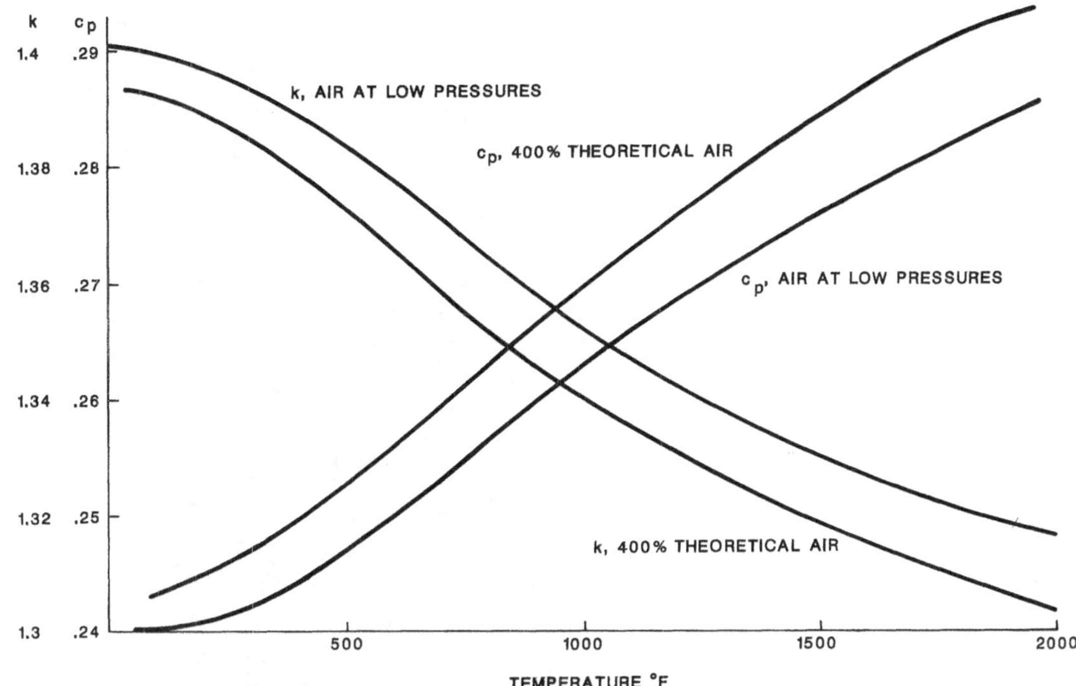

Figure 6.1 C_p and k as functions of temperature

a vital element for efficient power plant operation, but it is considered to be part of a particular airframe and not part of the engine. In a marine gas turbine unit the diffuser is replaced by a simple bell-mouth type of inlet, which is a device with practically zero loss for channeling the flow into the engine.

Although a well-manufactured bell mouth contributes zero losses, the flow ahead of the bell mouth may be so severely distorted, i.e., having large pressure/velocity variations, as to cause variations in the magnitude and direction of the velocity vectors that the local compressor blade sees, thus causing local stall (see below). Such distortion is caused by inadequate design of the downtake section or the intake duct (e.g., too small a duct cross-section, or sharp bends, or protruding joints and corners). Although intake distortion has been a recurring problem in aircraft engine-airframe design, it was initially assumed that a well made bell mouth would adequately take care of such problems in shipboard installations. Experience has shown that the duct design and installation may have to be integral considerations at the time of the turbine unit design.

In Section 3.11 it was shown that the basic gas turbine cycle efficiency depends only on the pressure ratio. Thus, efficient compression of large volumes of air is the key to an efficient gas turbine cycle. This can be achieved by two types of compressor, the centrifugal and the axial flow type (Figure 6.2). The fundamental principle here is that, be it a

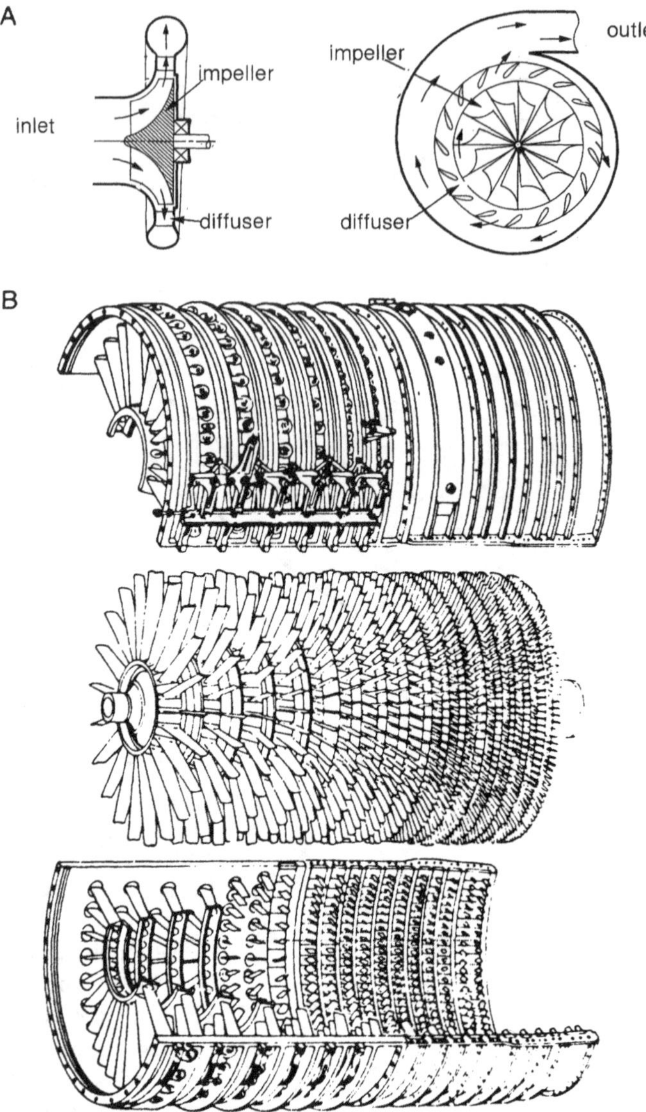

Figure 6.2 (A) Centrifugal compressor (B) Axial flow compressor

centrifugal or an axial type, the compressor adds kinetic energy to the air and then utilizes some sort of diffuser passage to decrease the velocity and convert the kinetic energy into a pressure rise.

A centrifugal compressor takes in air at the center of the impeller (the eye), and centrifugal force pushes it out radially at high speed (Figure 6.3) into the diffuser passage. Here the flow is decelerated, and the pressure rise is obtained. Rotor and diffuser blades have a variety of shapes, to be discussed further below, depending on pressure-velocity characteristics necessary for a particular type of application.

The axial flow compressors have blades similar to those of an axial flow turbine. The rotating blades speed up the flow, and the stationary blades, acting as diffusers, convert the velocity into the pressure increase. Thus, a compressor could be viewed approximately as the reverse of a reaction turbine.

The basic difficulty associated with either type of compressor is that the flow process in a compressor is one of deceleration, whereas the turbine process is acceleration. It is well known that a flow can be rapidly accelerated (nozzle, turbine) with very small losses, but the same is not true if the flow is rapidly decelerated. Then large losses would occur, as a result of stalling caused by rapidly rising pressure on a blade surface forming the diffuser passage. This is tantamount to saying that only a certain amount of flow can be pushed into a passage where pressure is rising.

In order to preserve smooth flow and to prevent stall, only a certain amount of pressure rise and turning can be allowed in a given blade row. Thus, for a given pressure ratio, axial flow compressors need to have many stages compared to an axial flow turbine, which needs only a few. This restricts the pressure rise at this time, per stage, typically to a value about equal to or less than 1.3. In more recently developed engines and in others now being developed this value is expected to increase to about 2. For example, the LM2500 compressor has 16 axial compressor stages with an overall pressure ratio of about 17. This indicates an average pressure rise per stage of 1.194 ($17 = 1.194^{16}$) (see equation 15, below). A centrifugal compressor also has flow and pressure rise limitations, with a maximum practical pressure ratio of about 5. In new engines this value is expected to increase beyond 10.

In order to consider the salient features and to provide useful comparisons, each of the compressor types needs to be discussed separately.

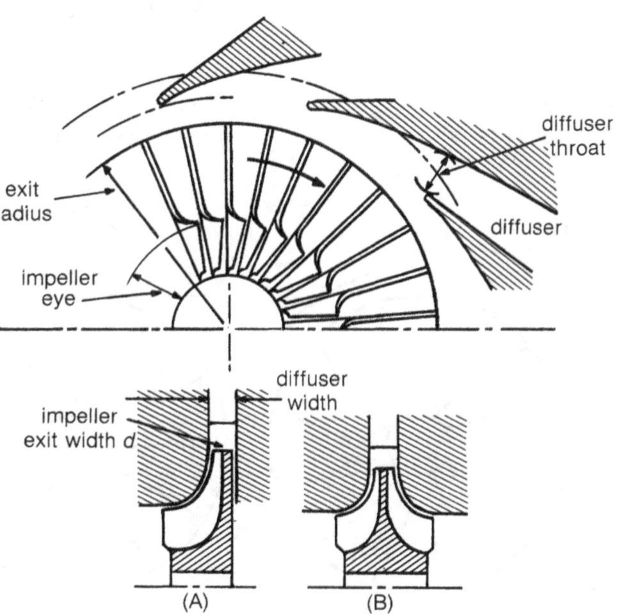

Figure 6.3 Centrifugal compressor arrangement. (A) Single entry (B) Double entry

The energy absorbed by the compressor can be determined by the conditions at the rotor inlet and exit, since no work is done in the diffuser. The basic computations can be carried out by using the velocity triangles at the respective locations, and by use of the Euler pump equation, equation (2), Chapter 4. Thus the work per pound of fluid is:

$$w_c = \frac{1}{g_c}(V_{b_2}V_{u_2} - V_{b_1}V_{u_1}) \tag{1}$$

The general velocity diagrams are shown in Figure 6.4. The tangential velocities V_{u_1}, V_{u_2} are also called whirl velocities, or simply whirl. Since the entry into the impeller is in most centrifugal compressors axial, the tangential velocity component V_{u_1} is zero or negligibly small. Thus, the following is obtained:

$$w_c = \frac{V_{b_2}V_{u_2}}{g_c} = H \tag{2}$$

where H is the ideal head or the total head rise across the compressor, or also a pump, excluding all internal losses. From the velocity diagram at the rotor exit one finds:

$$V_{u_2} = V_{b_2} - V_{r_2}\cos\beta$$

and with substitution into equation (2) the ideal head becomes:

$$H = \frac{V_{b_2}^2}{g_c} - \frac{V_{b_2}V_{r_2}\cos\beta,}{g_c} \tag{3}$$

Now, since the mass flow is given by the velocity normal to the rotor circumference ($V_r \sin\beta$):

$$\dot{m} = \rho A V_r \sin\beta$$

where A is the flow area ($\pi D d$), d being the width of the flow passage at the rotor rim (Figures 6.2a and 6.3). By

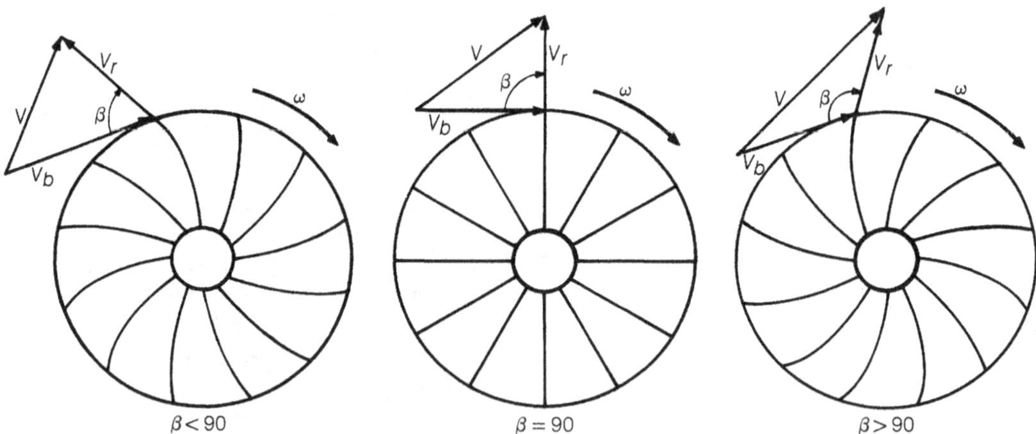

Figure 6.4 Centrifugal compressor pump velocity diagram

equation (3), Chapter 3, $\Delta p = \rho g H$, and upon substitution into equation (3):

$$\Delta p = \rho V_{b_2}^2 - \frac{\dot{m} V_{b_2}^2}{A} \cot \beta \tag{4}$$

which gives the ideal pressure rise in a centrifugal compressor or pump in terms of the rotational speed ($V_b = r\omega$) and the mass flow. Angle β is recognized as the rotor blade angle.

The results are shown graphically in Figure 6.5. The solid lines represent the ideal results of equation (4). The backward curved vanes ($\beta < 90°$) provide the smallest pressure rise, but require less power than the radial bladed wheel ($\beta = 90°$), which gives a constant Δp for all flows. The forward curved blades provide a large pressure rise (due to a large kinetic energy rise), but also require a larger amount of power. When compressor losses are taken into account, the dotted curves result. The points labeled S represent the margin of stable operation, and are located at the maximum points on the respective curves. The stable operation is toward the right. If the compressor is operated in the stable region and a disturbance or flow restriction occurs decreasing the flow, then the pressure tends to rise and in turn restore the flow. If the same happens at or near S, a pressure decrease will result, which causes the flow to decrease even further. This cycle then tends to build up and eventually leads to zero flow, or even to a reverse flow. These events occur in a very short time, leading to what is known as the compressor stall or surge. If the cause for such disturbance or flow reduction persists, a few such cycles can destroy the compressor.

It can also be seen from Figure 6.5 that the backward curved blades have the largest range of useful stable operation. Thus, this type of blade construction is found in most service pumps. In operations where a flow system requires a continuously high flow rate with a high head, forward curved blades are used. In gas turbine applications the radial curved blades are used as they represent the best compromise among the primary concerns of ease of manufacture, minimum blade stresses, and a relatively large stable range of operation with a fairly constant pressure rise.

Since equation (4) is more suitable for pump calculations where flow properties tend to remain constant, for air compressors a more realistic pressure rise can be obtained as follows:

Equation (18), Chapter 3, gives for compressor work, with kinetic energy terms included:

$$w_{12} = h_2 + \frac{V_2^2}{2g_c J} - h_1 - \frac{V_1^2}{2gH}$$

$$= h_{0_2} - h_{0_1} \tag{5}$$

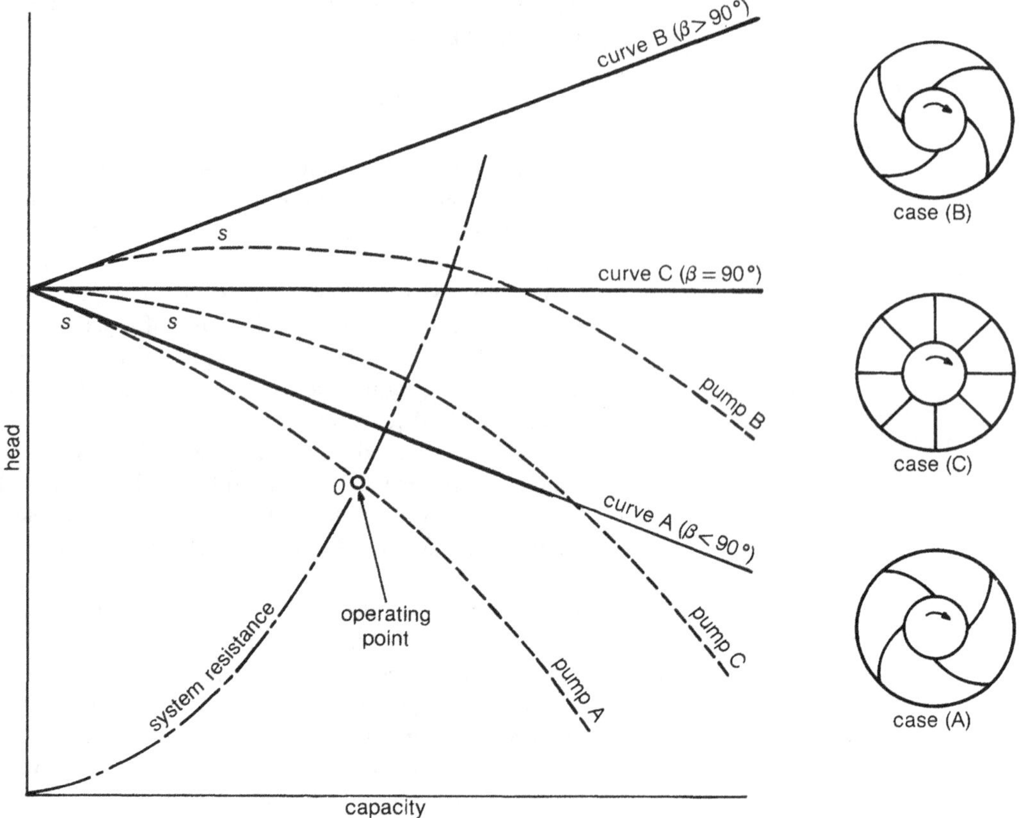

Figure 6.5 Centrifugal compressor characteristics

where h_0 is stagnation enthalpy. Thus, actual compressor work per pound of fluid, with an efficiency η_c, is with $h_0 = c_p T_0$, and by use of equation (17) Chapter 3:

$$w_c = \frac{h_{0_2} - h_{0_1}}{\eta_c} = \frac{c_p T_{0_1}}{\eta_c}\left[\frac{T_{0_2}}{T_{0_1}} - 1\right]$$

$$= \frac{c_p T_{0_1}}{\eta_c}\left[\left(\frac{p_{0_2}}{p_{0_1}}\right)^{(k-1)/k} - 1\right]$$

Solving this now for the pressure ratio gives:

$$\frac{p_{0_2}}{p_{0_1}} = \left[1 + \frac{\eta_c w_c}{c_p T_{0_1}}\right]^{k/(k-1)}$$

$$= \left[1 + \frac{\eta_c}{c_p T_{0_1} g_c}\left(V_{b_2} V_{u_2} - V_{b_1} V_{u_1}\right)\right]^{k/(k-1)} \qquad \textbf{(6)}$$

which is a general expression valid for both radial and axial compressors.

For radial blades $V_{u_2} = V_{b_2}$ and equation (6) is simplified for axial entry where $V_{b_1} = 0$;

$$\frac{p_{0_2}}{p_{0_1}} = \left[1 + \frac{\eta_c V_b^2}{c_p T_{0_1} g_c}\right]^{k/(k-1)} \qquad \textbf{(7)}$$

Temperature rise can be calculated from:

$$\frac{T_{0_2}}{T_{0_1}} = \left(\frac{p_{0_2}}{p_{0_1}}\right)^{(k-1)/k} \tag{8}$$

Example 6.1: A centrifugal compressor compresses 50 lb/sec air flow at a rotational speed of 10,000 rpm. Standard sea level air (60°F, 14.7 psi) enters the compressor axially. At the exit the wheel radius is 1 ft, the passage width is 3″, and the blade angle β is 80° (slightly curved backwards). Determine (a) the torque, (b) the power required to drive the compressor, (c) the pressure, and (d) temperature rise. An efficiency of .81 is assumed.

Solution:

Since entry is axial, the entry whirl, V_{u_1}, is zero. The blade speed at exit is:

$$V_{b_2} = \frac{2\pi r_2 N}{60} \quad N = 10{,}000 \text{ rpm}$$

$$= \frac{2\pi\, 10{,}000}{60} = 1047 \text{ ft/sec}$$

From the mass flow equation, the relative velocity can be calculated as:

$$V_{r_2} = \frac{\dot{m}}{\rho \pi D d \sin \beta}$$

$$= \frac{50}{(.075)2\pi \dfrac{3}{12} \sin 80°} = 430 \text{ ft/sec}$$

From the velocity diagram:

$$V_{u_2} = V_{b_2} - V_{r_2} \cos \beta = 1047 - 430(.174)$$
$$= 972 \text{ ft/sec}$$

(a) The torque is obtained from equation (1), Chapter 4:

$$\tau = \frac{\dot{m}}{g_c} r_2 V_{u_2}$$

$$= \frac{50}{32.2}(972)$$

$$= 1509 \text{ ft-lbf}$$

(b) The power required is then:

$$P = \frac{\tau \omega}{550} = \frac{1509(10{,}000)2\pi}{550(60)} = 2873 \text{ HP}$$

(c) The pressure rise can be determined from equation (7) with $k = 1.4$, $c_p = .24$ Btu/lb°R:

$$\frac{p_{0_2}}{p_{0_1}} = \left[1 + \frac{.81(1047)^2}{.24(520)32.2(778)}\right]^{3.5}$$

$$= [1 + .284]^{3.5} = 2.4$$

(d) The temperature ratio becomes, then, equation (8):

$$\frac{T_{0_2}}{T_{0_1}} = \left(\frac{p_{0_2}}{p_{0_1}}\right)^{(k-1)/k} = 2.4^{.286} = 1.28$$

or, the exit temperature is:

$$T_{0_2} = 1.28\ T_{0_1} = 1.28(520) = 666°R\ (206°F)$$

Successful compressor operation in the gas turbine depends on the ability of the compressor to supply working fluid to the rest of the system: combustor, turbine, and the exhaust ducting. These components then determine the requirements, or losses, for the flow and pressure, respectively, that the compressor must satisfy. In Figure 6.5 is shown the required pressure rise curve, which is the loss downstream of the compressor. For stable operation, the compressor must run at point O, where the pressure rise in the compressor exactly matches the pressure difference required by the rest of the system to move the flow. Operation to the right of this point is not possible, since the compressor pressure rise is inadequate to move the required flow. Operation to the left of point O requires additional losses or pressure drops in the rest of the system, which tend to move the operating point until the stall point is reached. Thus, for a given rpm value, there is only one stable operating point. This means that the compressor and the turbine (which provides most of the pressure drop) must be properly matched for stable operation below the surge line.

Complete compressor performance is represented on the compressor performance chart, Figure 6.6, where the pressure rise and also efficiency are given in terms of the non-dimensional mass flow and rpm variables to account for varying inlet pressure and temperature conditions. The corrected, or reduced, air flow is:

$$\frac{\dot{m}\sqrt{\theta}}{\delta} = \frac{\dot{m}\sqrt{T_{0_2}/520}}{p_{0_2}/14.7}$$

Figure 6.6 Typical centrifugal compressor performance plot

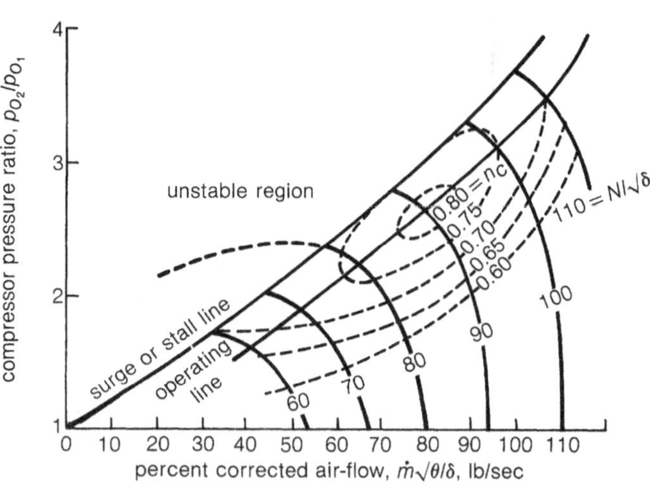

where T_{0_2} and p_{0_2} are total temperature and pressure at the compressor inlet, respectively. (See also Figure 6.20.) Similarly, the corrected rpm is:

$$\frac{N}{\sqrt{\theta}} = \frac{N}{\sqrt{T_{0_2}/520}}$$

The operating line represents the points of optimum combination of efficiency, pressure ratio, and mass flow rate; the attempt is made to match the turbine and operate the compressor as much as possible near the operating line.

Example 6.2: A gas turbine unit, designed for sea level operation at 9000 rpm and with a mass flow of 50 lb/sec, is equipped with a centrifugal compressor. The compressor performance map is given in Figure 6.6. What would be its mass flow and pressure ratio at the same rpm at an elevation of 6000 feet where the pressure is 11.78 psia and the temperature is 5°F?

Solution:

At 6000 feet:

$$\theta = \frac{T_{6000}}{520} = \frac{465}{520} = .894$$

$$\delta = \frac{P_{6000}}{14.7} = \frac{11.78}{14.7} = .8$$

Since at sea level the conditions 9000 rpm and 50 lb/sec represent 100% corrected rmp and 100% corrected flow, the corrected rpm at 6000 feet becomes:

$$\frac{\%N}{\sqrt{\theta}} = \frac{1}{\sqrt{.894}} \left[\frac{9000}{9000} \right] 100 = 105$$

At the intersection of the operating line and the percent corrected rpm of 105 (Figure 6.6):

$$\frac{p_{0_2}}{p_{0_1}} = 3.3$$

and:

$$\%\dot{m}\,\frac{\sqrt{\theta}}{\delta} = 103$$

The mass flow rate is calculated as:

$$\dot{m} = 103\,\dot{m}_{\text{S.L.}}\,\frac{\delta}{\sqrt{\theta}} = 103(50)\,\frac{.8}{\sqrt{.894}} = 43.6 \text{ lb/sec}$$

Axial Flow Compressor

A sketch of part of a cross section of an axial flow compressor is shown in Figure 6.7. A compressor stage is defined as a rotor row followed by a stator row; the inlet guide vanes are not considered to be part of the first compressor stage and must be treated separately. The rotor blades are fixed to the rotor drum, and the stator blades are attached to

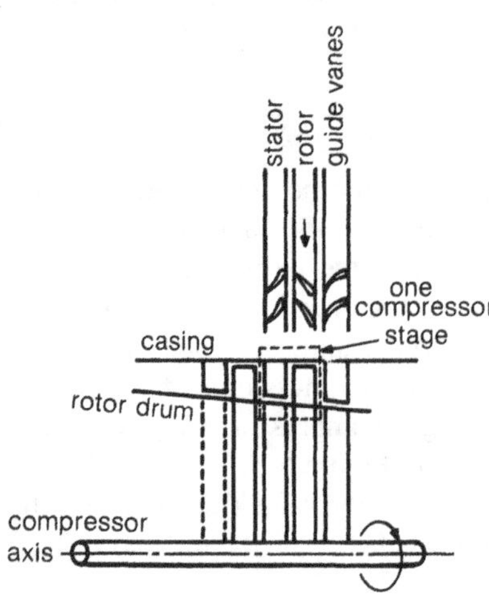

Figure 6.7 Axial flow compressor schematic

the casing. The rotating blades add kinetic energy to the fluid and increase the pressure slightly; then they discharge it at the proper angle to the stator row, where the pressure is increased further by a diffusion process. There is no pressure rise in the inlet vanes; they act more as a set of turbine nozzle vanes, since the flow is turned from the axial direction and speeded up for the proper entry angle into the rotating row. One stage is a compressor by itself. A multistage compressor is a row of single stage compressors stacked on a common shaft; the number of stages can exceed 20 in a high-pressure-ratio unit.

The general velocity diagram for an axial compressor is shown, with the pressure rise, in Figure 6.8. In the rotor row the pressure rise is obtained by diffusion of the relative velocity V_r, and the absolute velocity V is accelerated and turned for entry into the stator row. In the stator the diffusion (pressure rise) is obtained from deceleration and turning of absolute velocity V toward the axis. In most compressor designs, the exit angle, α_3, is equal to the entry angle into the stage, α_1, and the stage is then called a normal stage. The diagrams shown are for a mean radius (called also pitch radius) and average values of velocities and fluid properties. The turning and diffusion in a blade row are accomplished by relatively thin blades where the mean line between top and blade surfaces is curved (called camber). This blade shape and curvature are of great importance in exact performance analysis and prediction of axial compressor stages, called blade element theory, a subject beyond the scope of this book. The rest of the discussion here is restricted to the results obtainable from the momentum considerations used for turbines and the centrifugal compressor above.

The work of a stage per pound of fluid is, from equation (1) above:

$$w_c = \frac{1}{g_c}(V_{b_2}V_{u_2} - V_{b_1}V_{u_1})$$

$$= \frac{V_b}{g_c}(V_{u_2} - V_{u_1})$$

$$= \frac{V_b}{g_c}\Delta V_u$$

$$= \frac{1}{2g_c}(V_2^2 - V_1^2 + V_{r_1}^2 - V_{r_2}^2) \qquad (9)$$

since the radial effects are assumed to be negligible. The reaction is defined, similarly to a turbine:

$$R = \frac{h_2 - h_1}{h_3 - h_1}$$

Since the total enthalpy relative to stator rotor remains constant:

$$h_{0_{r_2}} = h_2 + \frac{V_{r_2}^2}{2} = h_1 + \frac{V_{r_1}^2}{2} = h_{0_{r_1}}$$

the reaction becomes:

$$R = \frac{V_{r_1}^2 - V_{r_2}^2}{2V_b(V_{u_2} - V_{u_1})}$$
$$= \frac{(V_{r_1} + V_{r_2})(V_{r_1} - V_{r_2})}{2V_b(V_{u_2} - V_{u_1})} \tag{10}$$

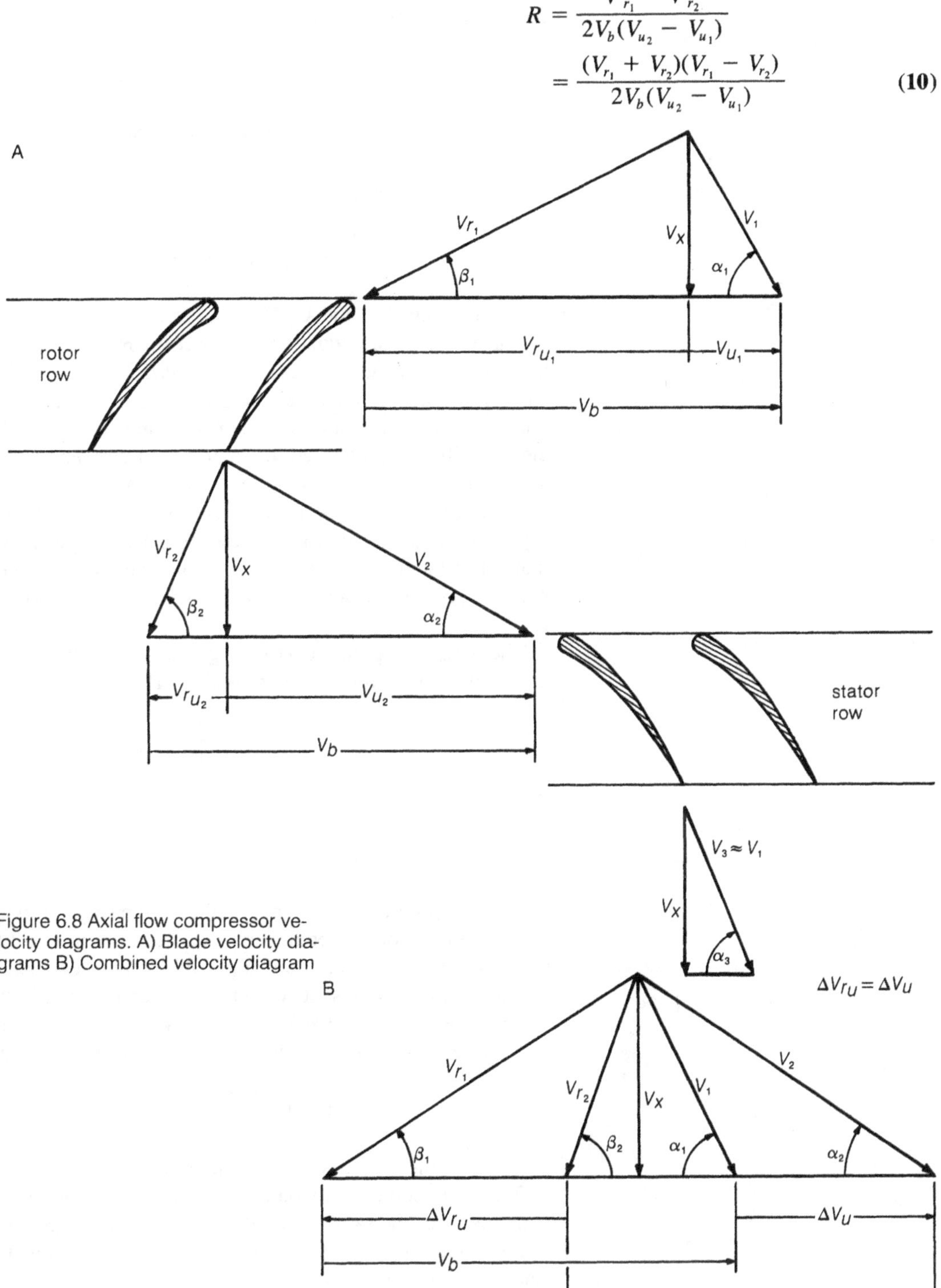

Figure 6.8 Axial flow compressor velocity diagrams. A) Blade velocity diagrams B) Combined velocity diagram

From the velocity diagram, Figure 6.8, one obtains $V_{u_2} = V_b - V_{r_{u_2}}$ and $V_{u_1} = V_b - V_{r_{u_1}}$, which give, since in an axial compressor $V_{b_1} = V_{b_2} = V_b$:

$$V_{u_2} - V_{u_1} = V_{r_{u_1}} - V_{r_{u_2}}$$

and, upon substitution, the reaction:

$$R = \frac{V_{r_{u_1}} + V_{r_{u_2}}}{2V_b} \tag{11}$$

Since $V_{r_{u_1}} = V_b - V_{u_1}$, the reaction becomes:

$$R = \frac{1}{2} + \frac{V_{r_{u_2}} - V_{u_1}}{2V_b}$$

$$= \frac{1}{2} + \frac{V_x}{2V_b}(\cot \beta_2 - \cot \alpha_1) \tag{12}$$

as $V_{r_{u_2}}/V_x = \cot \beta_2$, and $V_{u_1}/V_x = \cot \alpha_1$.

It can be seen from equation (12) that for $R = \frac{1}{2}$, the velocity diagram again is symmetrical and $\beta_2 = \alpha_1$, as in the case of turbines. In the axial compressor case, however, the reaction is a significant parameter affecting the stage efficiency. Fifty percent reaction staging is in common use in axial compressors, as turning and compression are accomplished equally in the rotor and stator rows, thus minimizing separation and stall in a given row. Experience has shown that for subsonic compressors, the efficiency and the temperature rise are at optimum values when a reaction of .5 is used.

The stage temperature rise is given by the thermodynamic expression for the compressor work:

$$h_{0_2} = h_{0_1} = w_c = \frac{V_b}{g_c} \Delta V_u$$

or:

$$\Delta T_0 = T_{0_2} - T_{0_1} = \frac{w_c}{c_p} \tag{13}$$

The subscripts 2 and 1 can apply either to the stage or to the entire compressor, depending on the meaning of w_c. The velocity triangle is useful and applicable only to one stage.

Equation (13) represents the ideal temperature rise in the stage if the stage could absorb all the work represented by w_c. The pressure ratio is obtained from equation (6) above:

$$\frac{p_{0_2}}{p_{0_1}} = \left[1 + \frac{\eta_c \Delta T_0}{T_{0_1}}\right]^{k/k - 1)} \tag{14}$$

where η_c is the compressor adiabatic efficiency.

From equation (14) it would appear that, for a multistage compressor of the same or similar blading in all stages, the overall pressure ratio can be obtained by multiplying the stage pressure ratios, or:

$$\left(\frac{p_{0_2}}{p_{0_1}}\right)_c = \left(\frac{p_{0_2}}{p_{0_1}}\right)^N \tag{15}$$

where N is the number of stages. This expression is only very approximate and, indeed, incorrect as the progressive temperature rise in the compressor stages gives a higher T_{0_1} for each stage in equation (14), thus reducing the pressure ratio in each successive stage. For a fixed stage efficiency and temperature rise, T_{0_s}, it will be assumed that the overall compressor efficiency is the same as the single stage efficiency, called also the polytropic or small stage efficiency. Then the overall pressure rise can be expressed as:

$$\left(\frac{p_{0_2}}{p_{0_1}}\right)_c = \left[1 + \frac{N\Delta T_{0_s}}{T_{0_1}}\right]^{n/(n-1)} \tag{16}$$

where $n/(n-1) = \eta_p \, (k/(k-1))$, η_p being polytropic efficiency.

Example 6.3: Estimate the LM2500 average single stage pressure ratio if the compressor polytropic efficiency is .84 (for a compressor it is always higher than the adiabatic efficiency). The inlet temperature is assumed to be 60°F (520°R).

Solution:

For an overall pressure ratio $(p_{0_2}/p_{0_1})_c = 17$, the stage temperature ratio is calculated from equation (16) for 16 stages of compression as:

$$\Delta T_0 = T_{0_1}\left(\frac{\left(\frac{p_{0_2}}{p_{0_1}}\right)^{(n-1)/n} - 1}{N}\right)$$

with:

$$\frac{n-1}{n} = \left(\frac{1}{\eta_p}\right)\left(\frac{k-1}{k}\right)$$

$$= \left(\frac{1}{.84}\right)\left(\frac{1.4-1}{1.4}\right) = .34$$

One obtains:

$$\Delta T_0 = 520\left(\frac{17^{.34}-1}{16}\right) = 52.7°F$$

The single stage pressure ratio is also obtained from equation (16), with $N = 1$:

$$\left(\frac{p_{0_2}}{p_{0_1}}\right)_{stage} = \left(\frac{520+52.7}{520}\right)^{1/.34} = 1.33$$

The axial compressor performance curves are very similar to those of a radial type, and a typical performance map is shown in Figure 6.9. The curves are plotted on the same nondimensional basis, with the pressure ratio plotted against the corrected mass flow for a set of fixed values

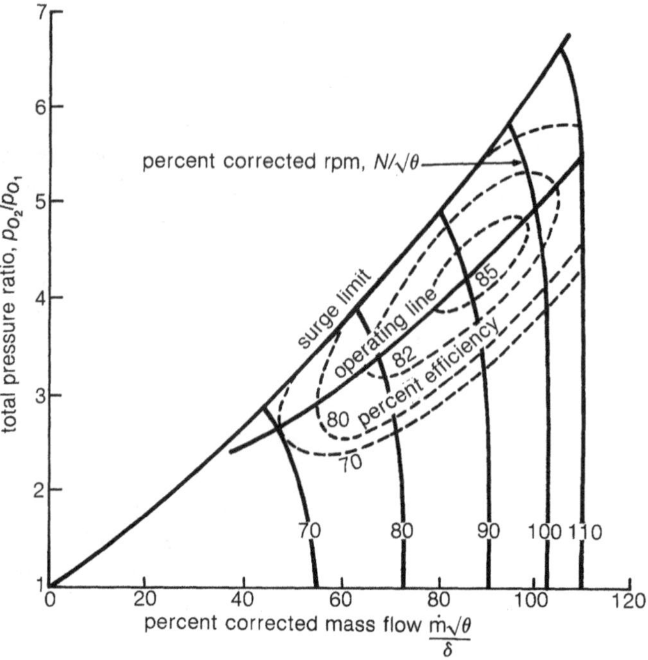

Figure 6.9 Typical axial flow compressor performance plot

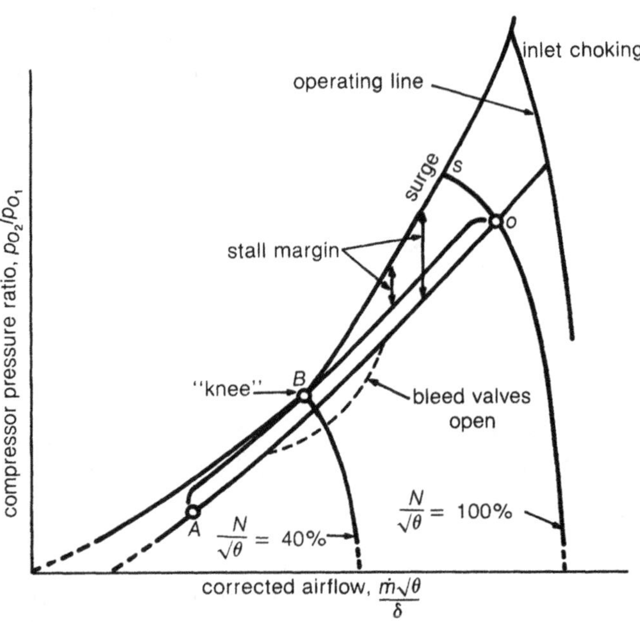

Figure 6.10 Axial flow compressor operation characteristics

of the nondimensional rpm. In comparison with the radial compressor, it is seen that the corrected rpm lines are much steeper on the axial flow compressor chart and cover a much narrower range of mass flow for a stable operation.

Two limiting conditions of compressor operations appear: stall and surge, which are indicated on Figure 6.10. Compressor stall can be explained briefly by reference to Figure 6.11. The design incidence angle is given by angle i. If the mass flow is decreased (or increased) for some reason, the incidence angle changes to i', and the back side of the blade will stall, with its ability to do work greatly diminished. Often a blade row does not stall together, as expected, but the blades stall in separate patches; in turn, these patches start rotating around the compressor annulus. This condition is known as rotating stall. Owing to the stalled back portion of the blade (in Figure 6.11), the blade to the left of the stalled blade has its incidence angle increased; being already close to the stall, this blade then will also stall out, and the stall starts propagating within the same rotating row. The rotating stall traveling around the blade row will load and unload blades at frequencies related to blade speed and the stall cells found in the blade row. If this loading frequency is close to the blade natural frequency, blade failure may result.

The mechanism of surging is very complex and not yet completely understood. It is not easy to separate surge from stall, as one event may initiate the other. However, it is possible to distinguish between rotating stall and surge by means of mass flow characteristics. A rotating stall does reduce the efficiency, but the total mass flow through the compressor is relatively steady. The stall patches only displace the local flow over the annulus. With a surging compressor, however, a strong oscillation or pulsation of the total mass coming from the machine is exhibited.

There are several reasons for mass flow change that can lead to eventual blade or rotating stall. At low speeds, and during start-up acceleration, A to B in Figure 6.10, the rear stages experience very high velocities due to lower than design pressure ratios, which reach sonic values and then limit the mass flow so that the choking conditions occur. As acceleration must take place at pressure ratios slightly higher than the normal operating line, the knee portion of the surge line may be transversed (Figure 6.10). To prevent this, the local pressure ratio is decreased, and the flow resistance is reduced by opening the bleed valves. As the rotor speeds

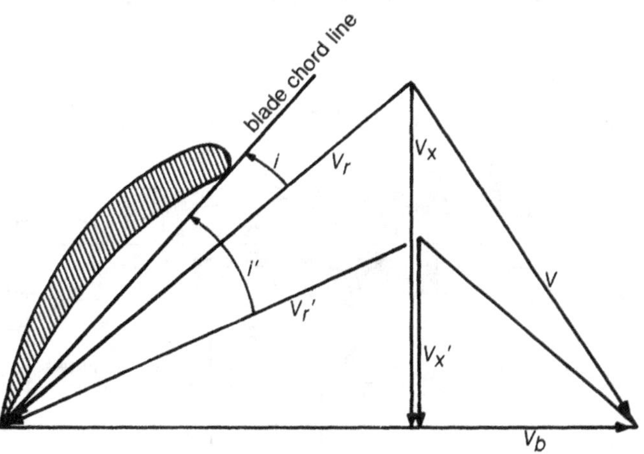

Figure 6.11 Effect of mass flow change on flow angles, axial flow compressor

pick up, the valves are again closed. As the rotor is speeded up past the design point, choking may occur at the inlet.

If the mass flow is decreased during operation at design rpm, point O in Figure 6.10—for example, because of low ambient inlet conditions—the operation tends to move toward point S near the surge line, since the density at the compressor exit is increased owing to higher pressure and the mass flow is decreased. The axial flow V_x is reduced to V_x' (Figure 6.11). This change increases the blade incidence angle from i to i', and the last stage blades may stall. If the rotor speed is reduced, from O toward B on the operating line, the mass flow falls off more rapidly at the inlet, and the first stage blades tend to stall for the same reason of reduced V_x. Concurrently, however, in the last stages pressure and density are decreased, resulting in a slight increase in velocity and a decrease in the incidence angle, leading to possible choking in the last stages.

Thus, summarizing, during operation at reduced corrected rpm, the incidence for the first stages is increased, while the last stage incidence is decreased. The effect is increased with increasing pressure ratio.

To alleviate and eliminate this matching problem and reduce wasteful bleed, two basic solutions have been utilized. Variable inlet guide vanes and stator blades are used in the first stages, which can be closed down at low rpm to increase the axial velocity and to keep the stage away from stall and near optimum operating conditions. The same effect can be achieved by decreasing the speed of the first stages and increasing it in the last stages. At a given flow rate, this effectively then corrects the incidence angles. This effect is obtained by dividing the compressor into two or more sections, each rotating at its own speed. Often a combination of variable stator vanes and twin spool compressors is used for most effective and flexible control of compressor operation (Figure 6.12).

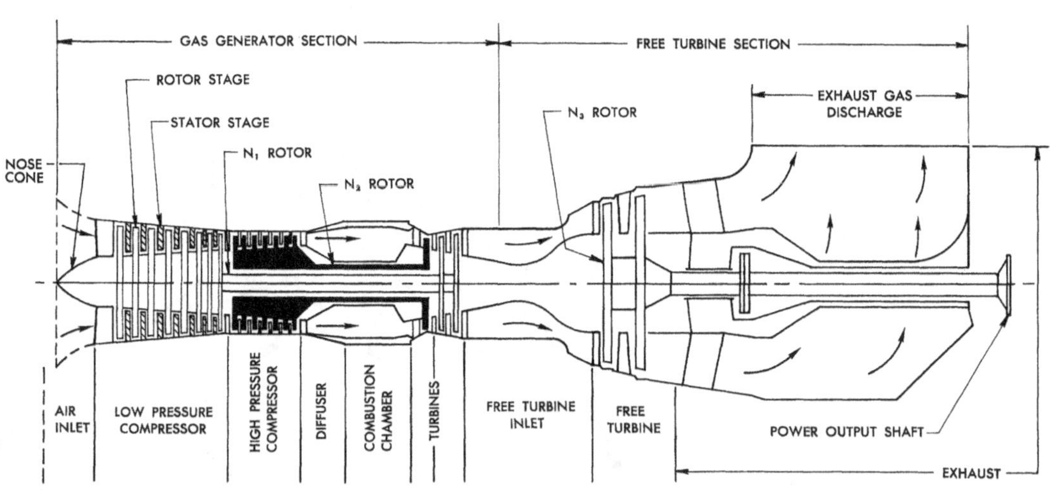

Figure 6.12 Split compressor turbine arrangement (Courtesy Pratt & Whitney)

The relative merits and some disadvantages of the two compressor types can be summarized as follows:

CENTRIFUGAL

Advantages	*Disadvantages*
High pressure ratio per stage.	More than two stages currently impractical due to increase in losses. Two stages used in some smaller units.
Good stable operating range.	
Maintains good efficiency over a wide range of operations.	
Simple to manufacture, thus lower cost.	Large cross-sectional area.
Lower weight.	

AXIAL

Higher peak efficiencies.	Has narrower operating range for good efficiency.
High pressure ratio due to many stages with low losses.	Higher cost.
Low cross-sectional area.	Higher weight.
Much higher mass flow.	Starting power can be high.

6.3 Combustors and Fuels

The purpose of the combustor in a gas turbine is to provide a steady flow of high energy fluid to the turbine sections. Since large gas turbine units use a mass flow of several hundred pounds per second, large quantities of energy must be added to the air in a very small space and in a very short time. To accomplish this, the fuel is burned directly in the air, resulting in large amounts of heat liberated in a relatively small space: over 20,000 BTU/ft^3/sec in a gas turbine as compared to a typical conventional steam plant where the working fuel is heated indirectly at rates of 7–100 BTU/ft^3/sec. Thus, the resulting gas temperatures are also much higher in gas turbines, reaching 3000°F and higher, at which point even high alloy steels melt unless measures are taken to protect the exposed surfaces.

One major effect of high pressure on heat release rates is as intuition suggests. The air, since it is compressed, requires substantially less volume. Hence, the heat release rate per unit mass is constant, while the heat release rate per unit volume increases dramatically. Another effect of pressure which is not intuitive, but has been determined empirically, is that combustion efficiency is better at higher pressures. Although it is not known precisely what accounts for the improved combustion, two important variables which affect combustion would have changed. The smaller volume in the combustion zone should facilitate mixing of the fuel and air at high pressures. Slightly more subtle is the effect of surface area. Although the surface to volume ratio increases, the important parameter of surface area to mass ratio decreases substantially. Surface area effects and poor mixing are known causes of incomplete combustion.

Although the combustor is perhaps the simplest part of a gas turbine unit, it has evolved from empirical experimenta-

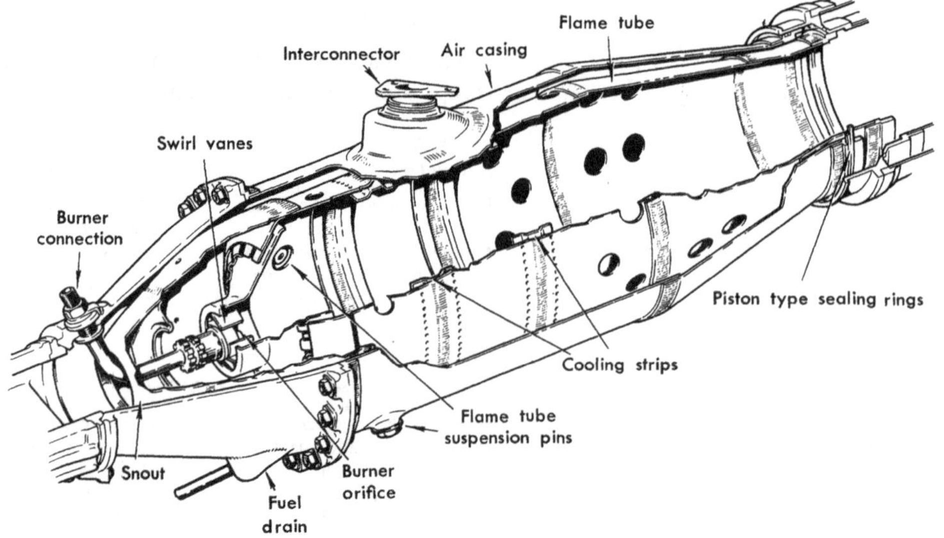

Figure 6.13 Tubular type combustor (Courtesy Rolls-Royce Ltd.)

Figure 6.14 Annular type combustor (Courtesy General Electric Co.)

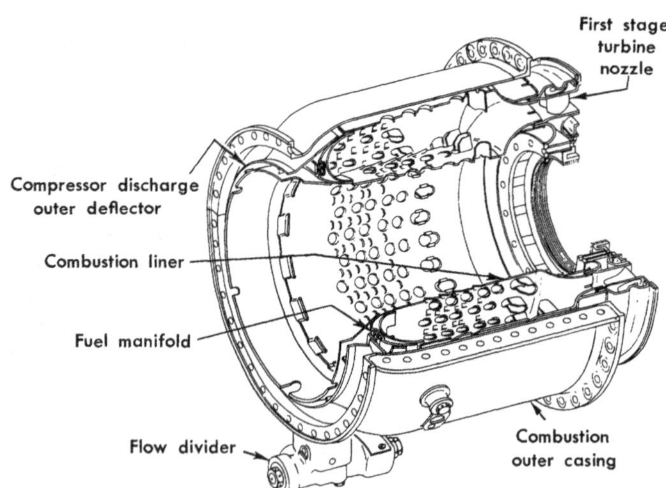

tion and testing, as it is not amenable to reliable theoretical treatment. Two basic types exist: the tubular or can type, Figure 6.13, and the annular type, Figure 6.14. They both contain the same basic parts: an outer casing, a perforated combustion liner, a flame tube, a fuel injection system, a primary combustion flame stabilization zone, and a means for initial ignition (usually by two spark plugs).

There is one consideration that dictates the basic geometry of conventional gas turbine combustors. The simplest form of combustion chamber would be a straight duct between the compressor and turbine. But the velocity of the air leaving the compressor and entering the combustor at about 500 ft/sec would be too high for combustion. The velocity is therefore slowed by the inlet section by a factor of about five as the air enters the combustor. The increasing area of the diffuser section can be seen in Figure 6.16. Even then,

Figure 6.15 Centrifugal compressor burner arrangement

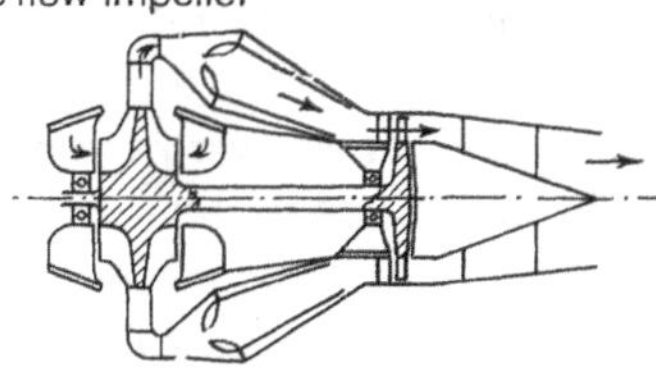

Figure 6.16 Annular combustion chamber. (A) JT9D combustor. (B) LM 2500 combustor liner (Courtesy General Electric Co.)

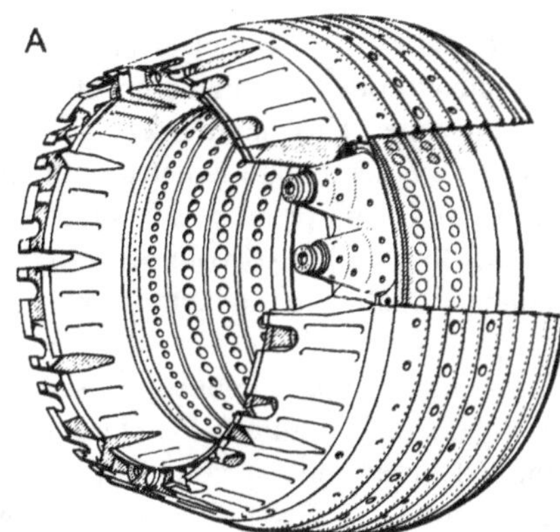

the velocity is too high for stable combustion. Hence, a flow reversal is created by baffles, louvers, and flame holders to provide a low velocity region where the flame can be sustained. A baffle may be used to bring burning gases back to mix with the fuel and air entering, in order to provide a stable source of ignition. The end of the flame tube, visible in Figure 6.13, and the fuel spray nozzle are effectively baffling the air from flowing into the center of the flame tube. A low pressure zone develops downstream of the baffle, and burning gases flow into the low pressure zone.

Can type combustors are used mostly with centrifugal compressors (not used in marine applications), where the air is divided when leaving the diffuser, and is ducted into individual combustion cans arranged around the compressor shaft section (Figure 6.15). Each can contains a fuel nozzle in its center. Primary air, entering at the nozzle, provides mixing and initial combustion. Secondary cooling air is introduced through and between the inner liner and the can case. It serves to cool the liner and to provide additional air to complete the combustion process. Before entering the turbine, the cooling air between the liner and can is mixed with combustion products from inside the liner to reduce the turbine inlet temperature. This type of combustor has excellent serviceability, as individual units can easily be removed and replaced. They tend to be somewhat longer, as their diameter must be held to a minimum for engine wraparound

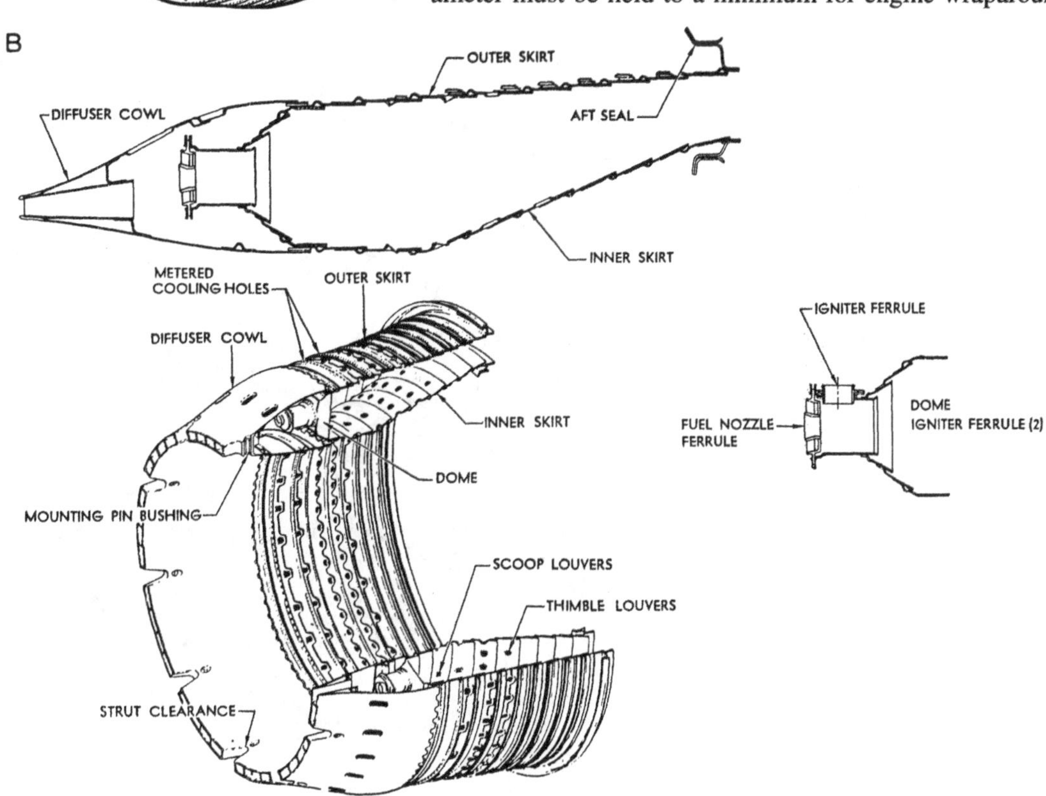

purposes. Their primary disadvantage lies in exposing turbine blades to severe temperature differentials if a fuel nozzle develops trouble or gets clogged. All can type combustors have two holes on opposite sides of the can near the primary combustion zone. These holes are connected to adjacent cans to enable the flame to cross over from one can to another during start-up operation.

The annular type of combustion chamber is in the form of an annulus arranged around the center of the machine. Fuel is introduced through a series of nozzles at the upstream end and is mixed and burned by primary and secondary air, as in the can type of combustor. In some burners (Figure 6.16a) the fuel is injected in a conical section upstream of the primary combustion zone to premix the fuel and to allow time for some vaporization. In another arrangement (Figure 6.16b) fuel is injected through atomizing nozzles directly into the combustion zone. This annular design has tended to correct the uneven exit temperature distribution of the earlier annular combustors. The primary advantage of this type of combustor is its low surface-to-volume ratio, so that it requires less secondary cooling air and a lower burner weight.

A combination of can-annular type combustors used in larger gas turbine units is shown in Figure 6.17. Here individual liners are assembled in an annular container. Each liner contains an array of fuel nozzles for better mixing and combustion. The annular outer chamber can easily be removed to allow access for individual liner inspection and replacement. The shorter length used in can-annular combustors serves also to reduce the pressure drop in the combustion system.

Figure 6.17 Turbo-annular combustor (Courtesy Rolls-Royce Ltd.)

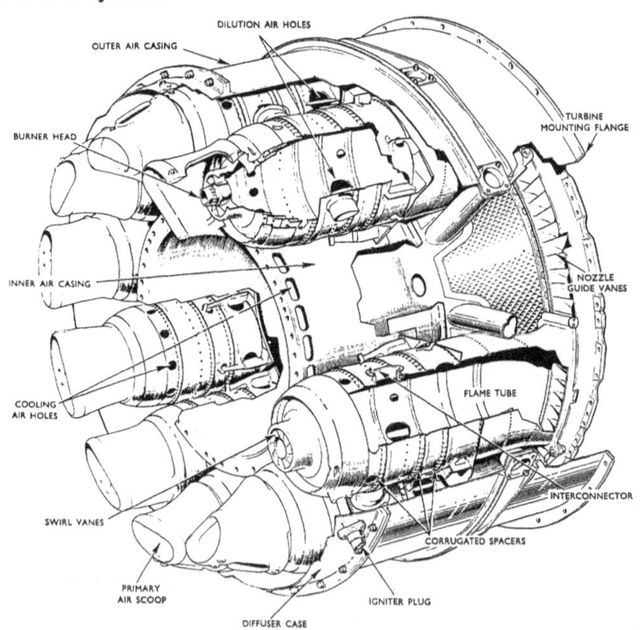

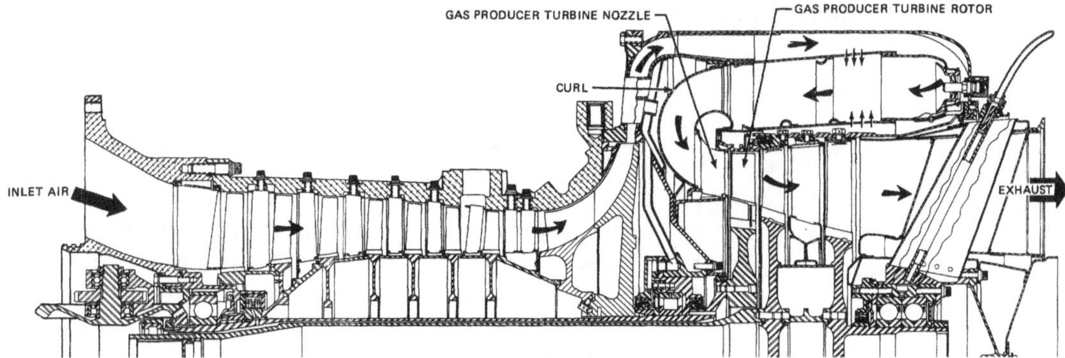

Figure 6.18 Reverse flow burner (Courtesy AVCO Corp.)

Another version of the combustor, the reverse flow annular type designed for compactness, is shown in Figure 6.18. The flow enters from the diffuser section and is split into two paths. The primary air enters the vaporizing tubes at the far end of the combustor liner, and burning takes place during motion toward the front of the engine. The secondary air shields the liner and the outer casing and enters the liner just before it starts the U-turn into the turbine, where it protects the annulus walls and cools the combustion products to allowable turbine entry levels.

For efficient combustion, engine design, and operation combustors should satisfy most of the following requirements:

1. They should be short and small.

2. There should be an even temperature distribution at the combustor exit.

3. Combustion should occur without deposits and visible exhaust.

4. There must be adequate time for burning.

5. There should be little stagnation pressure loss.

6. There should be no hot spots and no flame-outs.

7. There should be positive light-off to avoid hot starts (where the engine starts but the exhaust temperature is too high).

8. There should be good relight capability.

Obviously, several of these features are in conflict, and there must be compromise.

To provide reasonably stationary combustion in the burner where large quantities of air pass through, it is necessary to maintain the flow within certain limits. If the air flows too fast, the combustion flame will be blown out the exit. If the velocity is too low, the flame will travel upstream to the nozzle and will be extinguished. Maximum allowable exit velocity is about 400–500 ft/sec. To achieve this aim the primary and secondary air flows and flow areas must be carefully proportioned consistent with the cooling requirements.

The cooling requirements and air-fuel mixture proportioning arise from the high burner temperatures required

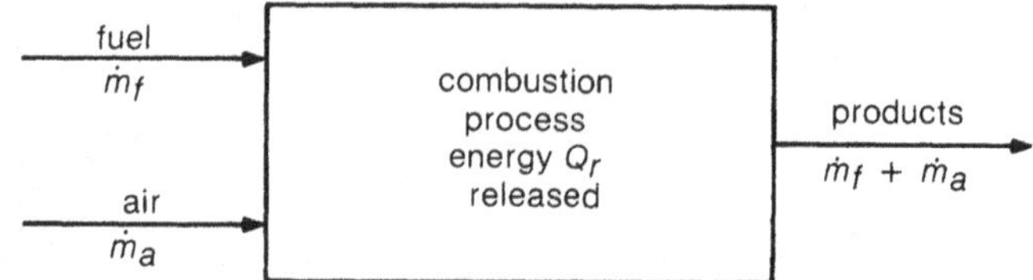

Figure 6.19 Combustion energy balance

for complete combustion and efficient use of fuel. Ideal combustion should take place as a stoichiometric process where all available oxygen chemically reacts with all available fuel. However, such burning occurs at impractically high temperatures.

The combustion process can be described by the first law of thermodynamics as follows:

The heat balance equation can be written approximately as (see Figure 6.19):

$$\eta_b \dot{m}_f Q = (\dot{m}_a + \dot{m}_f) c_p (T_{0_2} - T_{0_1}) \qquad (17)$$

where T_{0_2} = combustor outlet (turbine inlet) total temperature

T_{0_1} = combustor inlet (compressor outlet) total temperature

c_p = specific heat of the products mixture

η_b = burner efficiency or combustion efficiency (.95–.99)

$\dot{m}_a$ = air flow rate

$\dot{m}_f$ = fuel flow rate

The term Q is the lower heating value (LHV) of the fuel, defined as the heat available if the water vapor in combustion products is not condensed. For most hydrocarbon fuels Q may be approximated as:

$$Q = 15,900 + 15,800 \frac{H}{C} \text{ BTU/lb} \qquad (18)$$

where H/C is the hydrogen–carbon ratio:

$$H/C = \frac{1.008m}{12.01n} \text{ for a fuel represented by } C_n H_m$$

Most hydrocarbon fuels have a value near 18,500 BTU/lb. Since for most hydrocarbon fuels the stoichiometric fuel-air ratio is about .0667, it is easy to estimate the combustor exit temperature from equation (17), which can be rewritten:

$$f = \dot{m}_f / \dot{m}_a$$

$$\eta_b Q = \left(\frac{1}{f} + 1\right) c_p \left(T_{0_2} - T_{0_1}\right)$$

or:

$$.9(18500) = (15 + 1).26(T_{0_2} - 600)$$

where an average c_p = .30 BTU/lb°R has been assumed.

This gives $T_{0_2} = 4069°F$, which is too hot for present turbine inlet technology. In order to reduce the inlet temperature to a more manageable 1800–2200°F, it is necessary to operate the burner with a large quantity of excess air to provide adequate cooling. Equation (17) can also be used for this purpose when rewritten in a more convenient form:

$$f = \frac{1}{\dfrac{\eta_b Q}{c_p(T_{0_2} - T_{0_1})} - 1}$$

Suppose now that a turbine inlet temperature of 2000°F is desired with a combustor inlet temperature of 500°F and the same fuel properties as above. Then the required fuel air mixture is obtained:

$$f = \frac{1}{\dfrac{.9(18500)}{.26(2000 - 500)} - 1} = .024$$

which states that almost 300% excess air is required (as compared with the stoichiometric value of .0667).

Figure 6.20 shows the relationship between overall fuel/air ratio and the amount of temperature rise experienced for any reasonable compressor inlet temperature. The effects of temperature and non-air combustion products on specific heat have been taken into consideration. The calculations in Figure 6.20 used a lower heating value of 18,300 BTU/lb, which is representative of number two diesel fuel.

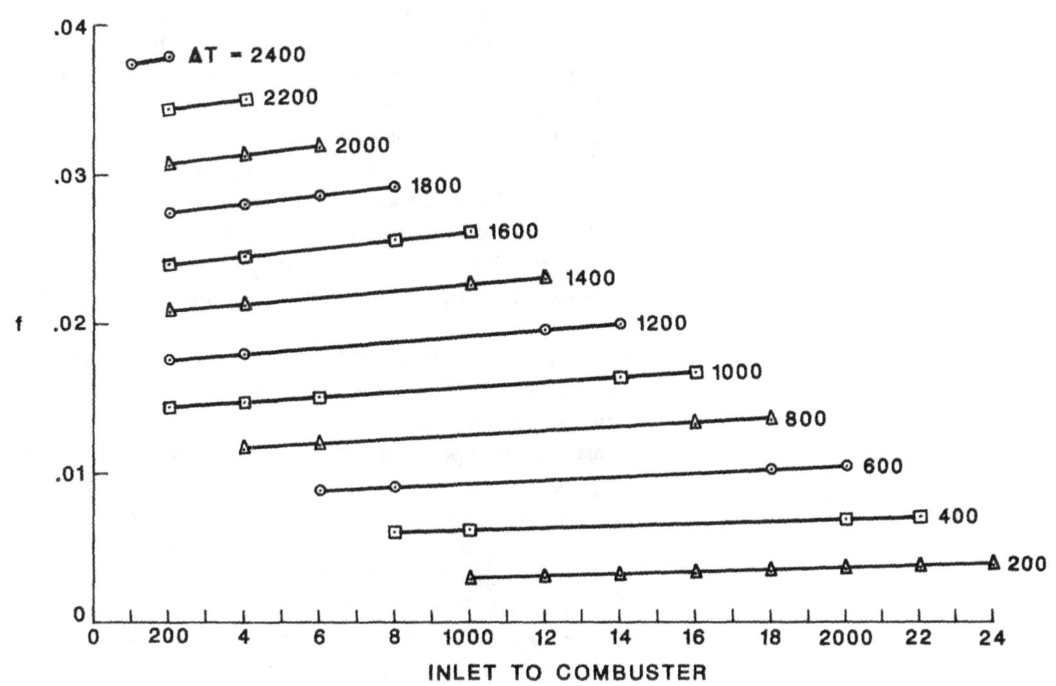

Figure 6.20 Fuel/air ratio f as a function of temperature rise in the combustor (stoichiometric f = .0667)

Fuel Limitations

Gas turbine combustors have fuel requirements very different from other internal combustion engines. There are no cetane or minimum octane rating requirements. The fuel may be either liquid or gas. Mostly the fuel requirements are dictated by considerations not directly related to the combustor. For example, although gases can be very good fuels for gas turbines, they are not likely to be found in marine applications owing to difficulty of storage. One fuel requirement almost universally applicable to gas turbines is the absence (or the removal) of ash, grit, or trace metals. Either ash or trace metals present in the fuel will contribute to high temperature corrosion in gas turbines. Sulfur is not too objectionable, if trace metals such as sodium are not present in the fuel or the air.

Air Requirements

If the fuel is not a determining factor in gas turbine combustor design, other factors do account for the similarities between the two types of gas turbine combustors, and their unique differences from other kinds of combustion chambers. The use of gas turbines for aircraft propulsion accounts for the compact, lightweight designs of most gas turbines. The use of a perforated liner to separate the primary combustion zone from the secondary air is strictly a combustion-related requirement. The overall air/fuel ratio of the mixture of gases leaving the combustor is far beyond the flammability limits of any fuel except hydrogen. Even within the limits of flammability, any fuel burns most rapidly near stoichiometric conditions. Speed of combustion and completeness of combustion are of primary importance in a gas turbine combustor. The perforated liner (Figure 6.14 is a good example) can be designed to admit the secondary air for dilution as required to protect the turbine blades from too high temperatures. A superior design will do so without excessive pressure drop through the perforations to put the correct amount of secondary air in the right places.

Mixing gases quickly, whether it is for combustion or for dilution, requires effective penetration of the air jet into the larger stream of gases. Intuition is correct in suggesting that a faster jet will penetrate further than a slower jet; pressure drop across the perforations gives the velocity needed for penetration. Owing mostly to momentum, a larger diameter jet (see the large holes in the flame tube in Figure 6.13) will penetrate further than a smaller jet. A larger jet is less effective, however, in mixing quickly. To allow adequate penetration by smaller jets, the distance to be penetrated can be minimized. There are two ways to accomplish this. Multiple jets comprise the first method. The main gas stream is usually penetrated from both sides in the annular design, and from the entire periphery in the tubular design. The other method is to keep the thickness of the main gas stream as small as possible. A single large combustor offers a more difficult mixing problem than several smaller tubes or the annular design; see Figure 6.17 for an extreme example of keeping the flow path thin.

Material Problems

Nickel alloys are usually specified as the material for gas turbine combustors, because they are resistant to oxidation at elevated temperatures. Strength at high temperatures also must be considered. The sheet metal used to make gas turbine combustors is very thin and easily formed. The importance of a thin wall is not so much weight savings as stress reduction, since combustors are often subjected to extreme thermal stress conditions.

Gas turbine combustors experience extremely rapid warm-up and cool-down across a very wide temperature range, which causes uneven thermal expansion. Stresses are produced from this uneven expansion. The thinner the cross section of the metal, the lower the stress levels during temperature changes. One end of the gas turbine combustor is always hotter than the other. This temperature gradient puts the combustor wall under continual stress in the axial direction, and this stress is the most common cause of combustor liner failure.

To accommodate expansion when hot, the combustor must be free to elongate.

The combustor wall is cooled by the secondary air before it enters the combustion zone; the air flow is designed to cool the wall as it penetrates the perforations in the liner. The film of air next to the combustor liner keeps the liner cool even after the secondary air has entered the combustion zone. The path of the secondary air can be seen in Figure 6.16. A major source of thermal stress occurs at the perforations. The wall is most effectively cooled at the edge of a hole, but the wall being thin cannot conduct heat readily to the edge where the cooling is occurring. The resulting temperature gradient produces extremely stressful conditions at the boundaries of perforations in the combustor liner which can easily lead to propagation of cracks.

Soot Formation

Liquid distillate fuels may form objectionable amounts of carbon particles—soot—in gas turbine combustors. Soot forms whenever hydrocarbons are burned in extremely rich mixtures. For example, natural gas has been used to produce carbon black commercially. In a gas turbine soot formation is unintentional and undesirable. If a liquid droplet were sprayed into the high temperature combustion zone of the combustor, the temperature in the primary combustion zone is so high that combustion of the droplet would occur before it would have the opportunity to vaporize and mix with enough air to support complete combustion, or even partial combustion to carbon monoxide. The free carbon radicals not oxidized would nucleate into submicron carbon particles, collectively referred to as carbon black or soot.

Soot particles once formed are resistant to combustion because the reactivity of carbon in the form of soot is low. Soot particles will oxidize, however, given enough time and enough oxygen for diffusion into the surface of the particle, and sufficient temperature to keep the kinetics high. It is better in the confines of a gas turbine combustor to avoid producing the soot particles. Avoiding soot production is effectively accomplished by prevaporizing the fuel. Advancing a step further, the fuel and air can be premixed and then ignited. If the mixture is lean, there is little possibility of fuel molecules remaining unburned.

6.4 The Turbine

The turbine, as a work extraction device, was treated in some detail in Chapter 4, where all of the theoretical considerations were of general applicability, but most of the discussions were preparatory for the analysis of the steam turbine. Although the basic principles and concepts discussed in Chapter 4 apply equally to gas turbines, there are some important differences that deserve further attention.

In steam turbines the densities and pressures are much higher than in gas turbines, and the blading and the casings must be sturdy and heavy enough to be able to carry the steam flow loads and to contain the steam safely. The turbine blades in a gas turbine unit can be much thinner and lighter, but their design is constrained by much higher temperatures which are necessary for efficient cycle operation (Section 3.11). In steam turbines inlet temperatures do not exceed 1050°F. In gas turbines the turbine inlet temperatures have been increasing steadily since the first unit was put into service (see Figure 6.21), and now exceed 2100°F.

Perhaps the most significant difference between the steam and gas turbines is how the energy is utilized once it has been extracted from the working fluid. The steam turbine converts all the energy into shaft work to drive a generator or a propeller shaft. Gas turbines extract the work first to drive the compressor and engine accessories (the compressor work w_c, Figure 3.12). In a turbojet unit the remaining energy is used to accelerate the flow to a high speed to get propulsive thrust. In a turboshaft or turboprop unit another turbine, called the power turbine, is placed in the flow to extract as much work as possible to turn a pro-

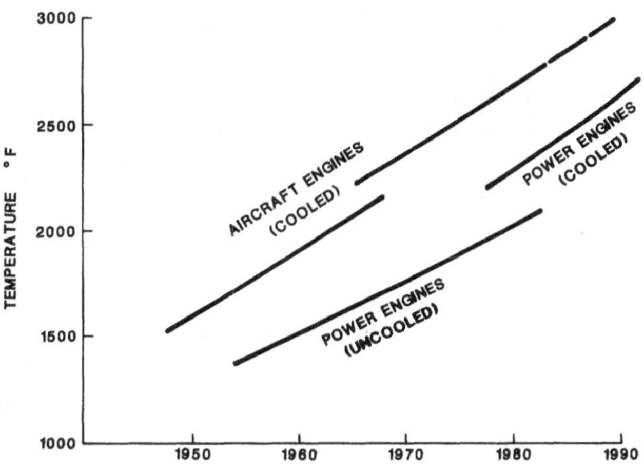

Figure 6.21 Typical turbine inlet temperature trends over time

Figure 6.22 Turbine blade construction. (A) nozzle vanes (Courtesy Rolls Royce Ltd.). (B) LM 2500 1st stage turbine bucket blade (Courtesy General Electric Co.)

peller. Since a very large portion of the energy ($\frac{1}{2}$–$\frac{3}{4}$ of the total) is required to drive the compressor, a very efficient power turbine is required to achieve good cycle efficiency.

Similarly to compressors, power turbines may be axial or radial types. Axial turbines can handle much larger flow, but at smaller flow the radial turbines can be more efficient and lighter. Multistaging, however, is much easier with the axial turbine. Thus, for marine propulsion applications requiring high power and high air flows, the favored type is the axial one.

Unlike compressors, both nozzle guide vanes—the ones between rotating blades in Figure 6.22—and the rotating turbine blades are highly curved to permit high flow expansion and large work extraction. Due to the expansive nature of the flow, the blades can be highly curved without undue losses. Most turbine rows have shrouds to reduce blade vibration (see Figure 6.23). Although shrouds add weight, they permit thinner blade sections, thus somewhat offset-

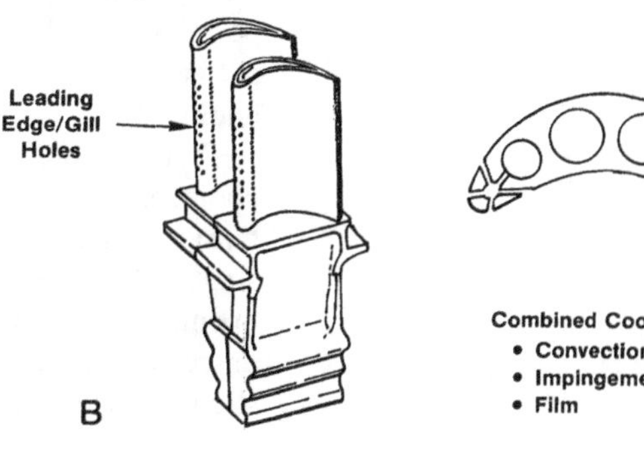

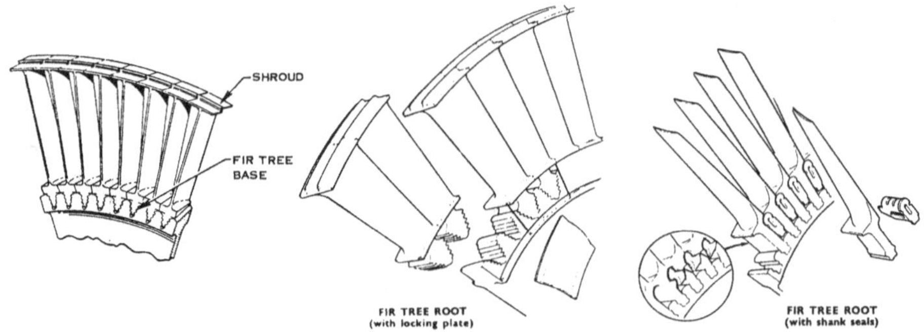

Figure 6.23 Blade shrouds and methods of attachment

ting the weight problem. The shrouds also help to control the leakage past the blade tips. The last figure also shows the fir tree blade-type attachment to the disc, which is now the most common way to affix the blades (see also Figure 5.27).

The turbine blades operate in a very severe environment. As pointed out above, they are exposed to very high temperatures; while glowing red they are rotating at high speed, thus being subjected to large centrifugal and axial loads. The result is that the blades will slightly twist and stretch, or elongate. The latter effect, called creep, is cumulative with time and its severity depends on the turbine load (rpm) and the temperature. As the design of the turbine blades and discs allows for only a given maximum amount of creep, exceeding the turbine inlet temperature or rpm can appreciably shorten the unit's service life.

Cyclic loading leading to fatigue, thermal shock, corrosion and sulfidation-oxidation also has a cumulative effect of shortening the engine life. This is the reason why cyclically operated aircraft and marine gas turbines must be inspected often, and usually must be overhauled after less than 10,000 hours of operation. However, (relatively) continuously loaded gas pipe line gas turbine units routinely achieve more than 30,000 hours' operating time between overhauls.

As efficient operation demands a high turbine inlet temperature, and since the latter is a limiting factor in most turbine designs, great effort is being spent on increasing the permissible inlet temperature. The most practical method is to actively cool the hottest and most sensitive parts of the engine.

Turbine cooling is achieved by bleeding the relatively cooler compressor airflow and feeding it through the internal passages to the first stage or high pressure turbine stages' discs and blades, as shown in Figure 6.24. Through openings in discs the coolant flow enters the blades and passes out through trailing edge slots or holes. Figure 6.25 shows typical blade passages used for nozzle and rotor cool-

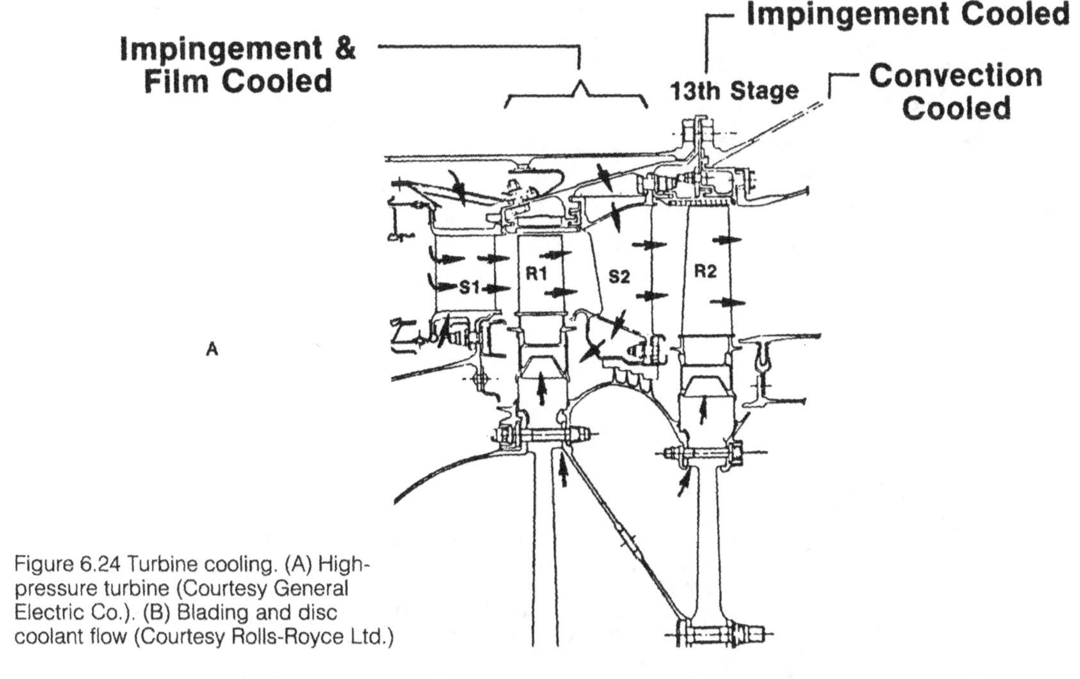

Figure 6.24 Turbine cooling. (A) High-pressure turbine (Courtesy General Electric Co.). (B) Blading and disc coolant flow (Courtesy Rolls-Royce Ltd.)

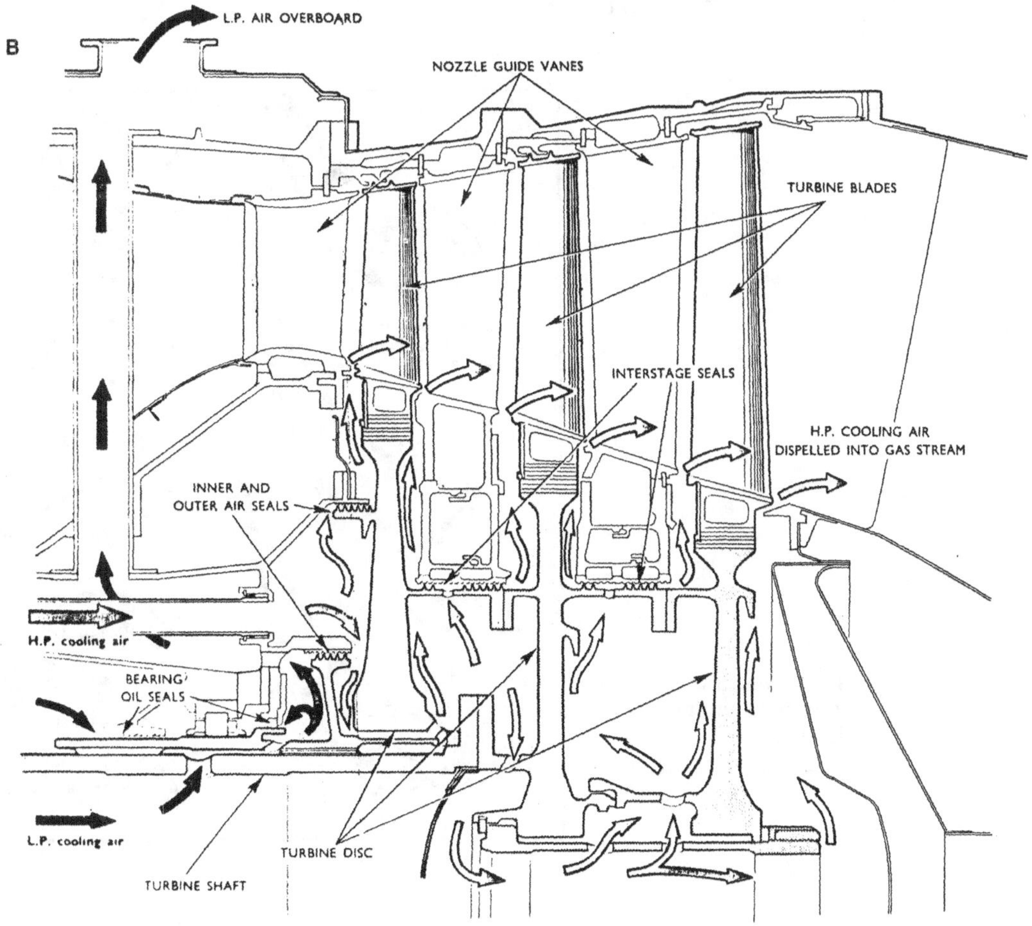

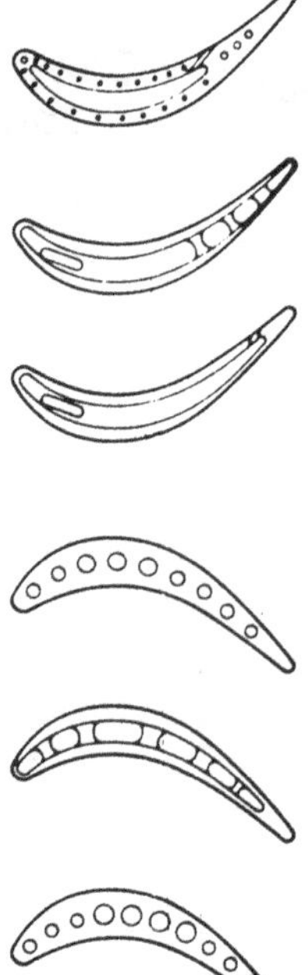

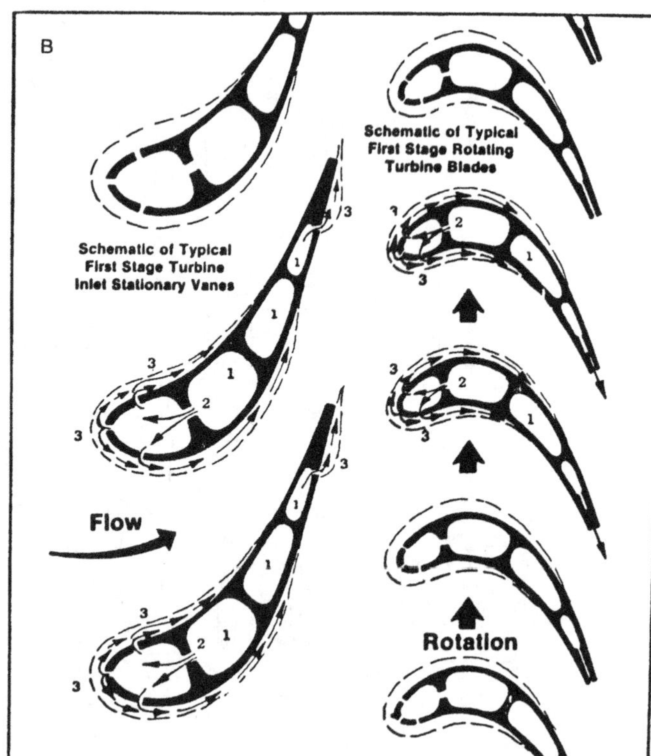

Figure 6.25. (A) Typical 1st stage turbine vane cooling passages (top) and blade cooling passages (bottom) for recent production engines (Courtesy Pratt & Whitney). (B) Types of air-cooling of turbine vanes and blades (Courtesy General Electric Co.)

A

1. Convection
2. Impingement
3. Film

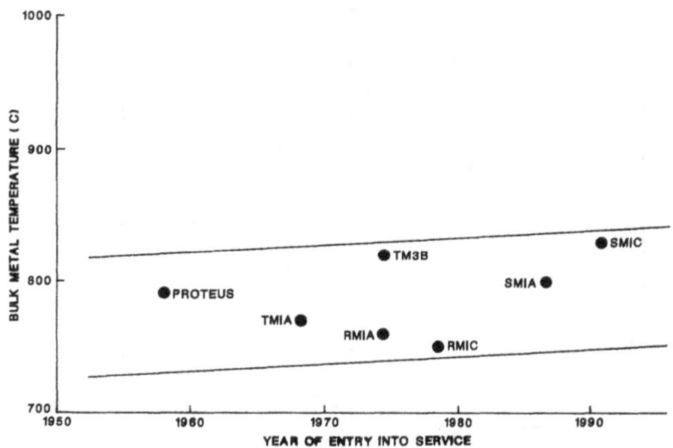

Figure 6.26 First stage turbine rotor blade bulk metal temperature (Courtesy Rolls-Royce Ltd.)

ing. A number of other configurations are shown in Section 6.6.

Figure 6.26 shows the blade temperatures that are obtained when active cooling is used. Theoretically, any given minimal blade temperature can be achieved if an adequate amount of cooling air can be pumped through the blades. Since the cooling air comes from the compressor and is effectively lost to the cycle, the cooling air represents another loss that must be minimized. Thus, 2% engine bleed is about the maximum that a designer is willing to allocate to coolant flow. Some of it will reenter the turbine with sufficient energy to be available for work in the low pressure section of the turbine.

6.5 Gas Turbine Performance

The purpose of this section is to consider the individual gas turbine components and their actual processes together in order to evaluate the overall unit performance. The process, as shown in Figure 6.27, begins at a with atmospheric air at h_a, p_a. The air enters the intake and proceeds through aerosol separators and compressor protective inlet screens to the engine inlet 1, where it enters an accelerating section to the compressor inlet 2. There is a small but definite inlet entropy rise and total pressure loss to 02. From 02 to 03 the air is compressed, with an increase in entropy due to compressor losses. State 03_s is the ideal compressor outlet state at the same pressure. Between stations 03 and 04 the compressor outlet air is mixed in the combustor with air, and combustion takes place with a small pressure loss from p_{0_3} to p_{0_4}.

From 04 to 05 expansion takes place in the high pressure turbine, providing power to drive the compressor. Thus, $h_{0_4} - h_{0_5} = h_{0_3} - h_{0_2}$. From high pressure turbine 05 the fluid enters the free, or power, turbine, which extracts useful external shaft work and expands to state 06. Both turbine processes exhibit losses, as ideal turbines would expand to

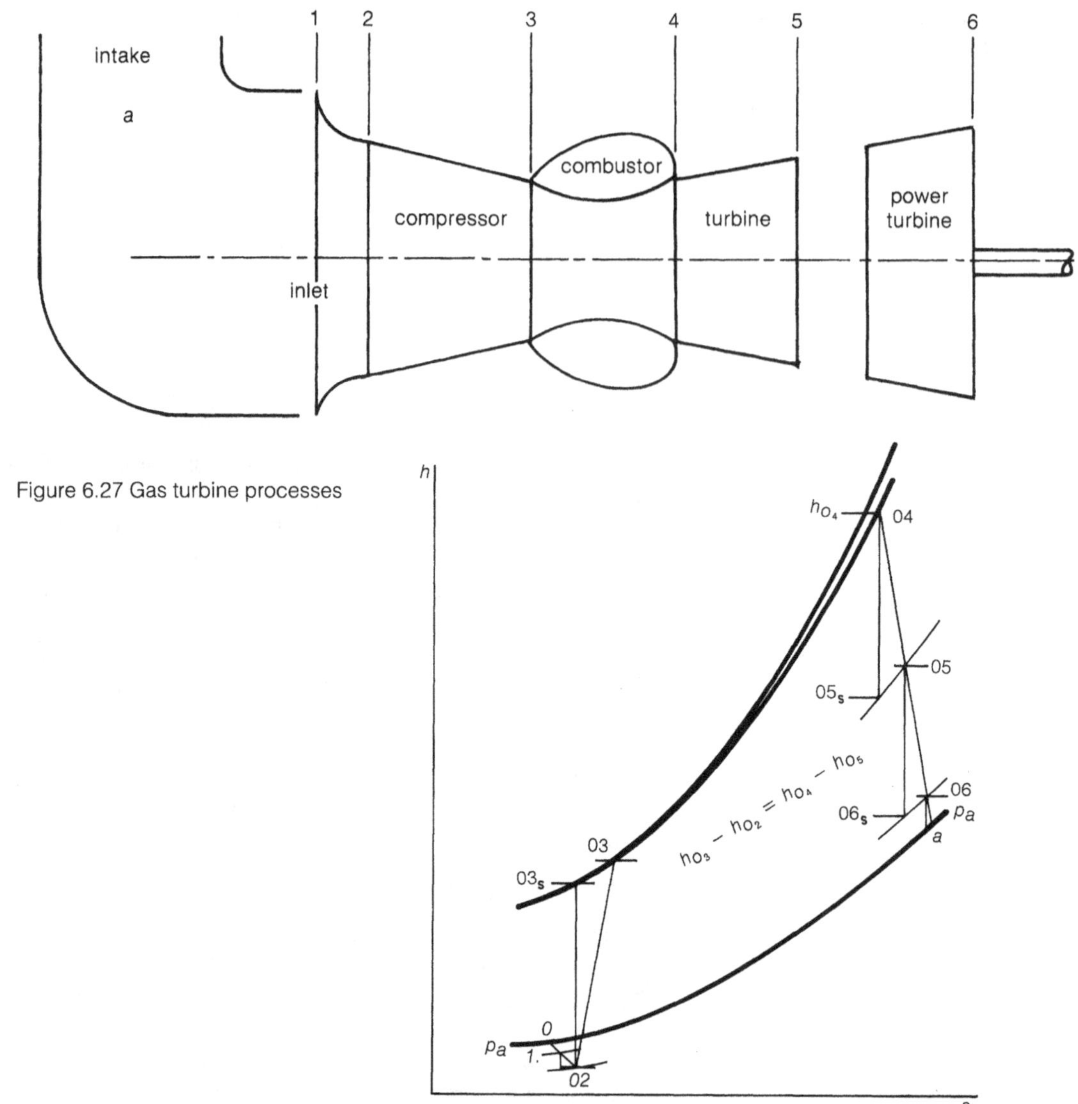

Figure 6.27 Gas turbine processes

05_s and 06_s, respectively. State 06 is at a higher pressure than the exhaust (uptake) exit at atmospheric pressure. The enthalpy drop between 06 and a represents kinetic energy loss and the losses through the uptake.

Since the total enthalpy and the total pressure changes are very small in the inlet area, but the pressure at the compressor inlet does differ from atmospheric, the pressure loss is usually given in terms of a Δp between atmospheric and the pressure obtained at station 02. Thus the pressure and enthalpy at 02 represent the starting point for a realistic calculation of turbine performance.

Compressor performance can be described by an adiabatic efficiency η_c, which is the ratio of the work required for an isentropic process to that for an actual process for the same stagnation pressure ratio. Thus:

$$\eta_c = \frac{w_{c_{ideal}}}{w_c} = \frac{h_{0_{3_s}} - h_{0_2}}{h_{0_3} - h_2}$$

with

$$\frac{T_{0_{3_s}}}{T_{0_2}} = \left(\frac{p_{0_{3_s}}}{p_{0_2}}\right)^{(k-1)/k} = \left(\frac{p_{0_3}}{p_{0_2}}\right)^{(k-1)/k} \tag{19}$$

There are two losses that need to be accounted for in the combustor. One is the pressure drop due to fluid friction and heat addition to a flowing fluid. It is given in terms of the combustor pressure ratio:

$$\frac{p_{0_4}}{p_{0_3}}$$

and is usually .95 or greater. The other is the combustion efficiency, η_b, or the fraction of the actual chemical energy released in the combustor.

Turbine adiabatic efficiency is given by:

$$\eta_{t_c} = \frac{w_{t_c}}{w_{t_{ideal}}} = \frac{h_{0_4} - h_{0_5}}{h_{0_4} - h_{0_{5_s}}}$$

with

$$\frac{p_{0_4}}{p_{0_5}} = \frac{p_{0_4}}{p_{0_{5_s}}} = \left(\frac{T_{0_4}}{T_{0_{5_s}}}\right)^{k/(k-1)} \tag{20a}$$

and:

$$\eta_{t_c} = \frac{w_{t_p}}{w_{t_{ideal}}} = \frac{h_{0_5} - h_{0_6}}{h_{0_5} - h_{0_{6_s}}}$$

with

$$\frac{p_{0_5}}{p_{0_6}} = \frac{p_{0_5}}{p_{0_{6_s}}} = \left(\frac{T_{0_5}}{T_{0_{6_s}}}\right)^{k/(k-1)} \tag{20b}$$

The uptake loss is usually expressed in terms of pressure difference between the power turbine exit and atmospheric exhaust pressures that define the turbine end state (or back pressure).

In performance calculations, the following determinations are necessary:

1. The compressor pressure ratio determines the specific work required, usually expressed per pound of mass flow.

2. The turbine inlet temperature, to be selected on the basis of materials and type of construction, determines the required heat addition.

3. High pressure turbine work is determined by the work required by the compression, i.e., $w_c = w_t$ or $h_{0_3} - h_{0_2} = h_{0_4} - h_{0_5}$. This determination permits calculation of turbine exit pressure.

4. The remaining enthalpy difference and the end state

06 determine the amount of shaft work available from the power turbine.

Unlike aircraft jet propulsion engines, marine engines lack a nozzle after the power turbine, and the turbine exhaust proceeds directly to the uptake.

In Chapter 3 the efficiency of an ideal simple Brayton cycle without heat exchangers and reheat was shown to be a function of only the cycle pressure ratio. With the state point nomenclature as shown in Figure 6.27 it becomes

$$\eta = 1 - \frac{1}{\left(\dfrac{P_{0_3}}{P_{0_2}}\right)^{(k-1)/k}}$$

and is shown plotted in Figure 6.28. This equation shows that the cycle efficiency increases with increasing pressure ratio. However, it does not tell the entire story, as the cycle operating temperatures also enter into consideration for an actual cycle. This was seen already in Section 3.11 when the specific work was calculated for an ideal cycle. In a real cycle, both the efficiency and the specific work depend on the pressure and cycle temperature ratios and on component losses. It is seen in Figure 6.28 that the efficiency has a different maximum for each temperature ratio. The drop in

Figure 6.28 Brayton cycle efficiency

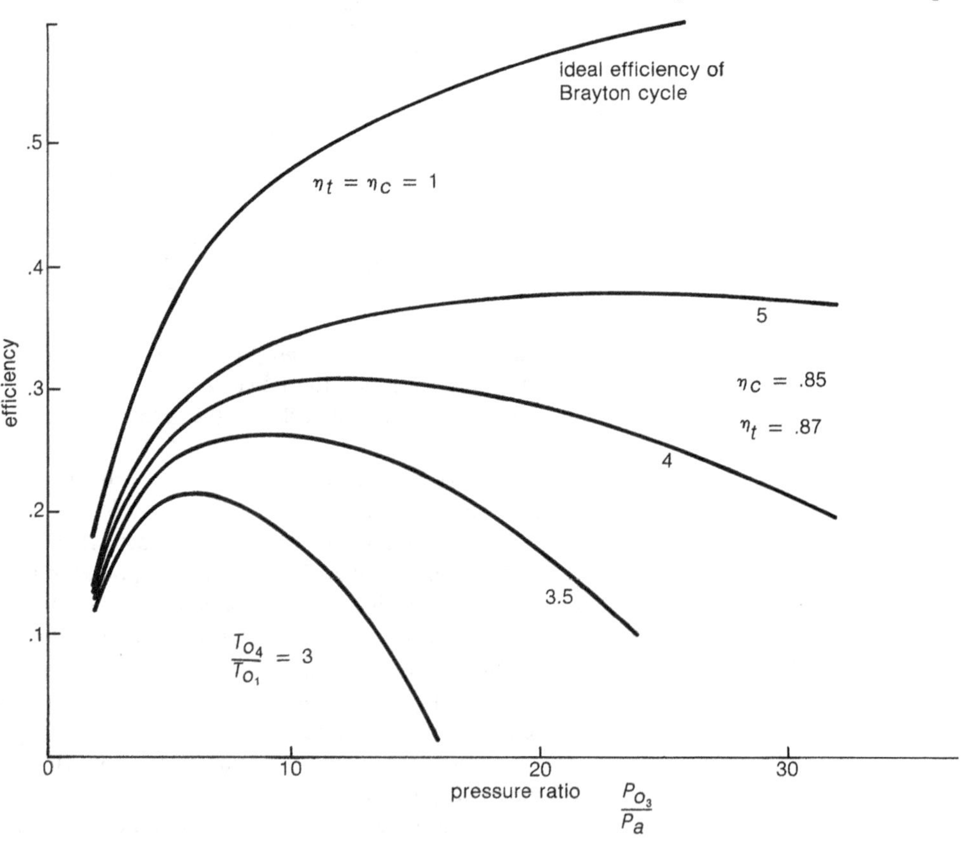

efficiency at higher pressure ratios is due to the increased work required to drive the compressor.

The effect of the pressure and temperature ratios and the component losses can be obtained if the net cycle work is calculated as it was done in Section 3.11. Rather than include all the engine losses, which would render the calculations rather cumbersome and would tend to obscure the main conclusions, only the principal losses due to the compressor and turbine are included. Moreover, the division of work, and possibly different efficiencies, of the compressor turbine and power turbine are not important to the discussion at hand.

Returning again to Figure 6.27, the net work can be expressed as (using temperatures rather than enthalpies, assuming c_p to be constant and neglecting pressure differences between inlet and exit):

$$w_n = w_T - w_c = c_p \left[T_{0_4} - T_{0_6} - (T_{0_3} - T_{0_2}) \right]$$

The temperature differences can be determined in terms of the cycle maximum temperature and pressure ratio by means of the component efficiencies defined above. Thus:

$$\eta_c = \frac{T_{0_{3_s}} - T_{0_2}}{T_{0_3} - T_{0_2}}$$

whence

$$T_{0_3} - T_{0_2} = \frac{T_{0_{3_s}} - T_{0_2}}{\eta_c}$$

and since

$$T_{0_{3_s}} = T_{0_2} \left(\frac{P_{0_3}}{P_{0_2}} \right)^{(k-1)/k} = T_{0_2} r$$

with

$$r \equiv (P_{0_3}/P_{0_2})^{(k-1)/k}$$

one obtains

$$T_{0_3} - T_{0_2} = T_{0_2} \left(\frac{r-1}{\eta_c} \right)$$

Carrying out a similar calculation for the turbine temperature difference, the net work becomes

$$\frac{w_n}{c_p} = T_{0_2} \left(\frac{1-r}{\eta_c} \right) + T_{0_4} \, \eta_t \left(1 - \frac{1}{r} \right)$$

To find the conditions for the cycle maximum net work, the standard optimization process can be carried out, which then yields

$$\frac{P_{0_3}}{P_{0_2}} = \left(\frac{T_{0_4}}{T_{0_2}} \eta_c \eta_t \right)^{k/2(k-1)}$$

which should be compared with equation 26 of Chapter 3, the corresponding result for the ideal cycle.

The actual cycle efficiency is obtained from

$$\eta = \frac{w_n}{c_p(T_{0_4} - T_{0_3})} = \frac{(1 - r)/\eta_c + t\eta_t(1 - 1/r)}{t - (1 + (r - 1)/\eta_c)}$$

where $t = T_{0_4}/T_{0_2}$.

This last expression is shown plotted in Figure 6.28 for several values of t with compressor and turbine efficiencies .85 and .87, respectively.

The net work for an ideal cycle was plotted earlier in Chapter 3 as Figure 3.14. The effect of component efficiencies is to slightly modify and lower the net work curves. The advantage of using a high turbine inlet temperature T_{0_4} is evident from both efficiency and net work curves. This was already seen in the discussions concerning the ideal cycle in Chapter 3. The increase in efficiency is not as marked as the increase in the output, which then permits construction of a smaller machine for a given application (total work output).

The choice of a design is a compromise among the lowest pressure ratio for required performance, the highest turbine inlet temperature from allowable materials or a cooling technology point of view, the shortest compressor blades possible from strength and mass flow considerations, and the maximum rpm and blade speed permissible for required output as the rotor size is decreased.

Another design compromise arises from the fact that, for given η_t, η_c, and t, the maximum turbine efficiency occurs at higher pressure ratio than the maximum net specific work. For example, Figure 6.28 shows that, at $t = 3.5$, maximum efficiency is obtained for a pressure ratio of about 9. The optimum pressure ratio for maximum work, however, is calculated as

$$P_{0_3}/P_{0_2} = (3.5 \times .85 \times .87)^{1.75} = 5.28$$

Thus, all other things being equal, unless a large efficiency loss is experienced the pressure ratio for higher output is the rule. The lower pressure ratio results in a lighter and smaller machine with fewer stages, in this case 3 fewer compressor stages.

The ambient temperature is a parameter that considerably affects gas turbine performance, and its increase reduces both cycle efficiency and the specific work. The efficiency is less affected, since the increased compressor work due to increased T_a is somewhat offset by a reduced combustion temperature rise, $T_{0_4} - T_{0_3}$, due to increased T_{0_3}. As Figure 3.14 shows, specific work is much more severely reduced.

On the other hand, if the power output is maintained constant, fuel consumption will increase.

Fuel consumption, for the shaft output type of gas turbine, is in general given by specific fuel consumption, defined as:

$$SFC = \frac{\dot{m}_f}{SHP} \frac{\text{lb}}{\text{HP-hr}} \left(\frac{\text{kg}}{\text{kW-hr}} \right) = \frac{f}{w_n}$$

where $\dot{m}_f$ is the fuel flow rate, lb/hr, and SHP is the horsepower produced by the shaft.

The following conversion factor will be of use:

$$1 \frac{\text{kg}}{\text{kW-hr}} = 1.644 \frac{\text{lb}}{\text{HP-hr}}$$

Specific fuel consumption can be expressed in terms of the ideal cycle parameters as:

$$SFC = \frac{3600 \dot{m}_f}{[(\dot{m}_a + \dot{m}_f)(h_3 - h_4) - \dot{m}_a(h_2 - h_1)]\dfrac{778}{550}}$$

or in terms of an actual cycle parameters with approximate efficiencies included in the end state points:

$$SFC = \frac{2545}{\left(\dfrac{\dot{m}_a}{\dot{m}_f} + 1 \right)(h_{0_3} - h_{0_4}) - \dfrac{\dot{m}_a}{\dot{m}_f}(h_{0_2} - h_{0_1})}$$

Example 6.4: Consider the gas turbine given in Example 3.5 with compressor and turbine efficiencies as follows:

sea level intake	$T_1 = 520°R$
compressor pressure ratio	$P_3/P_2 = 12$
turbine inlet temperature	$T_4 = 2200°R$
atmospheric exhaust	$P_6 = p_a = 15$ psi
compressor efficiency	$\eta_c = .85$
turbine efficiency	$\eta_t = .9$
specific heat and the ratio of specific heats	use average values from Figure 6.1

Calculate the work output, efficiency, and the specific fuel consumption.

Solution: Ignore kinetic effects and assume that total and static values are the same, i.e. $p_0 \equiv p$ and $T_0 = T$. In the compressor $k = 1.4$ may be assumed. $T_2 = T_a = T_0 = 520°R$.

The compressor ideal exit temperature is

$$T_3 = T_2 \left(\frac{P_3}{P_2} \right)^{(k-1)/k} = (520)12^{.286} = 1058°R$$

The average temperature is $(1058 + 520)/2 = 789°R$. Figure 6.1 gives the average c_p for the compressor as $c_p = .257$.

The compressor work is obtained as

$$w_c = \frac{c_p T_2}{\eta_c}[(P_2/P_1)^{(k-1)/k} - 1]$$

$$= \frac{.257}{.85}(520)[12^{.286} - 1] = 163 \text{ BTU/lb}$$

as compared to 129 BTU/lb for the ideal cycle. The turbine work is

$$w_T = c_p T_4 \eta_T \left[1 - \frac{1}{(P_3/P_2)^{(k-1)/k}}\right]$$

When average values of $c_p = .275$ and $k = 1.33$ are assumed, the turbine work becomes

$$w_T = .275(2200).9\left[1 - \frac{1}{12^{.248}}\right]$$

$$= 250\text{BTU/lb vs. 269 for the ideal cycle.}$$

The net work is

$$w_n = 250 - 163 = 87 \text{ BTU/lb vs. 140 for the ideal}$$
cycle.

For a mass flow of 50 lb/sec, only

$$w_n = \frac{87 \times 50}{.707} = 6153 \text{ HP vs. 9900 HP}$$

will be produced. Cycle efficiency is calculated from

$$\eta = \frac{(1 - 2.035)/.85 + 4.23(.9)(1 - 1/1.85)}{4.23 - (1 + (2.035 - 1)/.85)} = .268$$

To calculate the specific fuel consumption, one needs to determine the fuel air ratio from Figure 6.20. The combustor inlet temperature (compressor exit temperature) was found above as $T_3 = 1058°\text{R}$. The combustor temperature rise is $2200 - 1058 = 1142°\text{R}$. One finds then $f = .0185$. The specific fuel consumption is

$$\text{SFC} = \frac{f}{w_n} = \frac{3600 \times .0185}{87} = .77$$

which is not very good for this type of machine, especially since accessories' work and other losses were not taken into account. If the pressure ratio were lowered and/or turbine inlet temperature raised, the fuel consumption would be improved.

Example 6.5: A gas turbine with heat exchanger has the following specifications (similar to the LM2500)

Compressor pressure ratio	17
Turbine inlet temperature	1440 K
Ambient conditions, p_a, T_a	1 bar, 288 K
Compressor efficiency	.85
Turbine efficiency	.9

Heat exchanger efficiency   .8
Combustor efficiency   .98

Calculate the specific net work, cycle efficiency, and the specific fuel consumption.

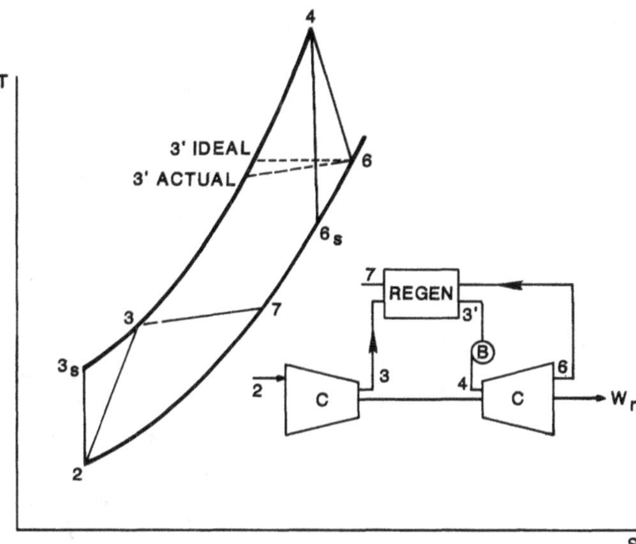

Solution: The ideal regenerative cycle was shown in Figure 3.15. The adjoining sketch shows the same real cycle with the nomenclature as used in Figure 6.27. As in the last example, it will be assumed that $p_a = p_1 = p_{0_2}$, same for the temperatures, and that the exit and heat exchanger losses can be neglected. In real cycles such losses do add up and must be accounted for in any realistic cycle evaluation.

The compressor exit temperature is calculated from the expression for compressor work (it is advantageous here to calculate individual temperatures first, as they are needed for heat exchanger calculations to evaluate T_3').

$$T_3 = T_2 + \frac{T_2}{\eta_c}\left[\left(\frac{P_3}{P_2}\right)^{(k-1)/k} - 1\right]$$

$$= 288 + \frac{288}{.85}[17^{.286} - 1] = 711 \text{ K}$$

Compressor work becomes

$$w_c = c_p(T_3 - T_2) = 1054(711 - 288) = 440 \text{ kJ/kg}$$

(see Section 3.7, conversion properties). Hence, an average $c_p = .252$ was used from Figure 6.1.

The turbine exit temperature is obtained from the expression for turbine total work between the same pressure ratio (in the actual case with *all* losses included the pressure ratio would be almost 10% less). Thus

$$T_6 = T_4 - T_4\eta_t\left[1 - \frac{1}{(P_4/P_6)^{(k-1)/k}}\right]$$

$$= 1440 - 1440(.9)\left[1 - \frac{1}{17^{.248}}\right] = 786 \text{ K}$$

Total turbine work is

$$w_T = c_p(T_4 - T_6) = 1.19(1440 - 786) = 778 \text{ kJ/kg}$$

and the net work becomes

$$w_n = w_T - w_c = 778 - 440 = 338 \text{ kJ/kg}$$

For a typical mass flow of 60 kg/s (132 lb/s), this gives a net work of

$$w_n = 60 \times 338 = 20,280 \text{ kW}$$
$$= 27,200 \text{ HP}$$

which, of course, is somewhat higher than expected due to some losses being neglected.

To find the fuel-air ratio from Figure 6.20, the combustion temperature rise $T_3' - T_4$, and the combustor inlet temperature T_3' are needed. These are derived from the heat exchanger information. The ideal heat exchanger would give $T_3' = T_6$ and $T_7 = T_3$. An actual heat exchanger would give a somewhat lower $T_3' = T_{3_a}'$. Thus, the heat exchanger efficiency is defined as

$$\eta = \frac{T_{3_a}' - T_3}{T_6 - T_3}$$

The combustor inlet temperature T_{3_a}' is then

$$T_{3_a}' = .8(786 - 711) + 711 = 771 \text{ K } (928°\text{F})$$

and the combustor temperature rise is

$$T_4 - T_{3_a}' = 1440 - 771 = 669 \text{ K } (1204°\text{F})$$

The fuel air ratio is obtained from Figure 6.20 as $f = .019$. The specific fuel consumption is therefore

$$SFC = \frac{f}{w_n} = \frac{3600 \times .019}{338} = .2024 \text{ kg/kWh}$$
$$= .333 \text{ lb/HPh}$$

The efficiency is obtained by assuming that the fuel energy content is

$$LHV = 18,400 \text{ BTU/lb} = 42,800 \text{ kJ/kg}$$

which gives for the efficiency

$$\eta = \frac{3600}{SFC \text{ } LHV} = \frac{3600}{.2024 \times 42,800} = .416$$

The SFC and efficiency values are both a bit more optimistic than values associated with real engines, partly due to omission of some losses, and partly due to inclusion of the heat exchanger (regeneration) in the cycle.

Without the regeneration, the last few values can easily be recalculated. The combustor inlet temperature is simply $T_3 = 711$ K, and the temperature rise is $T_4 - T_3 = 1440 - 711 = 729$ K. These values yield $f = .021$, and

$$SFC = .224 \text{ kg/kWh} = .368 \text{ lb/HPh}$$

and the efficiency

$$\eta = .376$$

which are still somewhat optimistic values, but closer to those actually obtainable.

Typical engine performance curves for three marine engines (LM2500, FT4, TF40) are shown in Figure 6.29a,b,c for various engine ratings. By rating is meant a particular power that a gas turbine unit produces at a given rpm and application. The top curve represents the maximum engine power available, but its extended use will lead to deterioration of the unit and a decreased interval between overhauls. The next curves—maximum continuous and normal—are just that. First is the manufacturer's recommended design curve for maximum use of continuous power with acceptable service life and time between overhauls. The normal rating is the recommended operation level for extended service life.

The rate of fuel consumption is also shown, either in terms of specific fuel consumption (SFC) or fuel flow rate (AVCO TF-40). For lower rpm and shaft output, SFC increases, owing to two effects: (1) As the power is reduced, so are the firing rate and temperatures of the gas generator and compressor turbine. This change also reduces the turbine inlet temperature and pressure, with an attendant decrease in the cycle efficiency. (2) Also important is the effect of the off-design operating condition of the turbine as the rpm and the flow angles are altered to suit the power demand. It is evident that the operation of a gas turbine unit at partial load is undesirable, as SFC drastically increases. Over a range of rpm's and power levels there is a flat bottom to the curve, suggesting an optimum operating rpm for the power turbine (Figure 6.29a,b, as stated in Figure 6.29c).

In order to improve the partial load operating efficiency, the turbine inlet area could be altered by means of controllable stator vanes that decrease the mass flow rate and effectively increase the combustor firing temperature. However, as seen from the compressor operating plot (Figure 6.6), such a change in compressor flow, at near-constant rpm,

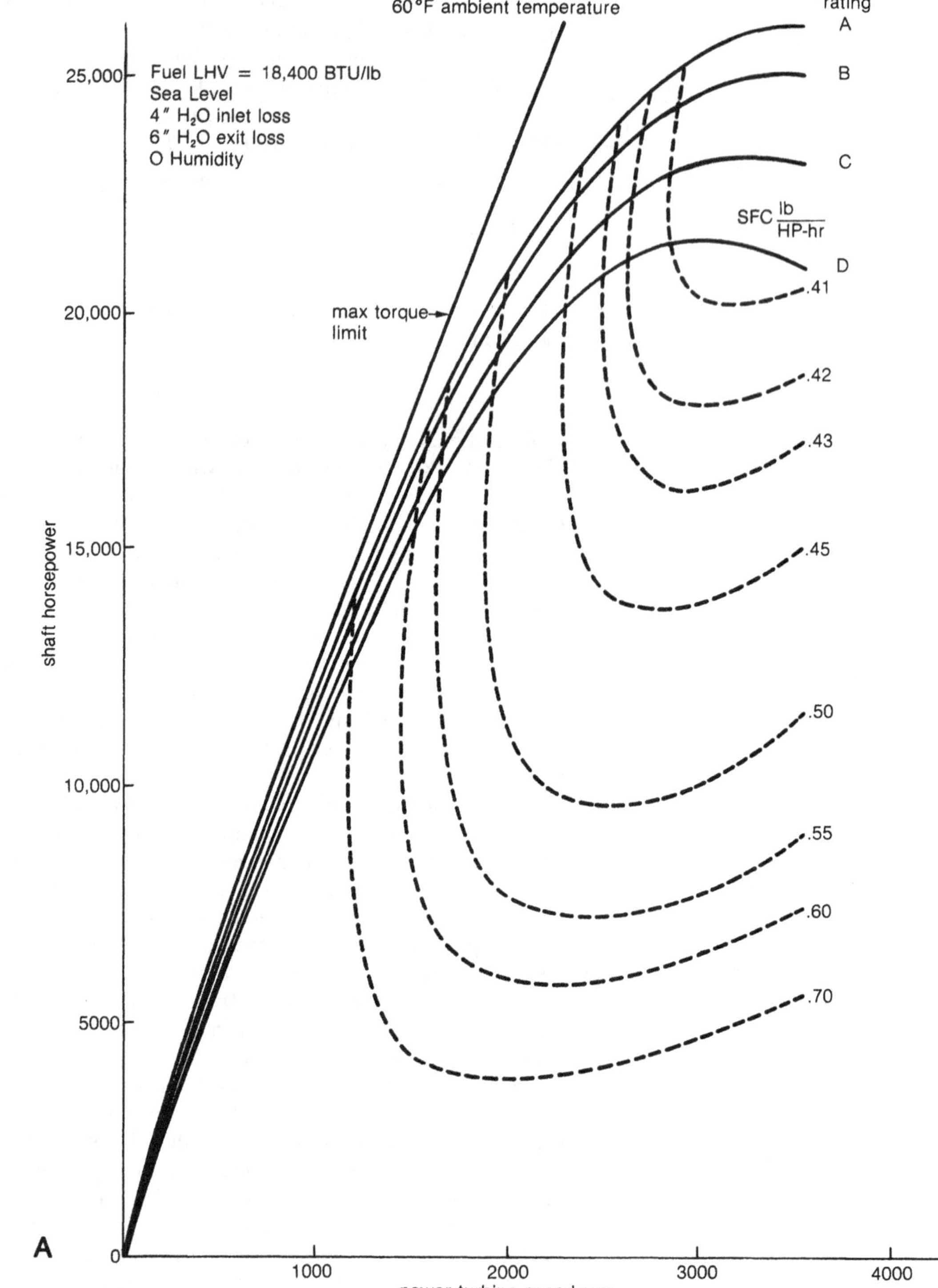

Figure 6.29 Engine performance characteristics. (A) LM 2500, General Electric Co. (B) FT4A-14, Pratt & Whitney. (C) TF-40, AVCO Lycoming

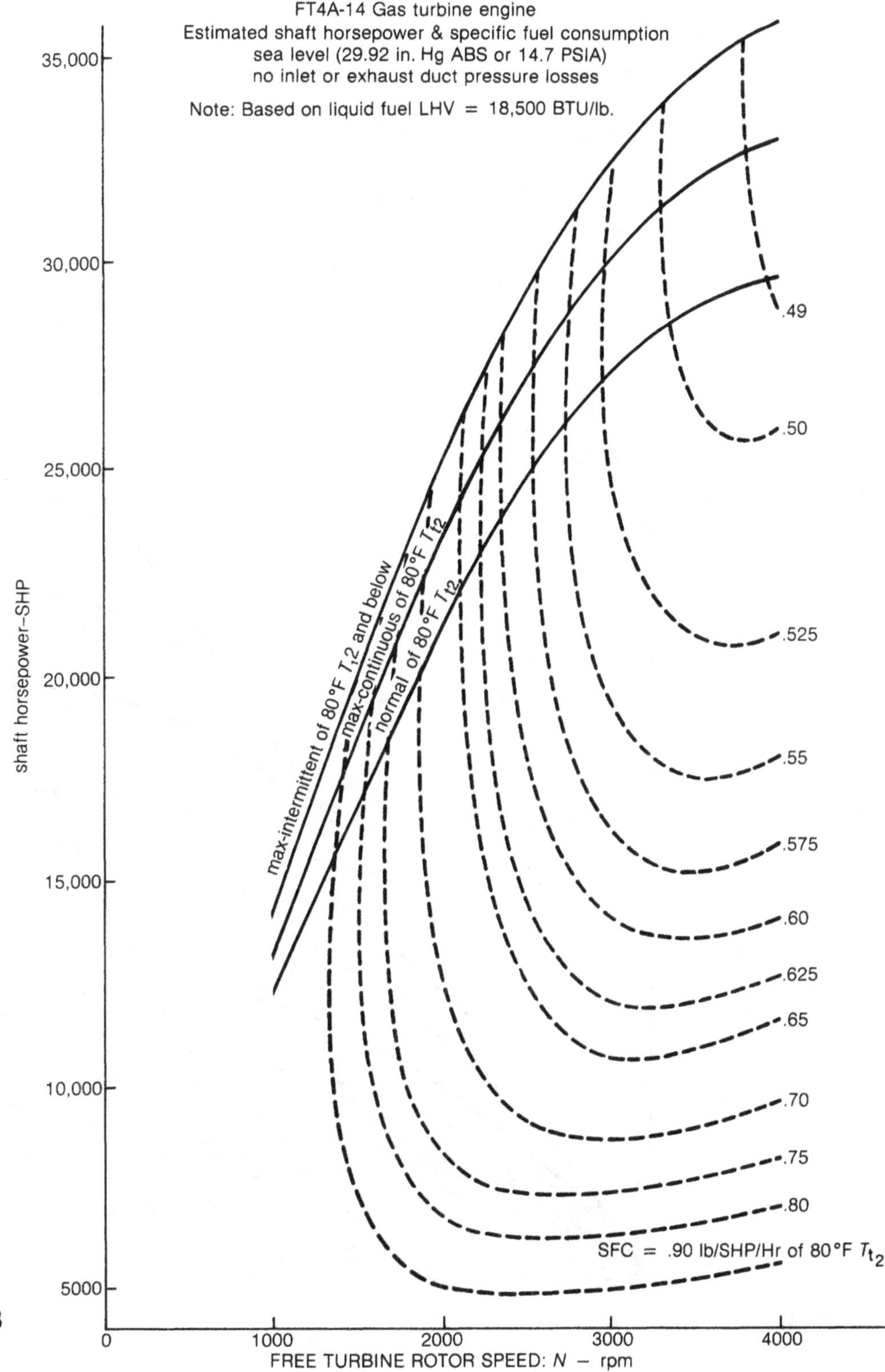

FT4A-14 Gas turbine engine
Estimated shaft horsepower & specific fuel consumption
sea level (29.92 in. Hg ABS or 14.7 PSIA)
no inlet or exhaust duct pressure losses
Note: Based on liquid fuel LHV = 18,500 BTU/lb.
shaft horsepower–SHP
35,000
30,000
25,000
20,000
15,000
10,000
5000
max-intermittent of 80°F T_{t2} and below
max-continuous of 80°F T_{t2}
normal of 80°F T_{t2}
.49
.50
.525
.55
.575
.60
.625
.65
.70
.75
.80
SFC = .90 lb/SHP/Hr of 80°F T_{t2}
0
1000
2000
3000
4000
FREE TURBINE ROTOR SPEED: N – rpm
B

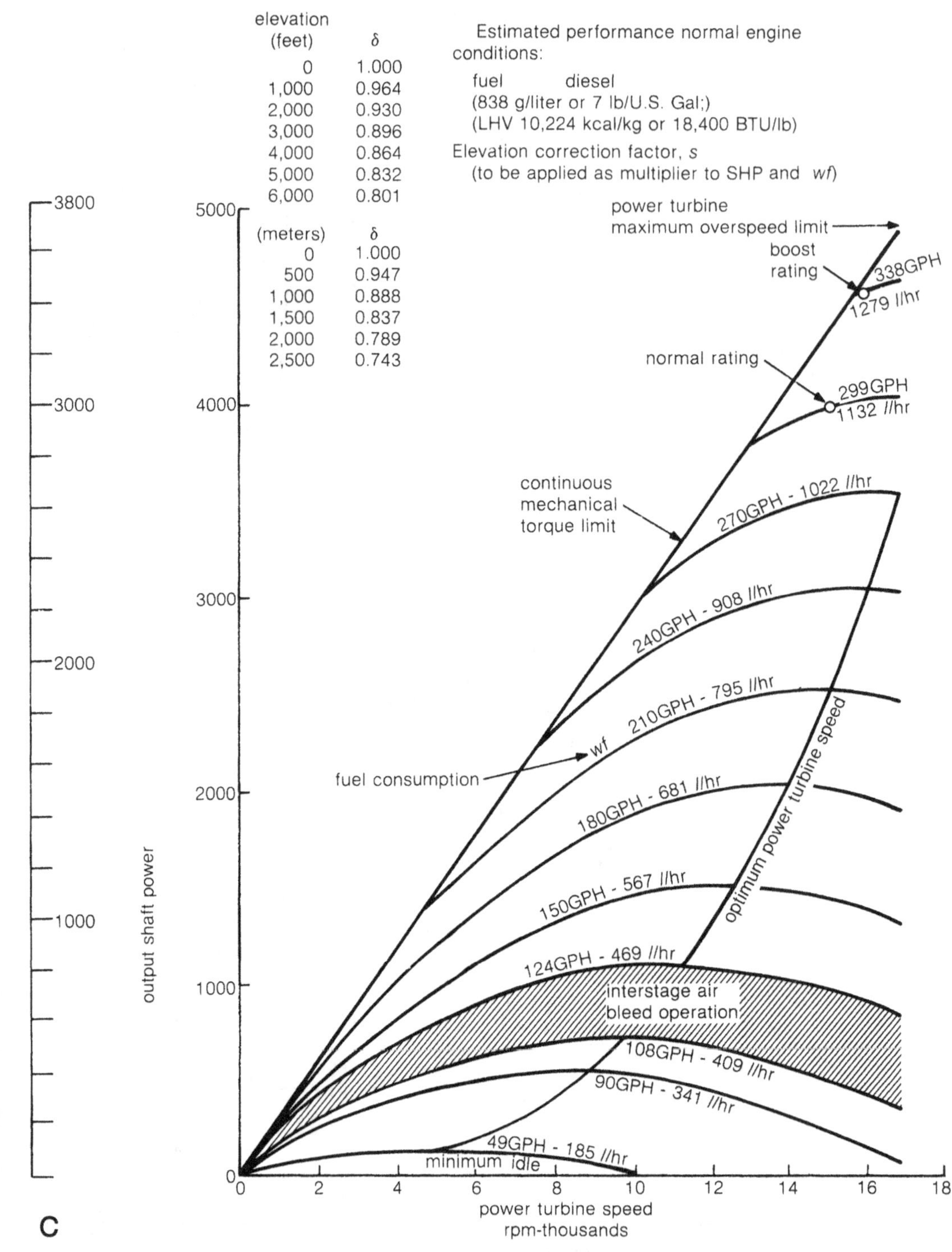

elevation
(feet)   δ
0   1.000
1,000   0.964
2,000   0.930
3,000   0.896
4,000   0.864
5,000   0.832
6,000   0.801

(meters)   δ
0   1.000
500   0.947
1,000   0.888
1,500   0.837
2,000   0.789
2,500   0.743

Estimated performance normal engine conditions:
fuel   diesel
(838 g/liter or 7 lb/U.S. Gal;)
(LHV 10,224 kcal/kg or 18,400 BTU/lb)
Elevation correction factor, s
(to be applied as multiplier to SHP and wf)

power turbine
maximum overspeed limit
boost rating
338GPH
1279 l/hr
normal rating
299GPH
1132 l/hr
continuous mechanical torque limit
270GPH - 1022 l/hr
240GPH - 908 l/hr
210GPH - 795 l/hr
wf
optimum power turbine speed
fuel consumption
180GPH - 681 l/hr
150GPH - 567 l/hr
124GPH - 469 l/hr
interstage air bleed operation
108GPH - 409 l/hr
90GPH - 341 l/hr
49GPH - 185 l/hr
minimum idle

output shaft power
3800
3000
2000
1000

5000
4000
3000
2000
1000
0

power turbine speed
rpm-thousands
0   2   4   6   8   10   12   14   16   18

C

will move the operating line close to the surge margin. Thus, the area variation effect is limited.

Another method for performance improvement lies in avoiding and eliminating partial load operation altogether by modifying the overall power plant cycle and using the gas turbine unit in conjunction with other kinds of power production modules in combinations like CODAG, CODOG, and so on. These will be discussed in more detail in the next chapter.

The torque supplied by the shaft is an important characteristic for selecting a gas turbine unit for a particular application. A single shaft engine, seldom used in marine applications, rotates at some multiple of the load speed that is fixed by the gearing ratio. Thus, a change in load speed will also change the compressor rpm. Then mass flow and torque also vary and are reduced as rpm is decreased (see Figure 6.30). The free power turbine torque curve is also shown in the figure. Free turbine speed is not a function of compressor rpm, which can provide an almost constant flow regardless of the free turbine speed down to zero rpm. The compressor speed may be cut down to a self-sustaining speed, which is the minimum idle condition. For a fixed gas generator condition, the reduction in output rpm yields an increase in torque. However, the load demand for marine propulsion tends to be small at low rpm and matching the compressor rpm and fuel flow leads to high *SFC*, low rpm, and the inefficient operation discussed above. Free turbine torque may be calculated from equation 2, Chapter 4, rewritten:

$$\tau = 5252\,\frac{P}{N}\ (\text{ft-lbf})$$

where P = power in horsepower
$\quad\ N$ = rpm

Another performance parameter, which is as symptomatic of the condition of the gas turbine as body temperature is for humans, is the exhaust gas temperature (EGT). This is one of the turbine performance parameters that is monitored to control the turbine and as a warning of possible malfunctioning, e.g., loose or clogged fuel nozzles, increased firing rate due to compressor fouling (see Section 6.7), and so on. The EGT is a function of the ambient temperature, turbine inlet temperature, and the amount of work extracted by the turbine, as already discussed above. For a given shaft power, ambient temperature changes affect the turbine inlet temperature and EGT more than other cycle parameters do. Also, the mass flow and fuel/air ratio will change due to the density changes with ambient temperature change (see also Figure 6.29). The fuel/air ratio change will

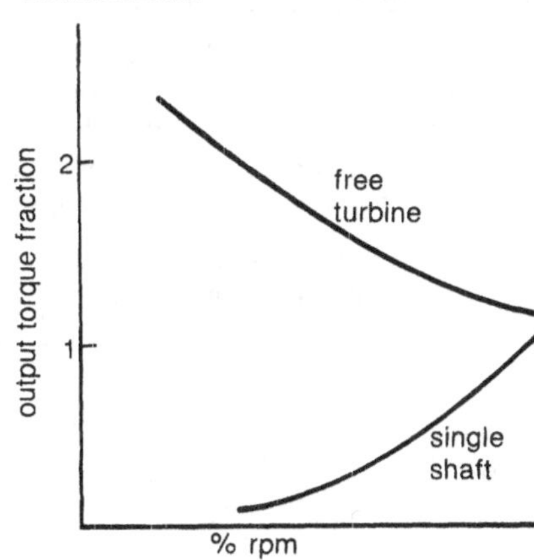

Figure 6.30 Gas turbine torque characteristics

then affect the flame temperature, with attendant change in EGT. These combined effects are shown in Figure 6.31.

Three additional parameters that influence gas turbine performance and which therefore must be included as corrections in accurate performance calculations are ambient humidity, and inlet and exhaust pressures. Since typical engine data (Figure 6.29) are given for zero humidity, ambient humidity must be determined and corrections applied as supplied with the engine instruction manual. Although the effect of humidity is to decrease the density of the incoming air, the net effect is to increase slightly both horsepower, at a given rpm, and the fuel flow. Since the increase in the fuel flow occurs at a higher rate than the SHP, the *SFC* (defined above) will also increase. The power turbine discharge temperature, in general, is not affected by humidity.

There are two types of pressure effects: change in ambient pressure, and the pressure change due to intake and exhaust losses. The ambient pressure change effect due to barometric variation for a given elevation is small and can be neglected. If a vessel moves through locks to a higher inland elevation, the pressure does drop, and this change will decrease both SHP and fuel flow rate, as indicated by the pressure ratio correction factor, $\delta = p_{\text{elevation}}/p_{\text{sea level}}$ as in Figure 6.29c. The intake and exhaust pressure losses are of

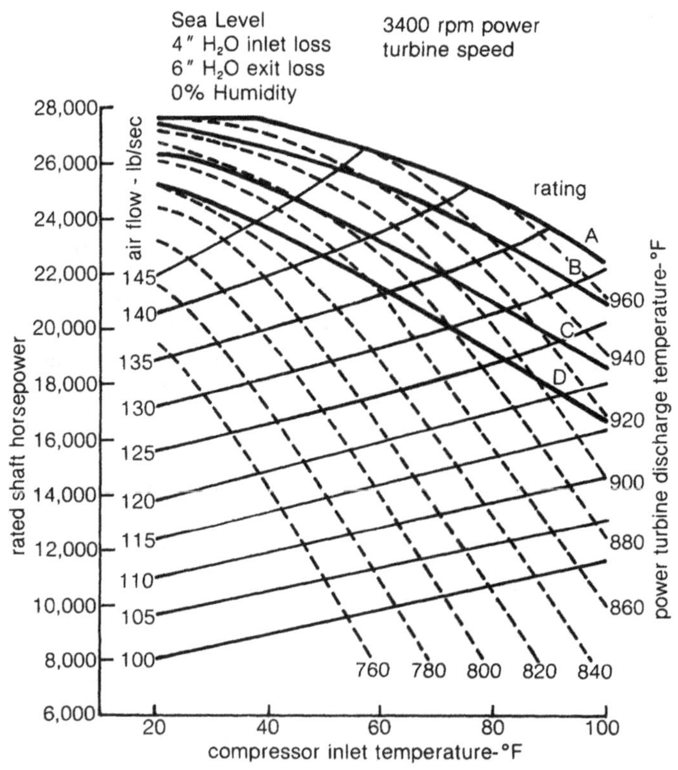

Figure 6.31 Inlet temperature effect on engine performance

a more serious nature and do affect the output and the cycle performance as discussed above with Figure 6.27. Figures 6.29a and d show the installed performance curves with standard Navy pressure allowance of 4″ H_2O intake and 6″ H_2O exit losses. Figures 6.29b and c give basic engine performance data where the direct losses must be specially compensated for according to the manufacturer's instruction manual.

Intake losses arise from pressure drops through aerosol separators and screening and from flow losses in the ducting between the duct intake and engine inlet. Intake ducting, depending on power plant location, may be practically nonexistent or may be tens of feet long. The next chapter will illustrate some typical gas turbine unit configurations and locations. Whatever the length, the gas turbine engine must expend power to suck air from the intake through filters and ducting into the compressor inlet. At the exhaust end, the unit must push the exhaust gas out through the uptake with its duct losses. For the same amount of pressure drop, the intake loss decreases SHP more than the exhaust loss, but the exhaust gas temperature is little changed.

6.6 Gas Turbine Components, Construction and Mechanical Features

The construction of a steam turbine depends largely on whether it is an impulse or a reaction turbine. Each has preferred ways of attaching blading to wheels or drums, and of providing adequate seals suitable for the expected pressure and temperature. A reaction turbine built in the 1950s does not differ much from one built now, with the exception of the use of improved alloys for high temperature application.

Gas turbines, on the other hand, represent a developing field with respect to metals, temperatures, pressures, compressor and turbine performance, methods of fabrication and assembly, and cycle applications. Thus, there are as many ways of putting a turbine together as there are turbine manufacturers, and to attempt to cover even most of the turbine construction details and component variations is clearly beyond the scope of this book. However, it is illuminating to review briefly the major components of the gas turbine unit, and some typical methods of construction and materials used for high temperature applications. In this study a comparison of construction principles retained and modified from the steam turbine should also prove useful.

There are essentially three ways of building and assembling a gas turbine unit for marine application—stacked, modular, and heavy duty (see Figures 6.32–6.34). Figure 6.32 represents the general aircraft derivative type of construction, emphasizing light-weight, highly modularized components. The individual frames, compressors, and turbines are assembled (stacked) on single or twin shafts (see also Figure 6.12). The frames can provide casings for combustor and turbines, or separate covers can be used, as for the compressor where stator blades are integral with the compressor casing.

A highly modular unit is shown in Figure 6.33, which is designed for compactness and for quick change-out of individual modules. Several unique features appear. The combustor is a reverse flow, folded type (see also Figure 6.18), which significantly reduces overall engine length and results in a shorter and more rigid drive shaft between the compressor turbine and the compressor. In contrast to the purely axial flow compressor employed in recent applications, the compressor is a mixed axial-centrifugal type (Figure 6.33b). The use of a centrifugal stage shortens the overall compressor by three to five stages and also serves to eliminate the very small vanes and blades that appear in compact units and are subject to corrosion and erosion damage in the hot final stages of a purely axial flow design. The power take-off is accomplished through the compressor end.

The third type of construction is found in heavy duty gas turbine units and is similar to steam turbine construction with heavy rotors, necessary support bearings, and also much heavier cast covers (Figure 6.34). This type of construction is used often with a regenerative cycle design, as the number of flanged connections in the compressor casings shows.

A fabrication technique which has been gaining in popularity is inertia welding. Shapes which are difficult to forge or even cast can be fabricated by joining two forged or rolled pieces. By holding one piece stationary and rotating the other, enough heat can be generated by friction to fuse the two sections. The weld may take place inside a shaft at a place impossible to reach with conventional welding techniques. Accessibility is of no import when friction is used to weld the pieces. The process is made feasible by the use of a flywheel to store the precise amount of energy required to fuse any two pieces of metal. The speed of the flywheel can be increased or decreased prior to contact between the surfaces to be welded, thereby establishing the amount of energy to be absorbed by the weld.

A typical feature of most aircraft derivative engines is the bellmouth-bulletnose inlet designed to provide smooth airflow to the compressor (see Figures 2.7, 6.32, and 6.35). To prevent icing, hot compressor bleed air is ducted through the intake support struts into the bulletnose and variable

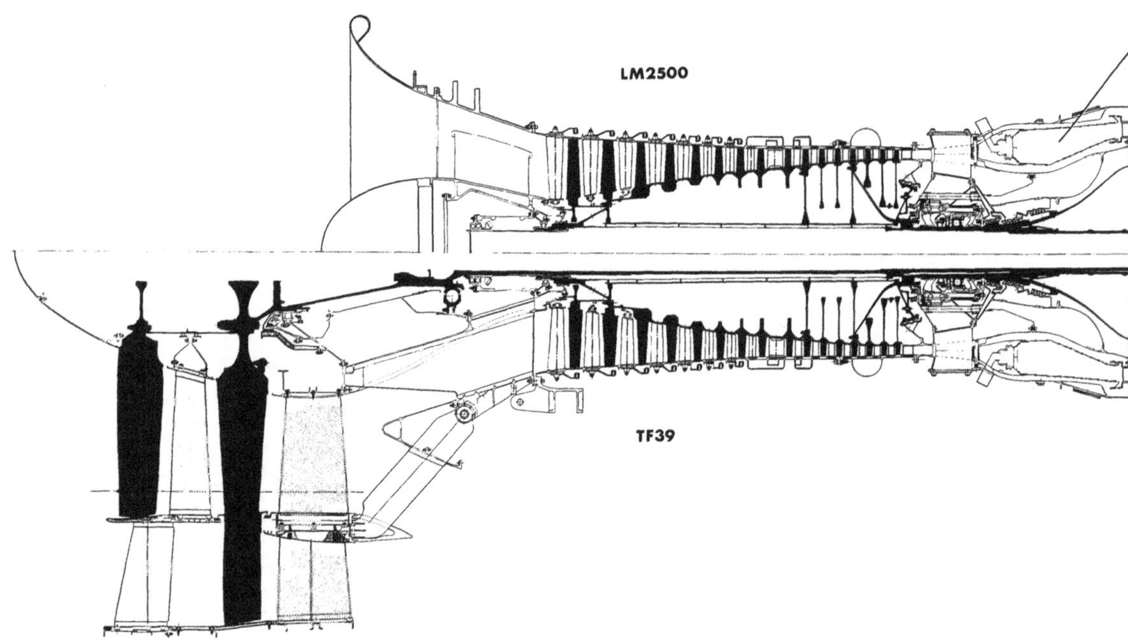

Figure 6.32 General Electric LM 2500
marine gas turbine

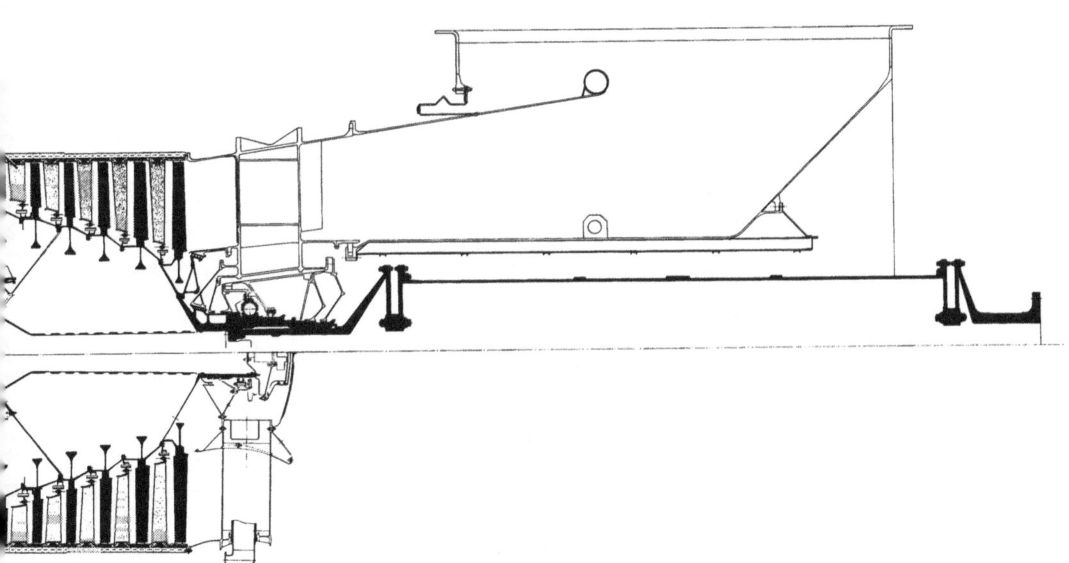

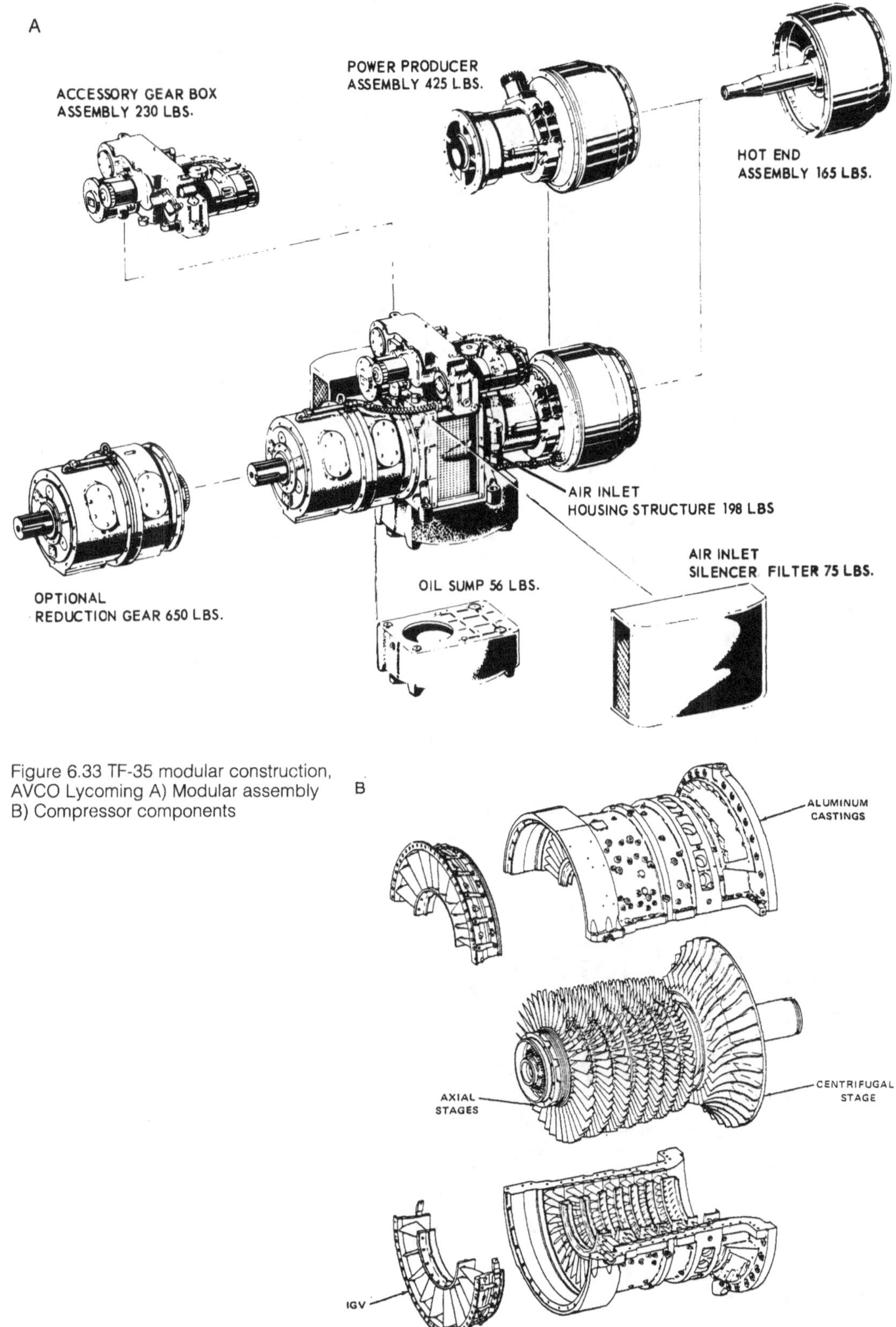

Figure 6.33 TF-35 modular construction, AVCO Lycoming A) Modular assembly B) Compressor components

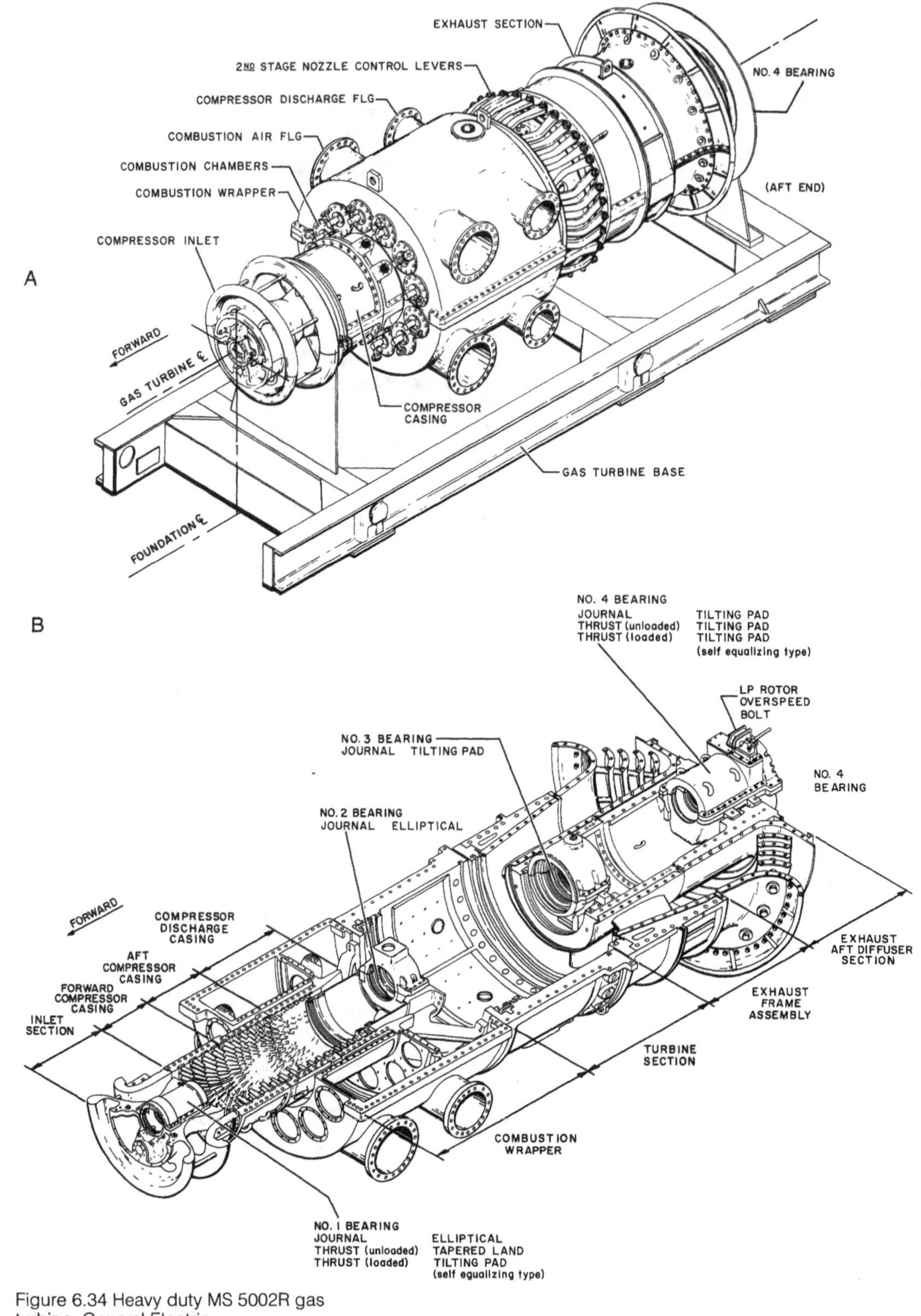

Figure 6.34 Heavy duty MS 5002R gas turbine, General Electric

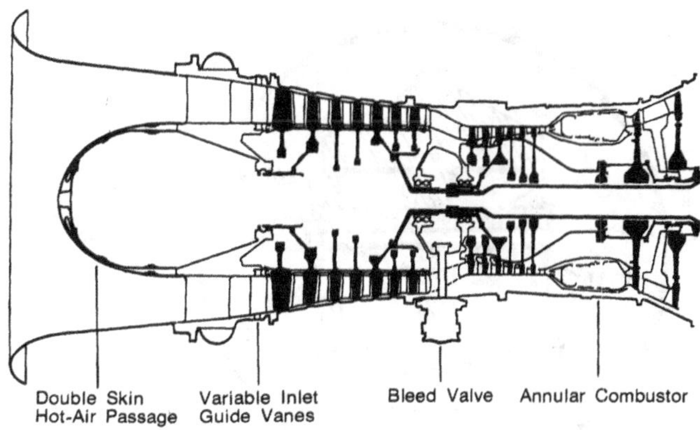

Figure 6.35 RB-211 arrangement, Rolls-Royce

inlet guide vanes. The bellmouth contains the necessary connections and spray rings for compressor washing to eliminate accumulated salt or other fouling matter. The bellmouth and bulletnose material is typically anodized aluminum for corrosion protection.

Figures 6.36–6.40 show typical construction details of compressors and turbines. The rotors are made up of forged wheels that are bolted together with tierods. The stator blades are attached to the compressor casing but are also supported by flanged rings running on rotor disc spacers. Two typical rotor blade fastening methods are shown in Figures 6.36, 6.23, and 6.37 with pinned and side entry dovetail fittings, respectively. Compressor sections and shafts usually operate at temperatures lower than 1200°F and thus fall in the cold-operating section category. Following the aircraft industry practice, the low pressure compressor section utilizes titanium alloys for weight savings. Titanium alloys are used for compressor discs to take advantage of their corrosion and oxidation resistance, and their high resistance to creep and fatigue. In low pressure compressor sections Ti-6Al-4V alloys have been used. In high pressure compressor hot sections Ti-6Al-2Sn-4Zr-6Mo and Inconel 718 (a nickel base superalloy) have been used, among others. Compressor blading has been made of stainless steel and titanium alloys (as above) for low pressure sections, with nickel base alloys used for the hot end (Inconel 718, Inconel X-750).

The combustors, discussed in Section 6.2 above, are subjected to the highest operating temperature (average metal temperature above 1700°F) and must possess a number of favorable characteristics: good erosion and corrosion properties, good thermal fatigue cracking resistance, and distortion resistance, for 8,000–10,000 hours of continuous or cycling operation. Hastelloy X and Haynes 188 in sheet

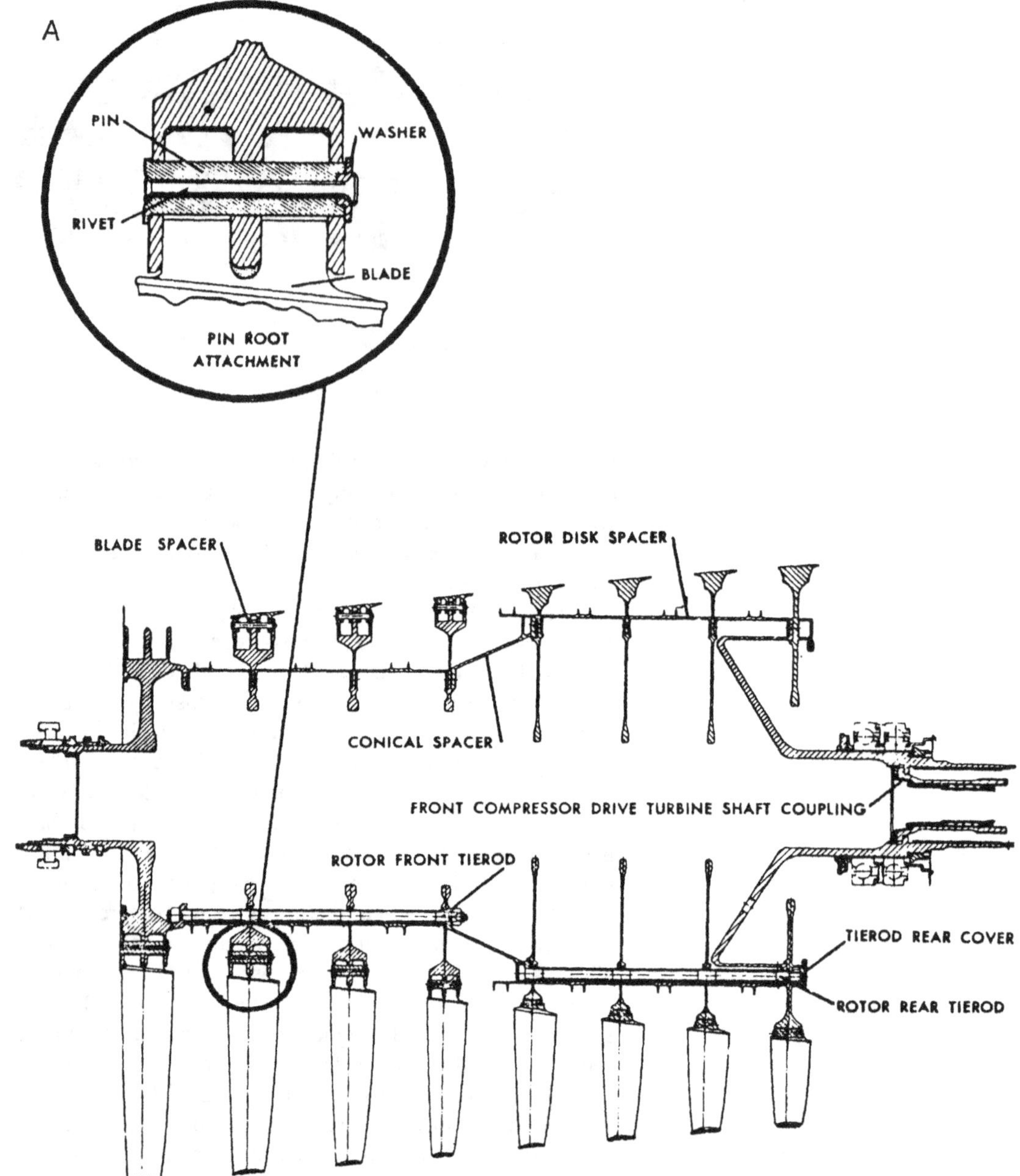

Figure 6.36 Low pressure compressor construction, FT4 engine (Courtesy Pratt & Whitney)

alloy form have been the most commonly used in annular combustors, which type are now most commonly used because of their even outlet temperature distribution, less cooling surface, and elimination of the flame tubes needed for can or cannular types. With increasing firing temperature TD-nickel or TD-NiCr (thorium dispersed nickel based superalloys) are coming in use, but they require protective coatings to resist oxidation and corrosion and for sulfidation resistance.

Resistance to oxidation, hot corrosion and erosion are of critical importance in the hot end blades and vanes of a gas turbine engine. Ceramics and super alloys have given way to new coatings. Special nickel alloys with high temperature strength and good creep resistance are protected by coatings. The combination of super alloys and coatings that are essentially ceramics makes use of the best properties of both materials.

Aluminum offers the best protection as a coating. The aluminum or aluminum compound oxidizes in service to alumina (Al O). Alumina is a refactory material and provides an excellent barrier to oxidation of the bare metal. However, since the coating gradually loses aluminum, and thick coatings are not possible, the aluminum will eventually be depleted and the base metal exposed.

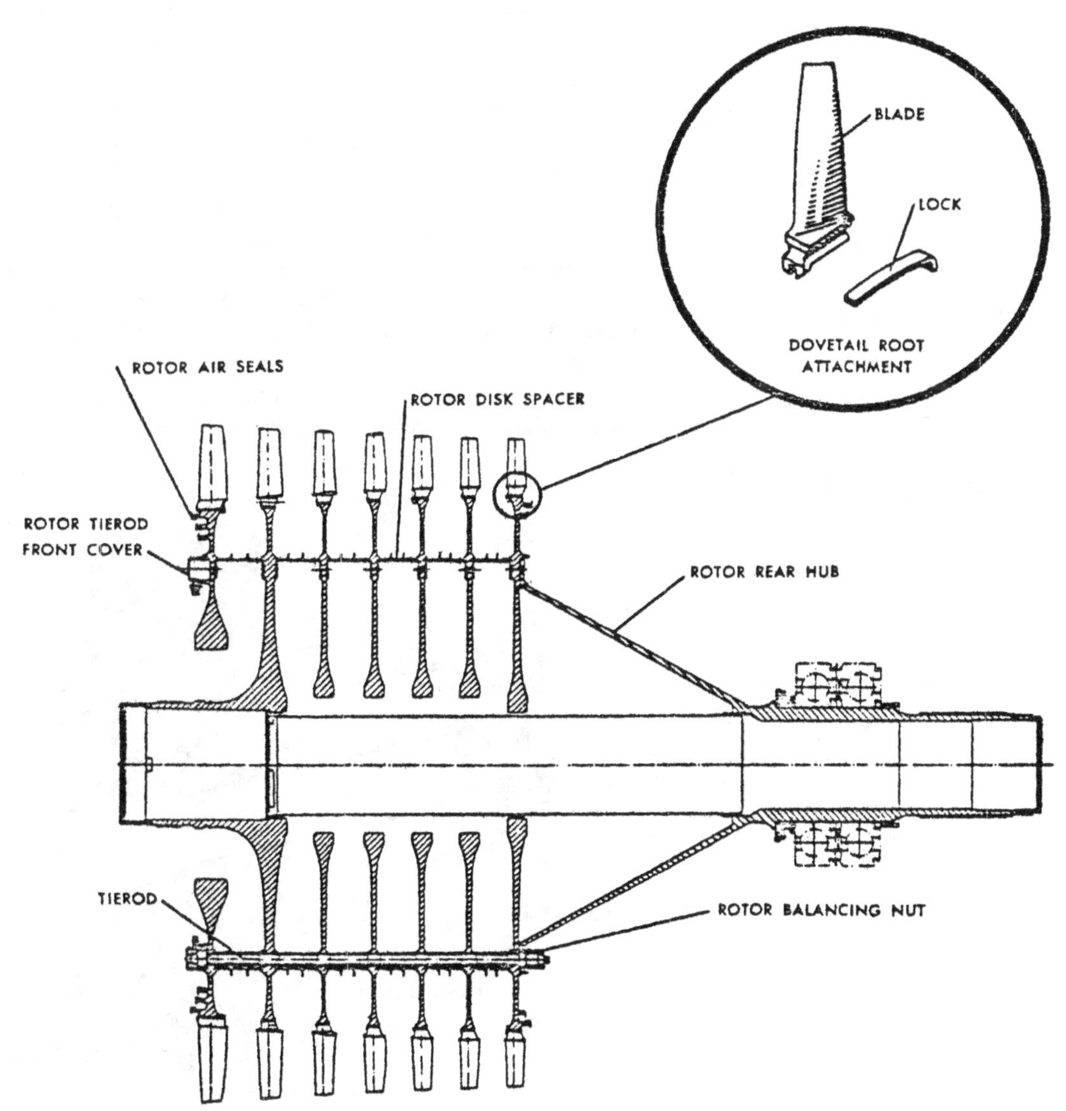

Figure 6.37 High pressure compressor construction details, FT4 (Courtesy Pratt & Whitney)

Other coating materials include chromium, titanium, yttrium, platinum, and tantalum. Each of these is more expensive than the preceding, but in certain applications the one to two years of added service life may justify the cost. At the elevated gas temperatures of modern gas turbine engines, with turbine inlet temperatures above 1600°F, use of alloys alone with no coating is impractical. Protective coatings are almost universally used for the first, second, and even third stage vanes and blades.

Typical turbine general construction details are shown in Figures 6.38–6.42. The first stage is the high pressure rotor (see Figures 6.35 and 6.38), which consists of the first stage wheel and bucket assembly. It is bolted to the rear compressor stubshaft. The low pressure compressor turbine consists of one or more stages (Figures 6.35, 6.41, and 6.42), and the wheel-bucket assembly is mounted on its own shaft, which is coupled to the front compressor. It should be re-emphasized that the examples shown here represent only a few of the numerous ways to couple drive turbines to the compressor. Typical power turbine arrangements are shown in Figure 6.40.

Typical bearing arrangements are shown in Figures 6.33, 6.40, and 6.41. Notable features are the duplex, or double bank, bearings, which serve as the unit's thrust bearings and in Figure 6.41 are grouped together in the rear compressor frame. This arrangement eliminates additional bearings required as twin spool intershaft bearings.

Similarly to steam turbine construction, Figure 6.39 shows balance plug holes and counterweights for initial bal-

Figure 6.38 LM 2500 high pressure turbine (Courtesy General Electric Co.)

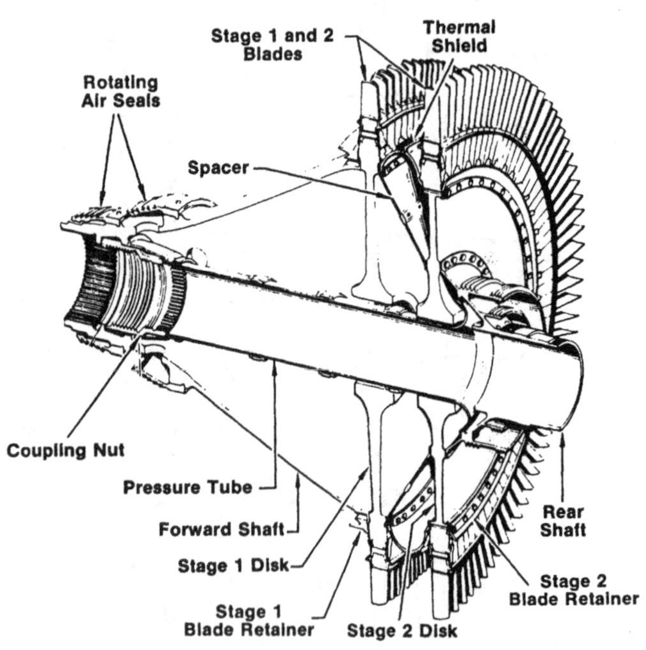

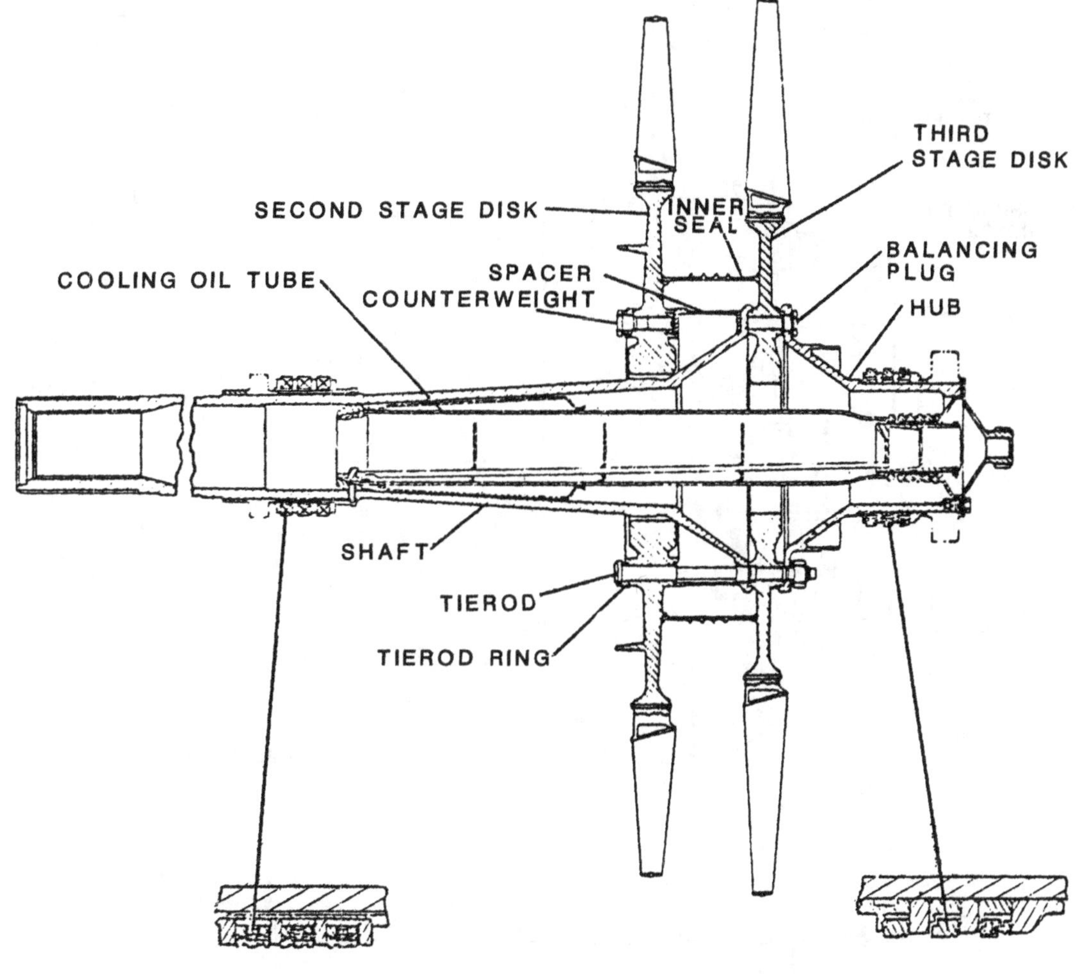

Figure 6.39 FT4 turbine (Courtesy Pratt & Whitney)

ancing and control of vibration during operation of the turbine. The gas turbine unit also develops axial thrust loads due to reaction blading employed in both turbines and compressors, and these loads are balanced by ducting compressor bleed air, or using special bleed orifices from last stages to the turbine wheel spaces. The vertical or conical spacers and seals act as dummy pistons (Figures 6.37, 6.38), and provide balancing thrust for the rotor. The balancing bleed is also used for cooling the turbine discs and the rotor.

Cooling of first stage turbine stator vanes and rotor buckets is one of the most demanding design requirements, for firing temperatures are now exceeding 2100°F and, most likely, will increase in the future to improve cycle efficiency. The first stage vanes (nozzle vanes) are subjected to serious thermal shock during start-up, and take the primary thermal impact during continuous operation. Since they are station-

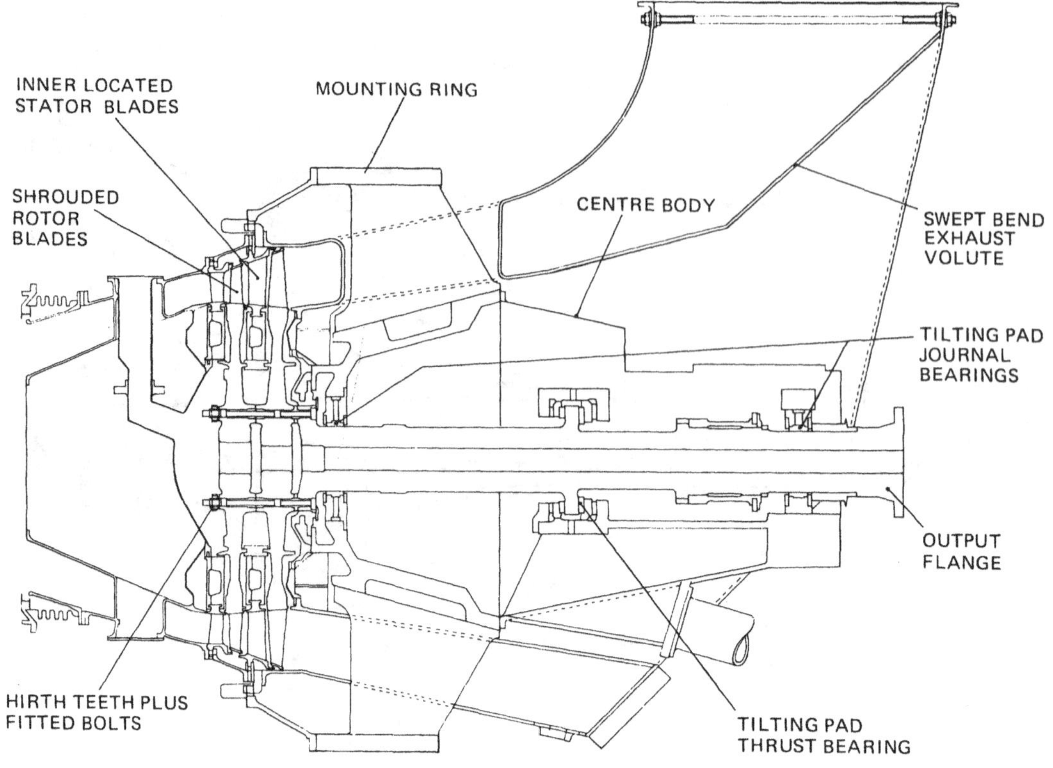

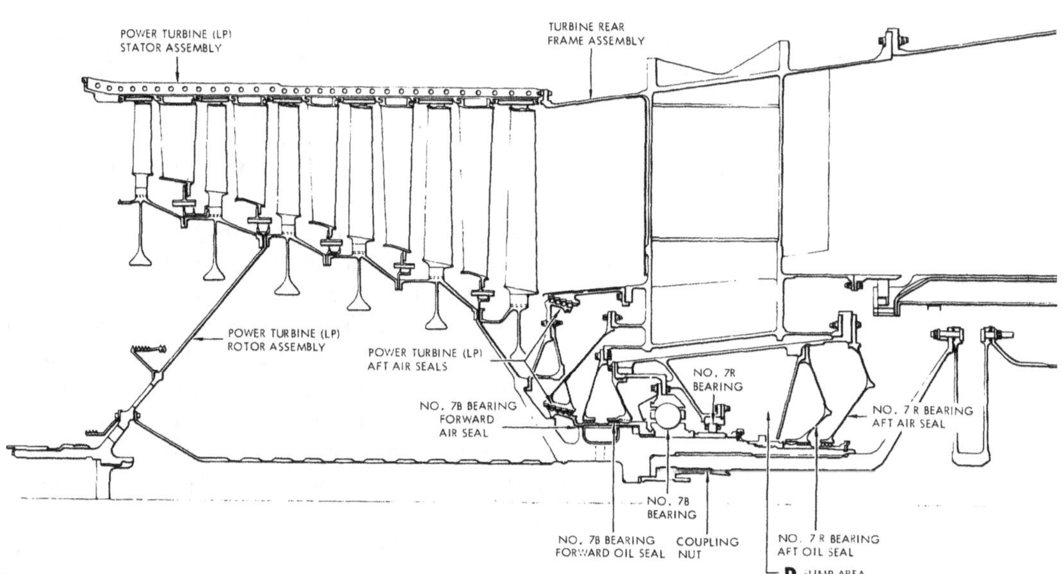

Figure 6.40 Turbine assembly nomenclature. (A) Marine Spey power turbine (Courtesy Rolls Royce). (B) General Electric LM 2500 power turbine

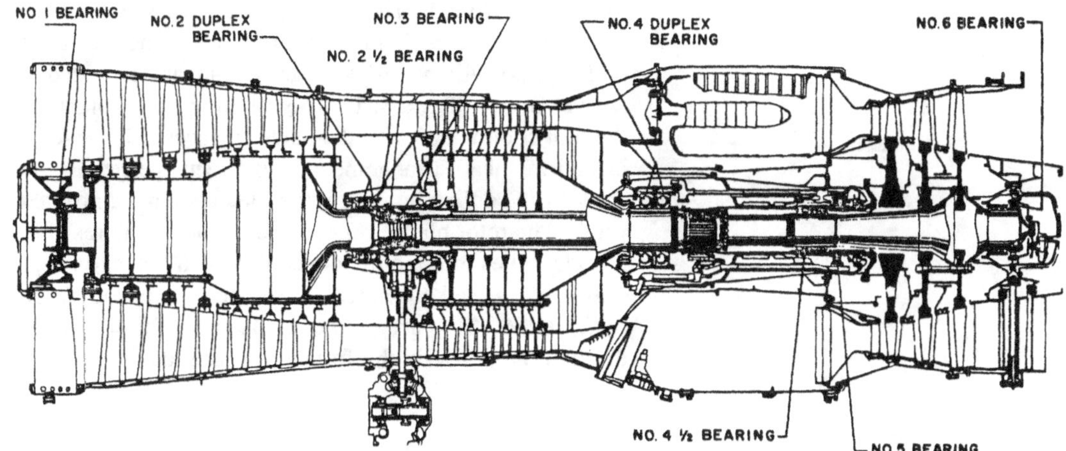

Figure 6.41 Gas turbine bearing arrangement (Courtesy Turbo Power & Marine Systems)

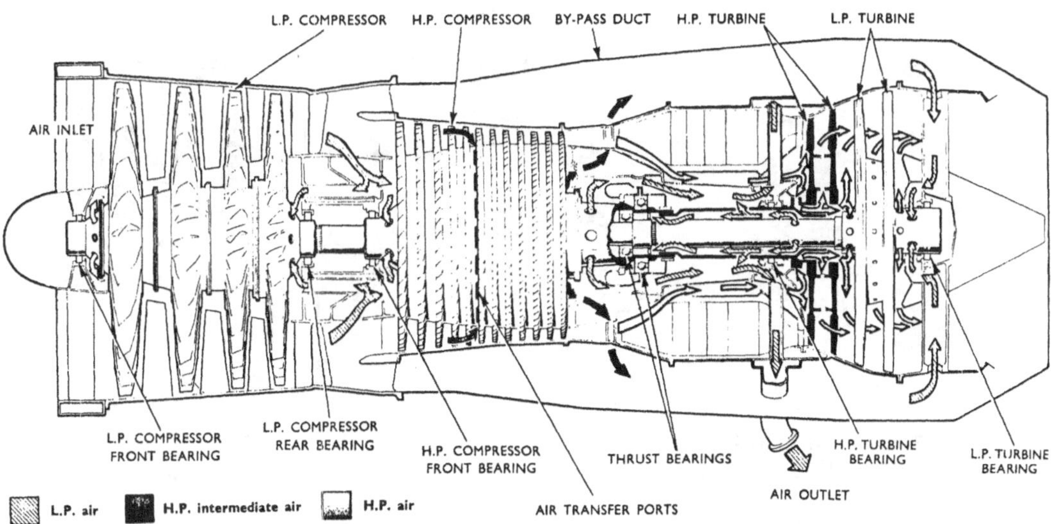

Figure 6.42 High pressure turbine cooling flow (Courtesy Rolls-Royce Ltd.)

ary and are not highly stressed, they tend to fail by erosion, oxidation, bending, or thermal fatigue cracking. The rotating buckets are subjected to a severe combination of high temperature and high stress, and special alloys combined with blade cooling provide sufficient creep-rupture resistance for running extended periods at high speeds and temperatures. The alloys are chosen to match the characteristics of fuels used, in order to minimize oxidation and sulfidation problems.

Some typical vane and bucket cooling arrangements are shown in Figures 6.43–6.47. Figures 6.43 and 6.44 show the LM5000 first stage nozzle vane film cooling arrangement, which reduces the blade metal temperature to less than 1560°F from a firing temperature of 2150°F. These cast blades are replaceable in groups of two (the next generation of LM2500 will be single shank), and are provided with special inserts that assist in distributing the coolant air to the film cooling holes on the blade surfaces. The nose holes or shower head arrangement helps keep the leading edge cool. Film cooling holes are provided also in the nozzle end walls. The rotor blade cooling arrangement features coolant air inlet holes through the dovetail, which helps to keep the critically stressed dovetail and the blade shank cool (Figures 6.42 and 6.43). Part of the air is used for film cooling through the holes on the forward portion of the blade; part

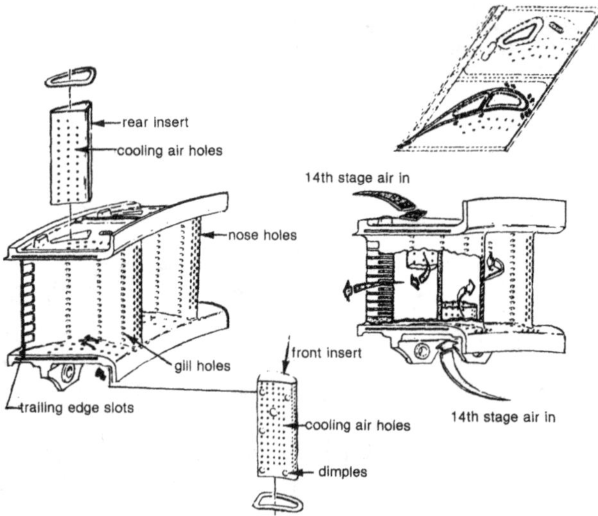

Figure 6.43 Turbine nozzle and film cooling (Courtesy General Electric Co.)

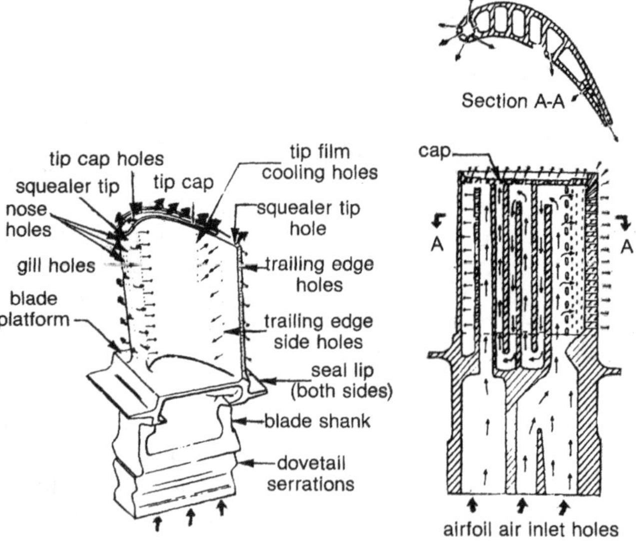

Figure 6.44 Turbine rotor cooling (Courtesy General Electric Co.)

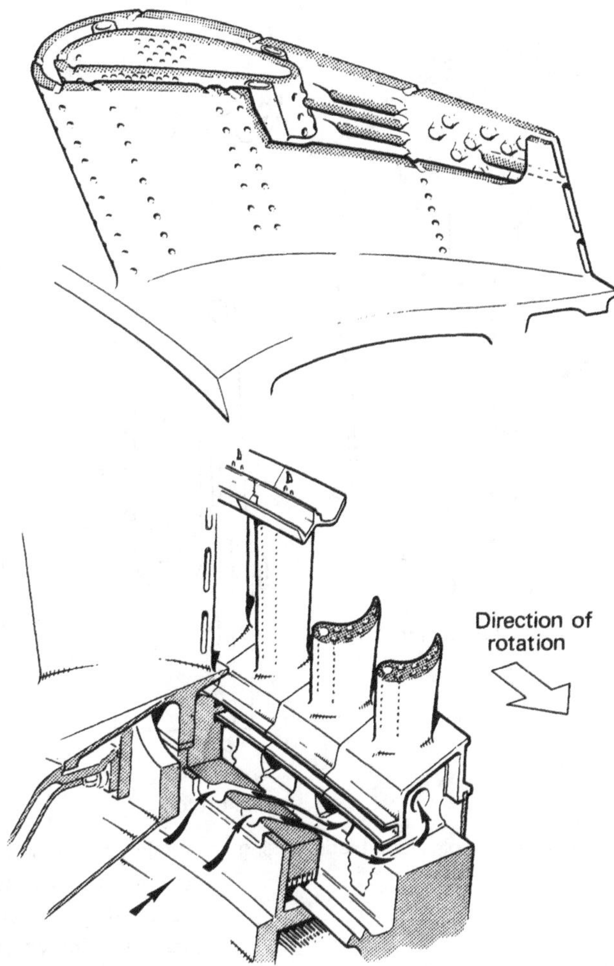

Figure 6.45 Rotor and blade cooling arrangement (Courtesy Rolls-Royce Ltd.)

of the air is recirculated and leaves through the trailing edge holes. The blade tip cap also contains film cooling holes to assist in cooling the clearance space and the tip of the trailing edge. In some designs the tip cap is not installed, and the coolant flows radially out into the tip clearance space. Figure 6.45 shows another arrangement for vane and rotor cooling. Rotor coolant air is supplied through nozzles that accelerate and cool the coolant air before it enters the blade root section. The leading edges and the convex surfaces are film-cooled; the rest of the blade is single-pass-convection cooled.

In the AVCO TF-40 the first two stages are air-cooled to below 1530°F from a firing temperature of 1950°F (Figure 6.46). The first sector vanes are internally cooled by impingement cooling. The second stage blade rows have a two-pass flow path for coolant air. The second stator vanes exhaust part of the air into the interior of the rotor to cool the turbine disc. Figure 6.47 shows the cooling features of a General Electric MS5002 heavy duty gas turbine first stage

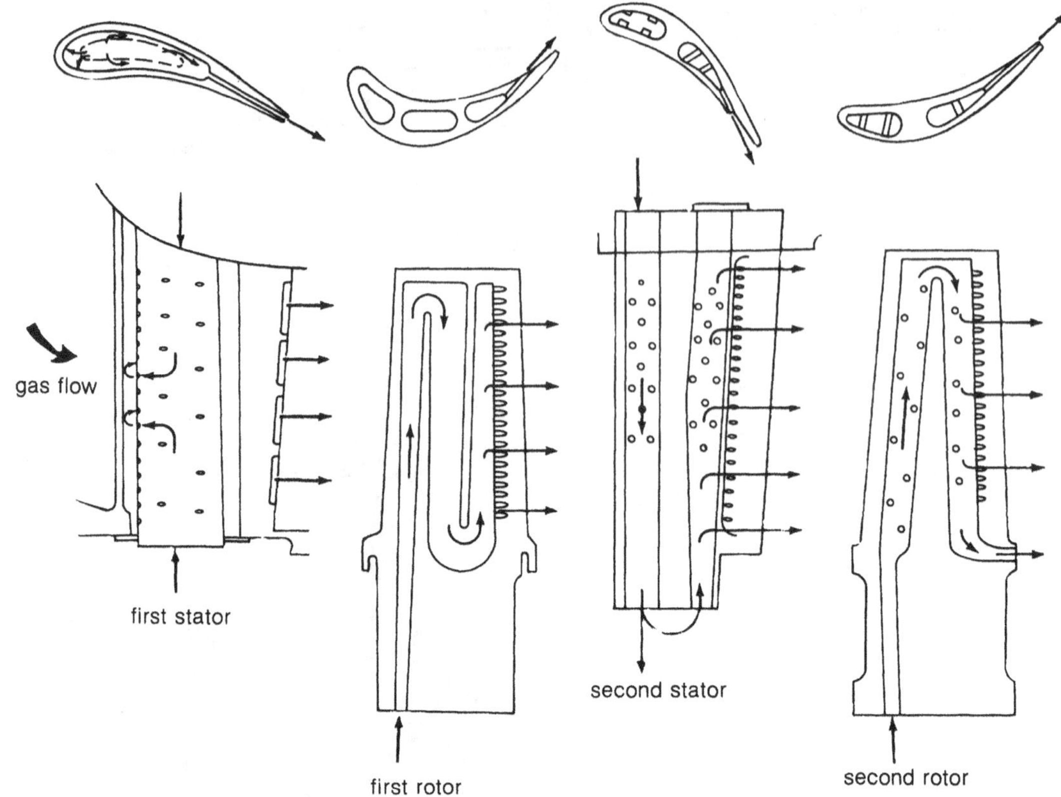

Figure 6.46 Multiple stage blade cooling (Courtesy AVCO Corp.)

nozzle section with leading edge impingement cooling and concave side trailing edge film cooling. Also shown is the second stage variable nozzle (for improved part load performance) arrangement (see also Figure 6.34). The first stage bucket (Figure 6.48) is an investment cast hollow two-section uncooled blade. The hollow vane section reduces the centrifugal stresses in the root of the vane, as well as in the bucket and wheel dovetails. The bucket has a long shank, which provides a substantial temperature drop between the hot bucket and the wheel dovetail. This arrangement reduces the temperature level and thermal stresses in the disc and dovetail area where high rotating stresses occur.

In general, the stator vanes are either cast or fabricated sheet metal hollow blades, which build up smaller thermal stresses. For casting, cobalt-based alloys are used, like Haynes Stellite 31 (also called X-40) (Co-25Cr-10Ni-7.5W), WI-52 (Co-21Cr-1Ni-11W-2Cb), and MAR-M509 (Co-21.5Cr-7W-3.5Ta-15Zr-.2Ti), which require marinizing coatings for oxidation and sulfidation resistance. Nickel-based cast alloys like U-700 (Ni-15Cr-18Co-5.2Mo-4.2Al-3.5Ti) are also used on stator vanes with coatings. Rene 41 (Ni-19Cr-10Co-3.1Ti-1.4Al) and A-286 (Fe-2.6Ni-15Cr-2.2Ti-1.2Mo) are forgeable alloys, easily formed

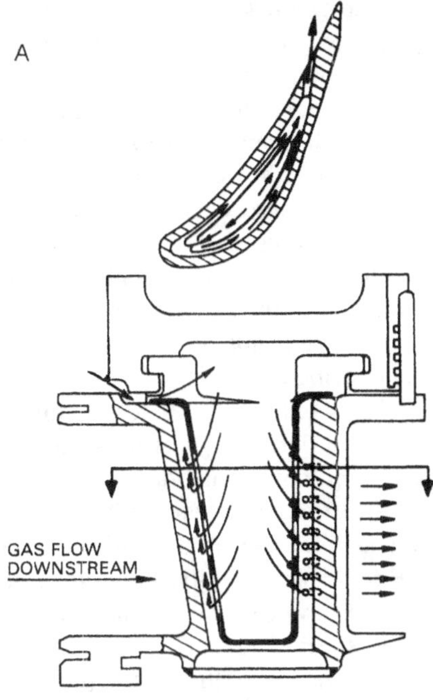

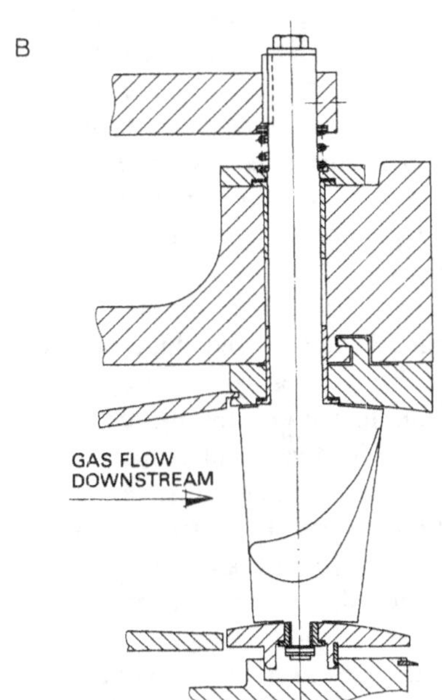

Figure 6.47 Heavy duty turbine blading. (A) First stage nozzle vane cooling. (B) 2nd stage variable nozzle vane (Courtesy General Electric Co.)

and molded, that have been used for sheet metal vane construction.

The turbine buckets, which are subjected to both thermal and centrifugal stresses, are made of nickel-based alloys or high-nickel stainless steel for cooler turbine stages, e.g., U-700, U-500 (Ni-19Cr-19Co-4Mo-2.9Ti-2.9Al), IN-100 (Ni-15Co-10Cr-5.5Al-4.75Ti-3Mo), or A-286.

Turbine discs are usually made from low alloy or stainless steels since higher alloying means a higher nickel base, and therefore, more difficult to forge materials. Materials used are A-286, Incoloy 901 (Fe-42.7Ni-13.5Cr-6.2Mo-2.5Ti), D-979 (Ni-27Fe-15Cr-4Mo-4W-3Ti-1Al), and Waspaloy (Ni-19.5Cr-15.5Co-4.3Mo-3Ti-1.4Al).

6.7 Marine Gas Turbine Environmental Protection

There are two aspects to be considered when discussing gas turbine engines and the environment. First there is the effect that the environment has on the operation of the gas turbine engine. Then there is the gas turbine's effect on the environment. Without precautions, either one can damage the other.

Protecting the Gas Turbine

A typical marine gas turbine must operate over a wide spectrum of environmental conditions: varying temperature, air quality, humidity, shock, vibration, and so on. The temperature effect has already been considered in the sections on compressors and unit performance; it is a significant parameter for which allowances must and can be made in the design phase. A more serious environmental problem is presented by airborne salt, potentially damaging particulate material and moisture, which lead to what are known as sulfidation, foreign object damage (FOD), corrosion, and fouling. Elimination of such damage is the purpose of marinizing efforts undertaken by all gas turbine manufacturers.

Sulfidation is defined as the combination of sodium (or any other metal, for that matter) with sulfur and oxygen during the combustion process to form a sulfate, and the resulting corrosive reaction with the parts subjected to the hot gas flow. Combustion liners, transition pieces, nozzle vanes, and turbine buckets are attacked and can be destroyed. Sources of metal impurities, primarily sodium and potassium, are contaminated fuel, or salt in the intake air. It is estimated that at sea level on a clear coastal day without spray or rain the air has a salt concentration of .05 ppm (parts per million). For a 20,000 HP unit, with a fuel-air ratio of about 1/50, this means that approximately 50 lb of salt could pass through the engine every 1000 hours of operation.

For a vessel operating in surf conditions, salt concentra-

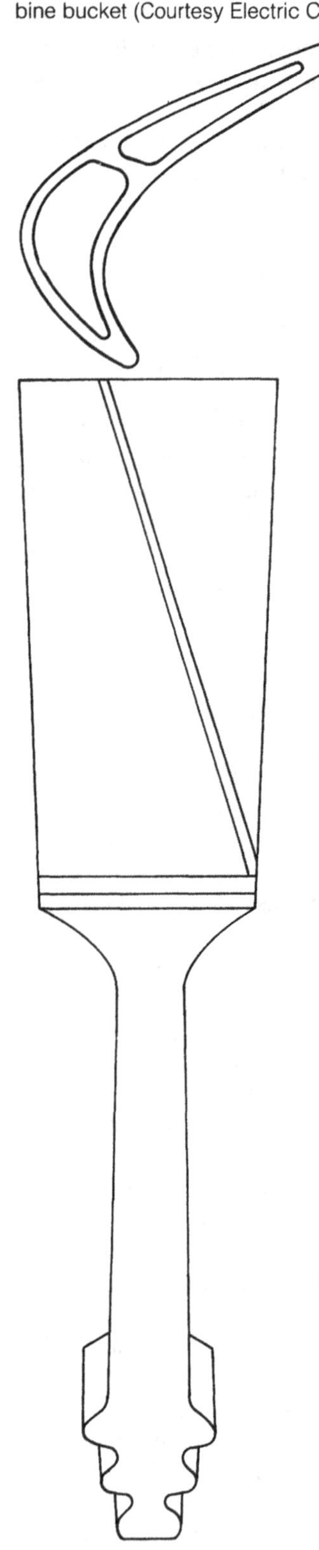

Figure 6.48 Long shank heavy duty turbine bucket (Courtesy Electric Co.)

tions in air can reach as high as 1000 ppm. When such a concentration passes through the air filtration system, the resulting deposits lead to rapid deterioration of the compressor pumping capacity and efficiency, resulting in loss of output power. Thirty minutes of operation under these conditions can reduce the power by as much as 20%. In addition, deposits may lead to exceeding the surge margin.

In any case, even very small amounts of salt can form caked particle deposits on compressor blades within a few hundred hours of operation. If these deposits are not removed by water or solvent washing, these caked particles break loose and are deposited on hot sections as highly concentrated "salt melt." Immediate conversion to sodium sulfate and pitting of the parts is the result. If salt is present in the fuel, it is more evenly distributed and is not deposited as a concentrate. However, since both forms of salt contamination cause rapid hot parts deterioration, it is important to eliminate and minimize the salt contamination.

A factor that seems to affect and aggravate sulfidation is firing temperature. With an increase of the turbine inlet temperature from 1600°F to as high as 2400°F, an increase in sulfidation has been found even when the fuel sodium content is within allowable specifications (less than 2 ppm when antisulfidation coatings/marine coatings are used) and the air salt content nil. In addition, it has been found that droplet size is as important a variable as concentration in compressor deterioration. The smaller particles evaporate more rapidly and pass through the compressor as dry crystals, but then cause sulfidation on turbine stages if the firing temperature is high. The larger droplets and higher salt concentrations tend to get deposited on rear compressor stages. The smaller droplets get deposited on forward stages if large enough not to be evaporated.

Compressor erosion may occur, due to droplet impingement and due to solid particle (dust, sand) ingestion. Erosion affects hot parts and the turbines as well. Both erosion and compressor fouling due to salt or other matter can lead to an 8–10% power loss in a relatively short time. To make up the lost power, the firing temperature must be increased with attendant deterioration of *SFC*. The effect of erosion is usually to eat away the blade tips with a corresponding increase in clearances between the blades and the casing; it is estimated that an increase of high pressure compressor blade clearance by .01 inch leads to an increase of the firing temperature by about 7° to make up lost power due to bypassed air.

The salt deposits and moisture cause rust and pitting of the blading and other parts. In addition, the high stress and corrosive environment of the rotating blades can lead to local stressing (stress corrosion) and reduction of the useful fatigue life of the rotating blades.

The immediate and most effective way to reduce, but not totally eliminate, sulfidation is to install aerosol separators to reduce the engine compartment concentration to a very low level of less than .005 ppm, on the average, for 95% of the operation time. Inertial separators (vane or centrifugal flow path) combined with woven pad filters (Figure 7.1) are now standard equipment on many installations. The inertial separator removes larger particles (10 microns or larger), and the filters remove most droplets and dust down to about 2 microns. To provide additional oxidation and sulfidation protection, improved corrosion-resistant alloys are used for compressor blades, anodized aluminum (vice lighter magnesium) is used on the housing parts, and sulfidation-resistant alloys and coatings are used on hot parts. Turbine buckets have been coated with aluminide, giving way now at high temperatures to platinum aluminide, or C101, or CoCrAlY electron beam deposited coatings. Alloys like U-700, IN-738, 601, and 617 have been used without coatings up to 2000°R with good service life.

In addition to coatings and inlet aerosol separators a standard procedure now is to provide additional protection by washing the engine periodically with fresh, preferably demineralized, warm water, or by special solvents. Regardless of the filter efficiency, over a long operating period deposits do build up, with long range deleterious effects and blocking of turbine cooling passages, control lines, and so on; and periodic plus immediate washing after a unit shutdown is recommended.

Engine parts corrosion can also be caused by the contaminants found in the fuel. Some of these contaminants are normal elements found in the crude oil, such as sulfur and vanadium; others, such as sodium, potassium, copper, and lead, originate from storage and transfer equipment, or are picked up from contamination by sea water if the fuel is stored in tanks that also see service as ballast tanks.

Sodium and potassium, abundant in sea water, are water-soluble compounds and can be removed by water washing or coalescer filtering. In the washing process fresh water may be added to the fuel; the water combines with sodium and potassium, and is then centrifuged out of the fuel. The coalescer is a series of filters. The first filter, of fiber glass construction, causes water droplets to coalesce on fibers; they are then drained into a sump. Follow-up filters are of the absorbing or membrane (separator) type.

Sulfur is the principal contaminant, which has lent its name to the corrosion-sulfidation process. It is naturally present in crude oils in varying amounts, and it is impossible, for all practical purposes, to remove it completely from the fuel. Thus, the sulfur is present in liberal amounts (up to 1%) as compared to other contaminants, and coatings must be used on parts exposed to hot gas flow. Next to sulfur, vanadium is the contaminant with the most serious corrosive effects on hot-gas-path parts. During combustion processes vanadium reacts with oxygen and sodium (or potassium), forming a liquid similar to the sulfur-sodium salt melt. It is deposited then on the first few stage blades with extremely corrosive effects. Thus, fuel specifications require that vanadium be limited to between .2 and .5 ppm. Fortunately, vanadium contamination is relatively easy and inexpensive to control by means of magnesium-containing inhibitors, which are directly added to the fuel (Epsom salts, for example), where occurrence of such contamination is regularly expected, as in the use of lower-grade fuels. The cost and complexity of vanadium control measures must be balanced against expected fuel cost savings or other anticipated benefits.

Table 6.1 summarizes the various materials used in gas turbine construction and Table 6.2 shows where some of the protective materials are used in the LM2500 gas turbine unit.

Protecting the Environment

The adverse effects that a gas turbine engine may have on the environment are similar to the effects from other engines and fossil-fuel–burning power plants. The high temperature exhaust must be directed safely away from everything it could damage. The gas turbine is a noise hazard, and should be insulated from the environment including muffling the intake air flow. The principal emissions, however, are combustion generated.

Carbon monoxide and unburned hydrocarbons are generated in gas turbine engines because residence time in the primary combustion zone is short and complete mixing of air and fuel under all operating conditions is very difficult. If the mixture of air and fuel in the primary zone is near stoichiometric, slightly lean, the resulting combustion will produce very low levels of hydrocarbon and carbon monoxide emissions. As the load on the engine changes, the amounts of both air and fuel entering the primary combustion zone change. If the air/fuel ratio moves away from stoichiometric either to rich or lean, the amount of hydrocarbon and carbon monoxide emissions will increase.

The formation of submicron solid carbon particles, carbon black or soot, has been a noticeable problem with gas turbines for years. The submicron particles are respirable and are therefore a potential health hazard. When the carbon particles are washed from the atmosphere, they make the rain water both dirty and acidic. The most effective solutions, as described in Section 6.4, involve combustor design. A lean, prevaporized, premixed combustion design

Table 6.1 Composition of Materials Used, Nominal Chemistry

Material	Fe	Ni	Cr	CO	Al	Ti	Mn	Si	Mo	C	Other	Trace
IRON BASE												
321 S.S.	68.5	9.5	18		0.2		2	0.7	0.5	0.08	0.5Cu	
347 S.S.	68.7	10	18				2	0.75		0.05	0.5Cb	
403 S.S.	85.6	0.5	12		0.5		0.5	0.5	0.2	0.1		Sn, P, B
410 S.S.	83.3	0.5	12		0.05		0.5	0.4	0.2	0.1		
15-7 PH	75.5	7	15		1.2		1	1	2.25	0.07		P, S
17-4 PH	72.4	4	16							0.05	4Cu; 4Cb	P, S
17-7 PH	73.7	7	17		1.15		0.7	0.4		0.07		
A286	52.8	26	15		0.3	2.1	1.5	0.7	1.3	0.04	0.3V	B
V57	52.5	27	14.75		0.22	2.95	0.35	0.57	1.25	0.08	0.3V	B
B5P5	97.2		1				0.55	0.3	0.55	0.45		
AM355	75.9	4.25	15.5				0.75	0.3	2.75	0.5		N
N155	49.7		21	20			1.5	1	3	0.12	1.0Cb; 2.5W	N
Nirralloy	92.9	3.5	1.15		1		0.55	0.3	0.25	0.23		N
Maraging S.	77.3	18			0.2		0.1		4.25	0.03		N
Chromoloy	97.7		1						1	0.2	0.1 Zr	
Greek Ascoloy	80.8	2	13				0.5	0.5		0.17	3 W	B, S
NICKEL BASE												
M252	5	51.5	19	10	1	2.5	0.5	0.5	9.75	0.15		B
SE1	1	50.8	15	22	4.35	2.35			4.4	0.08		B
SE1 15	.5	57.3	10.5	13.5	5.5	2.5			6.15	0.08	0.5 Cb, 1.5 W	B
TD Ni		97.7	0.01	0.1						0.01	2.2 THO$_2$	
TD Ni 1Or		76.2	21							0.1	2.7 THO$_2$	
Udimet 500	4	53.9	17.5	16.5	2.9	2.9	0.5	0.5	4	0.15		Cu, S
Udimet 700	4	48.8	15	18.5	4.5	3.25	0.15	0.2	5	0.1	0.15 Cu, 0.3 B	Zr
Inconel 100	7	69.7	15		0.45	0.55	0.5	0.5	3	0.15	3 W, Mg	P, B, S, Zr
Inconel 600	8	75.3	15				1	0.05		0.15	0.5 Cu	S
Inconel 601	14.1	58.7	23	1	1.35		0.5	0.25		0.1	1 Cu	
Inconel 625	5	58.0	21.5	1	0.4	0.4	0.5	0.5	9	0.1	3.6 Cb/Ta	S
Inconel 706	37.7	41	16		0.2	1.75	0.25	0.25		0.03	2.8 Cb	
Inconel 718	17.2	52.5	19		0.5	1	0.35	0.45	3	0.1	5.1 Cb/Ta, 0.75 S	B
Inconel 722	27	67.6			0.7	2.4	1	0.7		0.08	0.5 P	S
Inconel 901	34.7	43	13			2.8	0.5	0.22	5.7	0.05		
Rene 41	5	49.9	19	11	1.5	3.15	0.1	0.5	9.75	0.12		B
Rene 63	0.5	54.3	14	15	3.8	2.5	0.1	0.2	6	0.06	3.5 W	B
Rene 77	0.5	58	14.17	15	4.3	3.35	0.15	0.2	4.2	0.07		B, S, Zr
Rene 80	0.2	55.6	14	9.5	3	5	0.2	2	4	0.17	4 W	B, S, Zr
Rene 85		58.2	9.5	15	5.25	3.25			3	0.25	5.5 W	B, Zr
Rene 95		60.9	14	8	3.5	2.5	0.15	0.2	3.5	0.15	3.5 Cb, 3.5 W	P, B, S, Zr
Rene 100	1	60.0	9.5	15	5.5	4.2	0.5		3	0.18	1 V	B, S, Zr
Rene 120		59.5	9.25	10	4.25	4			2	0.17	7 W, 3.75 Ta	B, Zr
Waspaloy	2	55.1	19.5	13.5	1.25	3	0.5	0.75	4.25	0.1		P, B, S, Zr
Astroloy	0.5	54.5	15	17	4	3.5	0.15	0.2	5	0.06		B, S, Zr
Hastelloy X	18.5	46.3	22	1.5			1	1	9	0.1	0.6 W	
Hastelloy B	5	57.4	1	8.5					28	0.1		
COBALT BASE												
X40	2	10.5	25.5	51.9			1	1		0.5	7.5 W	P, S
L605	3	10	20	49.4			1.5	1		0.1	15 W	
BS 188	3.5	22	22	34.4			2	0.4		0.1	15.5 W	La
Marage 509	2	10	23.5	53.5		0.2	0.1	0.1		0.1	3 Ta, 7 W	Cu, P, B, S, Zr

(continued)

Table 6.1 Composition of Materials Used, Nominal Chemistry (*continued*)

Material	Fe	Ni	Cr	CO	Al	Ti	Mn	Si	Mo	C	Other	Trace
TITANIUM BASE												
6A1-4V	0.3				6.1	89.2				0.1	4 V	N, O, H
5A1-2.5 Sn					5	94.7				0.1	2.5 Sn	
Ti 55						100						
6A1-2Sn-4Zr-2Mo	0.12				6.125	85.8			2	0.1	2 Sn, 4Zr	
6A1-6V-2Sn	1				5.5	85.7				0.05	5.5 V, 2 Sn	O, H
8A1-1Mo-1V	0.3		4		8	89.4			1	0.08	1 V	O, H
Rene 17					5	83			4		2 Sn, 2Zr	
ALUMINUM BASE												
606116	0.7		0.25		96.7	0.15	0.15	0.6			1 Mg, 0.25 Cu, 0.2 Zn	
C355	0.2				92.4	0.1	0.5	5			0.5 Mg, 1.25 Cu, 0.1 Zn	
7005			0.15		93.3		0.45				1.4 Mg, 4.6 Zn, 0.13 Br	
7075	0.7		0.3		88.4	0.2	0.3	0.5			2.5 Mg, 1.5 Cu, 5.6 Zn	
MAGNESIUM BASE												
EZ33						(3.0 rare earth)					94 Mg, 0.5 Zr, 2.5 Zn	

Table 6.2 Use of Corrosion Resistant Materials in LM 2500

Bellmouth, bulletnose	AMS-4026 AMS-4218
Compressor inlet, front frame	17-4PH M-50 S2111
Compressor frames, casings (rear), discs, shafts	INCO 718
Compressor casing (front), front blades and vanes	Ti-6A1-4V
Compressor aft blades and vanes	A286
High pressure compressor blades	Ti-6A1-2Sn-4Zr-2Mo
Combustor	Hastelloy x HS188
High pressure turbine 1st stage vanes	X-40 HS188
High pressure turbine blades, second vanes	Rene 80
Turbine discs, casings	INCO 718
Power turbine blades, front vanes	Rene 77
Power turbine rear vanes, supports	Rene 41

Codep B and C (Ti diffussion coating) and BC21 (Y vapor deposition coating) used on high pressure turbine blades and vanes.

effectively prevents carbon particles from being formed in the primary combustion zone.

There are two other gaseous pollutants which contribute to acid rain. The amount of total sulfur emitted is mostly a function of the organic sulfur content of the fuel. Sulfur ends up as sulfurous or sulfuric acid. In a gas turbine, little can be done to reduce total sulfur except adding expensive clean-up devices such as scrubbers to the stack. At present, however, the low sulfur tolerance of gas turbines restricts them to fuels so low in sulfur that emissions are not a problem in most places.

Oxides of nitrogen, NOx, on the other hand, are generated primarily as a function of combustion temperatures if oxygen is available. Fuels bearing organic nitrogen, though not commonly used in gas turbines, cause a dramatic increase in nitric oxide, NO, production. In all combustors that generate oxides of nitrogen, the oxide produced is nitric oxide, NO. As the gases cool nitric oxide will oxidize to nitrogen dioxide, NO_2, and may oxidize to other oxides of nitrogen. All of the oxides of nitrogen, NOx, which are formed must be considered to get an accurate measure of the total pollutant emitted. Nitrogen dioxide is bad because it enters into the photo chemistry smog cycle. First, it is one of the few colored gases. It is brown and visible as smog. Sunlight reduces the nitrogen dioxide, NO_2, to nitric oxide, NO, and generates ozone in the process. Ozone is so reactive as to be a health hazard. Oxides of nitrogen also form acids in water and contribute to acid rain. The formation of nitric oxide, NO, can be minimized by proper combustor design.

Combustor design for reduced nitric oxide, NO, formation can take advantage of either the need for oxygen or the high temperatures required to produce nitric oxide. In fuel rich mixtures, the fuel uses all of the oxygen molecules, thereby denying the nitrogen access to oxygen. In stratified charge combustors, the fuel can be burned in a rich mixture to yield low nitric oxide formation followed by completion of combustion in a very lean mixture. In the lean mixture oxygen is present; however, the temperature of the mixture has been reduced by the dilution. In a gas turbine, the gases leave the combustor much too lean to support combustion. The lean combustion step would have to preceed the addition of the bulk of the secondary air.

The fuel could be burned initially at a very lean mixture to minimize nitric oxide formation, except for two reasons. Very lean mixtures require much more ignition energy than do rich mixtures. Also, lean mixtures burn more slowly than richer mixtures. Combustor design would have to accommodate the need for more ignition energy by returning more hot gases to stabilize the flame. Furthermore, the combustor would have to slow the velocities through the combustor or lengthen the path to tolerate slow flame speeds. Stratified combustion is the most likely to be found in gas turbine combustors.

Chapter Seven

Comparison of Steam and Gas Turbines

7.1 General

The preceding chapters have covered the basic underlying principles, main construction methods, and essential operating features of both steam and gas turbine units. The fundamentals set forth in Chapter 4 were used to develop the basic characteristics of both types of units, and a number of similarities have already been noted and discussed. In this chapter, a more extensive comparison of similarities, differences, advantages, and disadvantages will be made, with a section devoted to the common problem of transmitting the power to the shaft via the gearing, and another section highlighting the versatility of the gas turbine unit.

A quick survey of the factors involved makes it apparent why the gas turbine has been readily adopted for marine and naval applications. Aside from cost considerations (initial acquisition, installation, fuel, and maintenance), the advantages seem to be many, and the disadvantages only a few. However, it is expected that the gas turbine will not replace the other modes of propulsion, and that it will continue to coexist with steam and diesel engines, mainly because of current inherent drawbacks of the gas turbine at low speed and low power. Moreover, there always will be some missions and applications in which the steam turbine is superior to the gas turbine propulsion plant.

Studies also indicate that increasing fuel and oil prices will not have a drastic effect on the selection of the machinery type for use in commercial shipping. There capital and related costs (financing, insurance, and so forth), decreasing fuel quality with increase in prices, and mission needs will be more important factors in determining the type of power plants. For naval applications, although the fuel cost will always be a factor, the mission and logistic requirements tend to be the prime considerations in determining the type of machinery selected.

7.2 Comparison of Construction and Mechanical Features
Simplicity, Weight, Size

A number of the advantages of gas turbine (GT) units derive from their simplicity. Steam turbines (ST) with their associated auxiliaries such as blowers, pumps, piping, vacuum equipment, condensers, and so on, require great amounts of space, are heavier, and in combatant ships are more vulnerable to shock and battle damage. The whole GT

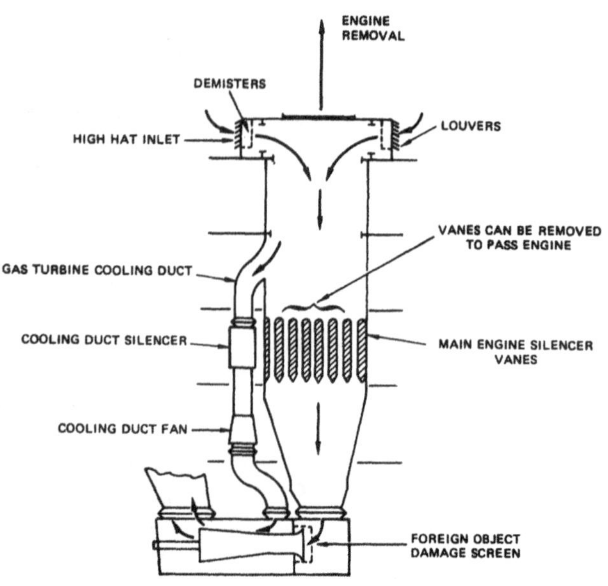

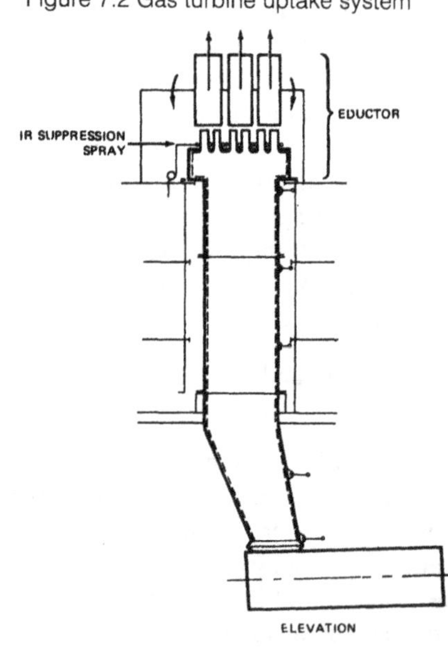

unit itself is about the same size as an HP or LP steam turbine or its condenser alone. Thus, both weight and space savings are considerable for the GT unit as compared to an ST plant. A comparison based on the same horsepower shows that a GT plant weighs about one-third as much as an ST plant and about one-half to one-fifth as much as a diesel engine, depending on the power. Aircraft derivative engines, having initially been developed for lightweight applications, tend to weigh less than the GT units developed for marine applications. The specific installed weight of a GT is 5–30 lb/SHP; for an ST plant it is 40–60 lb/SHP.

Weight saving is not of great importance for a ship like a destroyer, but its low weight and compactness are the factors that make use of the GT very desirable in surface effect ships, hydrofoils, and patrol boats. In destroyers and other larger ships, on the other hand, the low weight can lead to stability problems, as the overall power plant weight plays a significant role in the stability of the ship. Moreover, as Figures 7.1 and 7.2 show, the GT unit's intake and exhaust ducting are large and thus add weight topsides, which further tends to reduce stability.

One aspect of simplicity and compactness is implicit in Figure 7.1. The size of the unit and that of the duct are such that if a major failure occurs, the entire unit can be removed through the intake duct (see Figure 7.3), and another can be put in its place and be ready to go within 24 hours. The same is also true of the auxiliaries. Most GT auxiliaries, if spares are on board, can be replaced by the crew in hours. In the aircraft derivative or modularly designed units, e.g., Figure 6.33, the ship's crew can also replace hot sections, combustor units, controls, and so forth. By contrast, little major overhaul can be accomplished on an ST while under way, except possibly auxiliaries' component repair and replacement. On the other hand, the main turbines and the gears in a steam turbine unit rarely require major maintenance, and overhaul cycles are very long.

Reliability, Operating Experience, Maintenance

Both ST and GT plants exhibit good reliability. For steam turbines, which are mechanically quite simple, much operating experience has been amassed over many years, leading to a long life cycle and MTBO (mean time between overhauls). Keeping the steam temperature below 1000°F has contributed greatly toward smooth and safe operations. GT units show good dependability for a shorter MTBO, which is expected to reach 20,000 hours in the not too distant future. The shorter MTBO is essentially due to high temperatures associated with the aircraft engine cycle, which lead to hot section deterioration after a certain time even with the best of marine coatings. One essential difference that should be noted is that GT failures, when they occur, take place rapidly within seconds, as compared to ST failures, barring catastrophic occurrences or battle damage. There is no real protection for either type from such damage, but redundancy, as in a twin screw multi-unit GT plant,

Table 7.1 Marine Gas Turbine Engine Comparisons

Turbine	Manu-facturer	Base Load 1000 HP	Stages Comp.	Pressure Ratio	Combustor Shape	Fuel Air	Thermal Efficiency
IM5000	2/6	45.0	19	29.3	Annular	.01606	37.4
FT4C-3F	3	41.0	16	14.0	Can-annular	.01677	31.9
MS5002B	2	34.0	16	8.6	Cannular reverse flow	.01738	28.2
Marine RB211	1	33.0	13	19.1	Annular	.01844	35.5
Olympus TM3C	1	32.5	12	11.0	Can-annular	.01850	29.1
LM2500-PE	2/7	29.5	16	19.1	Annular	.02080	36.8
IM2500	2/6	28.3	16	17.1	Annular	.02075	35.9
Olympus TM3B	1	28.0	12	11.0	Can-annular	.01606	28.7
MS5002A	2	25.6	15	6.9	Cannular reverse flow	.01779	26.0
LM2500-PD	2	17.6	16	15.0	Annular	.01566	34.5
Spey SM1A	1	17.1	16	18.5	Cannular	.01554	34.9
IM2000	2/6	17.0	16	15.0	Annular	.01556	33.5
MS3002J	2	14.6	15	7.1	Cannular reverse flow	.01821	26.4
IM1500	2/6	13.6	17	11.2	Cannular	.01301	26.2
571-KF	4	7.7	13	12.7	Annular	.02026	33.9
570-KF	4	6.5	13	12.0	Annular	.02022	30.0
LM500	2	5.5	14	14.5	Annular	.01959	31.0
Tyne RM1C	1	5.3	15	12.5	Ruboannular	.01516	29.3
501-KF	4	4.3	14	9.5	Can-annular	.01800	27.0
CA-3	4/7	4.3	14	9.5	Can-annular	.01800	27.5
TF40	5	4.0	7ax+1cf	8.4	Annular reverse flow	.02088	26.0
TF25	5	2.5	7ax+1cf	6.9	Annular reverse flow	.01985	23.0

Manufacturers: (1) Rolls-Royce, Ltd.; (2) General Electric Co.; (3) Pratt & Whitney; (4) Allison Gas Turbines; (5) AVCO Lycoming; (6) Ishikawajima-Harima; (7) Creusot-Loire; (8) Fiat Avia Zione

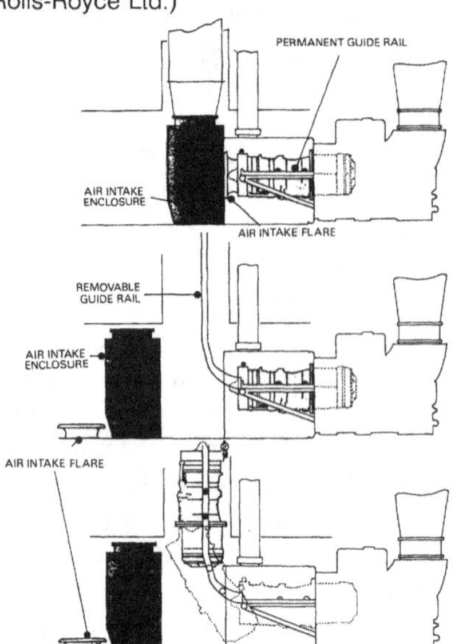

Figure 7.3 Steps in removal of a gas turbine through the intake duct (Courtesy Rolls-Royce Ltd.)

can alleviate some of the effects of this. It should be pointed out that since a gas turbine (gas generator–power turbine) comprises the major portion of a GT plant, any significant failure or damage to the unit is potentially more disruptive to the entire plant operation than in the case of an ST plant, where the turbine unit is a relatively smaller fraction of the entire plant.

The high reliability and shorter MTBO of the GT have resulted in a fundamental change in maintenance philosophy from that of the ST, derived from aircraft turbine maintenance experience accumulated by the airlines and military aviation community. Specifically, the shorter MTBO is dealt with by simply removing the unit and quickly replacing it as required, rather than repairing and/or overhauling it in place. The units that have been removed are not subjected to wholesale parts and module replacement. Improvement in materials, fabrication techniques, and nondestructive testing methods permit selectively changing out only those components that are determined to be worn out or defective. On-board maintenance is relatively simple and is limited essentially to periodic checks, compressor cleaning, and module replacement, if necessary. This modular maintenance concept permits on-board maintenance with a minimum of specialized tools, as well as minimum training and mechanical skill requirements. Furthermore, a minimum spares inventory is required. Thus, although a higher degree of knowledge is required of the crew, the maintenance is simpler and requires less training and specialization.

Parts, Cost

In GT construction mass production techniques provide interchangeable parts, components, and modules. ST units, on the other hand, tend to be one-of-a-kind machines, requiring a much more complex logistics support system. Whether aircraft derivative engines or heavy duty, heavyweight regenerative machines are used, the number of spare parts required and/or practical to have on hand is relatively small for the GT, and increasing modularization greatly simplifies maintenance and reduces the operating cost. Although cost is an ever varying factor, the GT units, especially the aircraft derivatives, have been cheaper to acquire and to install than steam or diesel plants of the same SHP, even taking into account the increased cost due to the controllable-pitch propeller or reversible type of reduction gear needed in the interim while the reversible rotation GT unit is being developed.

Intake, Uptake

The intake and exhaust systems have already been mentioned as part of the weight-stability problem unique to GT

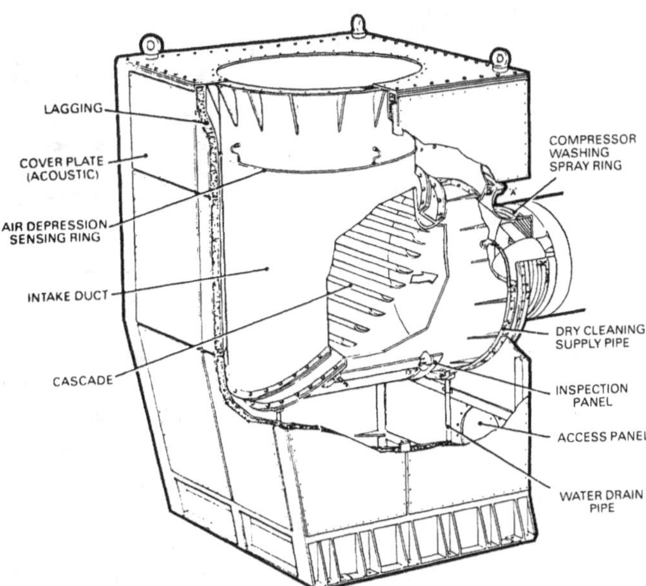

Figure 7.4 Gas turbine inlet compartment (Courtesy Rolls-Royce Ltd.)

plants. Figure 7.1 shows intake sound silencing and the engine cooling duct. As the hot section skin temperature may reach 1000°F, it must be cooled to maintain safe temperatures within the engine enclosure and to protect engine auxiliaries. Since a great portion of the engine noise originates at the inlet, sound suppression is necessary to provide for crew comfort and safety (See Figure 7.4). The intake duct is the sole external source for catastrophic engine failures. If a nut, bolt, or some other foreign object breaks loose and falls into the intake, it could be ingested by the compressor, an event which may, but will not always, wipe out an engine in a second. Thus, extreme care must be exercised in demister and screen maintenance. A direct counterpart to the inlet screens in a GT unit can be found in every HP, cruising, and auxiliary steam turbine, in the form of a fine screen emplaced usually in the last elbow in the steam supply line, to prevent foreign object damage to the turbine (see also Figures 5.43 and 5.44).

7.3 Operational Considerations

SFC, Speed, Response

Near the rated power setting, the aircraft derivative gas turbines now have a specific fuel consumption approaching or less than .4 (see Figures 6.29 and 7.5), and are competitive with diesel and ST plants at high power. The steam turbine average *SFC* approaches .6. The heavy duty GT units, designed to burn heavy distillate at about 1600°–1700°F vs. 2100°F for aircraft derivatives, have a realistic *SFC* between .5 and .6. At a reduced power level, the GT *SFC* increases rapidly because it is a full admission variable tem-

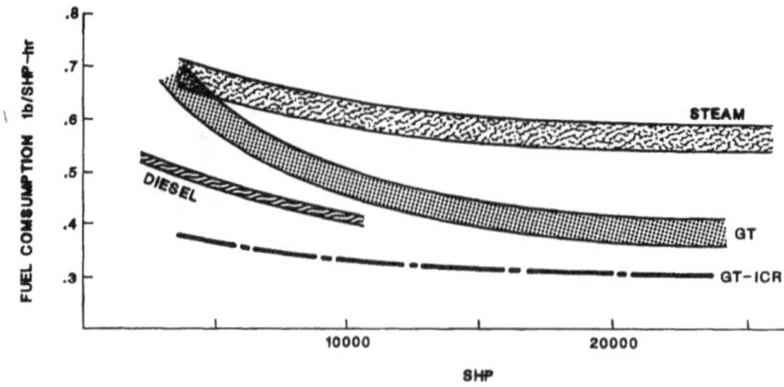

Figure 7.5 Fuel usage rate comparison, steam/gas turbine/diesel plants

perature turbine, while ST consumption increases at a much smaller rate because of its flow path maximizing measures such as partial admission, bypass, series-parallel arrangements, cruising turbines, and so forth. Although this problem represents perhaps the greatest disadvantage of a GT unit, it is overcome and eliminated by combining the gas turbine with various other types of power plants, including paralleling with other GTs of equal or different performance levels, diesel engines, and steam turbines, an arrangement which will be discussed further below. Also shown is the anticipated performance for the intercooled regenerative (ICR) engines now being developed.

Although gas turbines have not yet reached 100,000 SHP levels—and it is not certain that it will be necessary for them to do so, owing to the ease with which two much lighter gas turbines can be combined to give the same SHP rating—they have proved themselves by providing high speed (in excess of 40 knots) and good operational response characteristics. Steam turbines provide smooth continuous power but slower large power addition and deceleration capabilities. Gas turbines have a quick response and possess burst power capability (Figure 6.29). For example, the FT4A marine gas turbine was timed at 8 seconds between idle and 20,000 HP, and demonstrated repeatedly its ability to generate 25,000 HP in one minute from a cold start.

Response and flexibility comparisons are shown by the following statistics. A cold iron ST requires up to 48 hours to get under way. In emergency procedures 3–4 hours has been the minimum. With fire in the boilers, one-half hour is the minimum, while 3 hours is the norm, before full power can be achieved. A GT ship can go to 20 knots in 2–10 minutes, and from a stopped condition with engines idling to full speed in less than 2 minutes. At the same time, a crash astern maneuver can stop a ship going full speed in little more than one ship length. Performance like this is unattainable using current ST technology.

Manning Requirements, Remote Control

Manning requirements for GT-powered ships seem to be lower than for ST counterparts. Watchstanding duties are simpler, due to various automatic data display, logging, and control features, although failures in the latter usually present the most difficult maintenance problems requiring the most skill on the part of the crew to repair.

The size of the engineering staff on a GT ship is about 50–60% of the equivalent staff on an ST-powered ship. The reduced maintenance requirement alluded to above is one of the reasons for reduction in crew size. Reduction in watchstanding requirements is another factor. An engineering watch only 25% of the size required for an ST ship is needed for GT ships of comparable size. Another reason for reduced crew size is the adaptability of a GT plant to remote control. In new ships the philosophy of the "24 hour unmanned engine room," except for the roving watch, permits the operation of the entire propulsion plant by four to five men. For short periods three people will suffice. The automated remote control systems, unlike the situation on ST ships, permit starting, monitoring, and securing the entire plant from a single centralized location like the bridge, the engineering control center, or, of course, the engine room.

Vibration, Noise, Shock

Although the unattenuated GT intake and exhaust noise is extremely high, the sound suppression systems installed in the uptake and around the engine enclosure have effectively reduced the noise to practical and safe levels. As a result, the noise radiated by a GT ship is now less than that for steam and diesel power plants of comparable size. Also, the structural vibration is lower than in ST ships, and flow-induced noise is almost nonexistent, owing to the presence of fewer pumps and auxiliaries. A GT ship can be run with as clear a stack as one powered with steam or diesel units. One disadvantage of a GT combatant ship, however, stems from the high exhaust temperature (800°–1000°F), which provides a strong infrared signature and may require special exhaust system cooling measures.

Fuel

Although it has been stated that a gas turbine can be designed to run on almost anything burnable from margarine to natural gas, the marinizing requirements, discussed in Chapter 6, dictate high-firing-temperature units that require special care in selection and handling of fuels. Thus, fuels like JP-4, JP-5, and diesel fuels, which are costlier than the heavy distillates used for ST plants, are preferred for aircraft derivative engines. However, heavy duty GT plants now being designed with marine applications in mind are capable of using both No. 2 diesel and residual (bunker C) fuels.

Unlike the steam turbine, the gas turbine is highly susceptible to variations in ambient inlet conditions, especially inlet temperature. Due to the direct heating of the working fluid, high inlet temperature and sea spray ingestion (clogged filters) will degrade power output. Although the aerosol separator system can be monitored and maintained, the other ambient parameters are beyond the control of the operator. To balance the picture on the steam side, the only serious source of contamination of feed water and steam (besides oxygen, which can be removed by the DFT) is the salt water used for cooling purposes. The combustion superchargers of a pressure-forced boiler are also sensitive to reduced performance from inlet salt deposits. However, if contamination occurs, before it can leave any noticeable deposits in the turbine it will have ruined the boiler. Thus, one important task of the engine watch in ST ships is to monitor the feed water, for which there is no counterpart in GT power plants.

Thus we complete the direct comparison of GT and ST powered vessels. Table 7.2 summarizes briefly the various

Table 7.2 Comparison Summary of Gas and Steam Turbine Plants

	Gas Turbine	*Steam Turbine*
Simplicity	Simple, compact.	Mechanically simple.
Weight	Lighter, 10–30 lb/SHP.	More auxiliaries, 40–60 lb/SHP.
Space	Less.	Does not need large intake, uptake ducting. Needs separate space for boilers, DFT, and all condensate, feed water, and boiler auxiliaries.
Maintenance	Little while under way. Unit modular replacement. Quick unit replacement. Small spares inventory. Mass production/interchangeability of parts.	Continuing effort with respect to boilers, pumps, and auxiliaries. Nil with respect to main engine and gears.
Reliability	Good, short MTBO. Aircraft operations experience useful.	Good, high MTBO. Long operating experience.
Manning	40–60% less engineering force, 75% smaller watch. Unmanned engine room. Remote control; automatic data, control, safety features.	Broader watchstanding requirements.
SFC	Lower, .4 to .5 at full power. Poor at low power. Improve by COGAG, CODAG.* Mechanical compounding.	.5 to .6 over a relatively wide power range. Uses fluid compounding.
Response	Higher rapidity, fast cold and hot starts, good acceleration.	Smooth continuous power, slow addition of large blocks of power unless all boilers are on line. Requires boiler and engine "warm-up."
Vibration	More engine vibration but less structural vibration at all powers.	More pumps, auxiliaries.
Noise	Less, with sound suppression.	Can be quiet with special measures as in submarine operation.
Susceptibility to ambient conditions	High compressor inlet temperature, salt ingestion degrade SHP.	Relatively not susceptible, supercharger compressors for pressure-fixed boilers suffer performance degradation with salt ingestion.
Fuel	Needs fuel selection, treatment; sulfidation problems, marine coatings, susceptible to bacteria contamination.	Cheaper, no special treatment within limited range of petroleum fuels for which a steam ship is outfitted with tank heating coils, pumps, in-line heaters, and sprayer plates.

*For definitions of COGAG and CODAG see Section 7.5.

points made above. It should be emphasized that the comparisons are rather broad and are based on the assumptions that the plants produce the same SHP and are used in similar ships. For very small or extremely large vessels the comparisons may be less applicable on the basis of unsuitability or as yet undetermined scale or mission factors. The following sections offer a brief description of the gearings used in both ST and GT ships and a summary of the many and varied ways that gas turbines are put to use in basic propulsion schemes.

7.4 Gearing

From the discussion of turbines it may be recalled that higher rotor speeds lead to more energy conversion per turbine stage and therefore more powerful turbines for a given size. However, the mechanical energy developed by the turbine must be converted to thrust delivered by the propellers; and propellers work efficiently only at much lower speeds. From 1914 to the mid-1960s, when it was again applied in a nuclear submarine, the direct drive turbine was not in use on large ships. Some means of coupling the higher turbine speed with the desired lower propeller speed must be provided. Historically, this has been done in two principal ways: by electric drive and by mechanical reduction gears. Electric drive initially found favor in many large naval installations as the turbine began to supplant the reciprocating engine.

In electric drive (Figure 5.8) the turbine is used to propel a high speed generator, which in turn furnishes power to a lower speed motor. By varying motor pole combinations for various speed steps, the speed range of the turbine can be kept within relatively narrow limits, an advantage in turbine operation. Another advantage to certain types of ships is that full astern power can potentially be applied to the shaft by reversing electrical connections while maintaining constant turbine direction (a substantial heat exchanger is also required for rapid reversing). Still another advantage is the possibility of connecting several generators, each driven by a relatively small unit, to one large motor driving a single shaft. These advantages of electric drive for high power installations are offset by the larger size and complexity of the electric systems as compared to a mechanical reduction gear, and some decrease in efficiency of power transmission as opposed to direct mechanical means.

The development of facilities and techniques for construction of large mechanical gears in this country has given their use a decided edge, particularly in large high-powered combatant ships. Such gears are cut under the most exacting conditions of temperature control by machines that obtain tolerances of one forty-thousandth of an inch while working on gears of diameters in the order of ten feet and more. This accuracy has resulted in a very high degree of operational reliability and a high efficiency of power transmission. For double reduction gears, the efficiency varies somewhat with speed and load, but remains in the range of 96.5–98%.

Reduction Gear Principles

When two gears are in mesh, the ratio of the angular speed of the follower to the angular speed of the driver is called the train value, the symbol for which is r. The linear velocity of the pitch circles is the same for both gears, and is related to the angular velocity as follows:

$$V = R\omega = 2\pi RN = \pi DN$$

where V = linear velocity, fps
ω = angular velocity, radians per second
R = radius of pitch circle, feet
N = angular velocity of gear, revolutions per second
D = pitch diameter, feet

Then, for two gears, a and b, in mesh:

$$V_a = V_b = R_a\omega_a = R_b\omega_b = 2\pi R_a N_a = 2\pi R_b N_b$$

whence:

$$r = \frac{\omega_b}{\omega_a} = \frac{N_b}{N_a} = \frac{R_a}{R_b} = \frac{D_a}{D_b}$$

But the number of teeth, T, on each gear is proportional to its radius; therefore:

$$r = \frac{T_a}{T_b}$$

Modern HP steam turbine speeds are in the order of 6000 to 9000 rpm; LP turbine speeds are about 5000 rpm. Gas turbine power turbine seeds approach 20,000 rpm. Maximum propeller speeds are about 200 rpm for large ships, and around 350 to 400 rpm for smaller ships. Maximum propeller speeds in the order of 95 to 110 rpm are common in commercial shipping. Assuming a 300-rpm propeller, a 9000-rpm HP turbine, and a reasonable turbine gear diameter of about 12 inches, the diameter of a single large reduction gear attached to the shaft would have to be:

$$D_b = \frac{D_a N_a}{N_b} = \frac{12'' \times 9000}{300} = 360''$$

or 30 feet.

In order to reduce the size and weight, a double reduction gear is therefore universally used. Figure 7.6 shows a common type. Figures 7.7 and 7.8 show alternative arrangements. These are known as locked train, double reduction gears. The term "locked train" will be explained later. Note

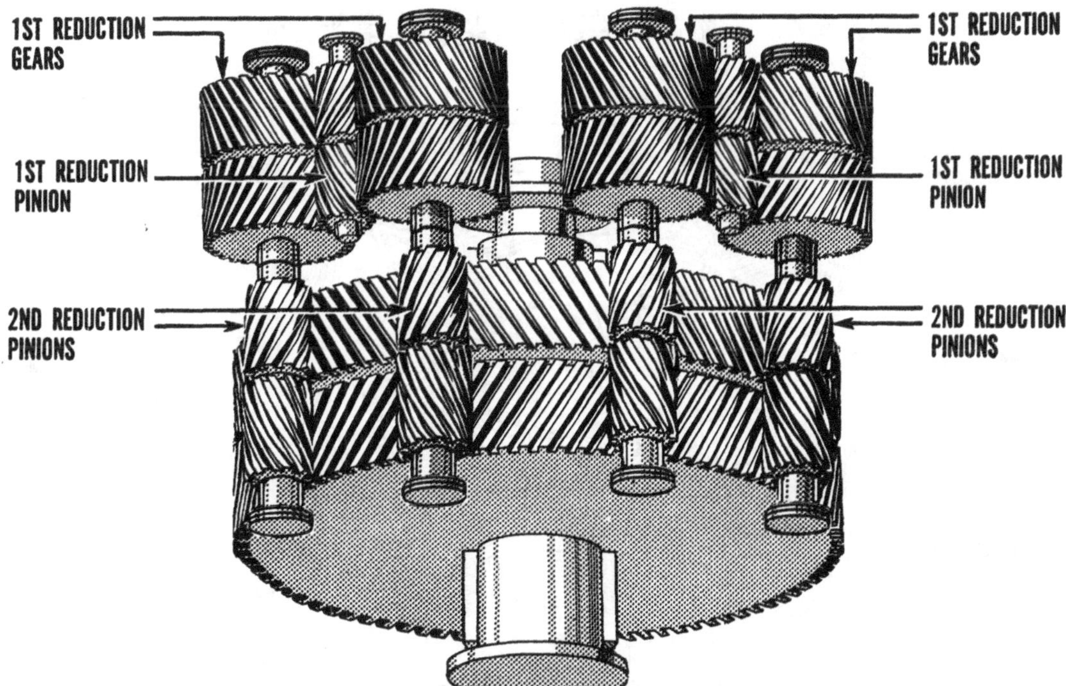

Figure 7.6 Locked train double reduction gear

that each turbine is connected to a first reduction pinion. These are often called simply HP pinions or LP pinions, to denote the turbine that drives them. Each pinion is meshed with two gears called the first reduction gears. They are mounted on quill shafts, each of which carries a smaller gear termed the second reduction pinion. These pinions drive the second reduction gear, commonly referred to as the "bull" gear. The arrangement of dividing the load to the four pinions permits lower tooth loads for a given power.

The force exerted on one gear by another is not tangential to the pitch circles at the point of contact (Figure 7.9), but acts along a line of action which lies at an angle with the tangent, called the pressure angle. The product of this force, the radius of the gear, and the cosine of the pressure angle, is the torque transmitted by the gear. This force must be opposed by an equal and opposite force acting at the center of the gear. The opposition force is transmitted to the bearing. Thus, bearings that support shafts carrying gears must carry not only the load imposed by the weight of the gear and its shaft, but also an additional load which is a function of the torque transmitted. In general, these two loads act in different directions, and their resultant is the vector sum of the two loads.

In Figure 7.9a, the small gear, A, is driving gear B. It exerts a force F_G alone the line of action, which is opposed by the reaction force F_R, acting at the center of gear A. Figure 7.9b shows F_R added to the weight of gear A and its shaft to determine the magnitude and direction of the resul-

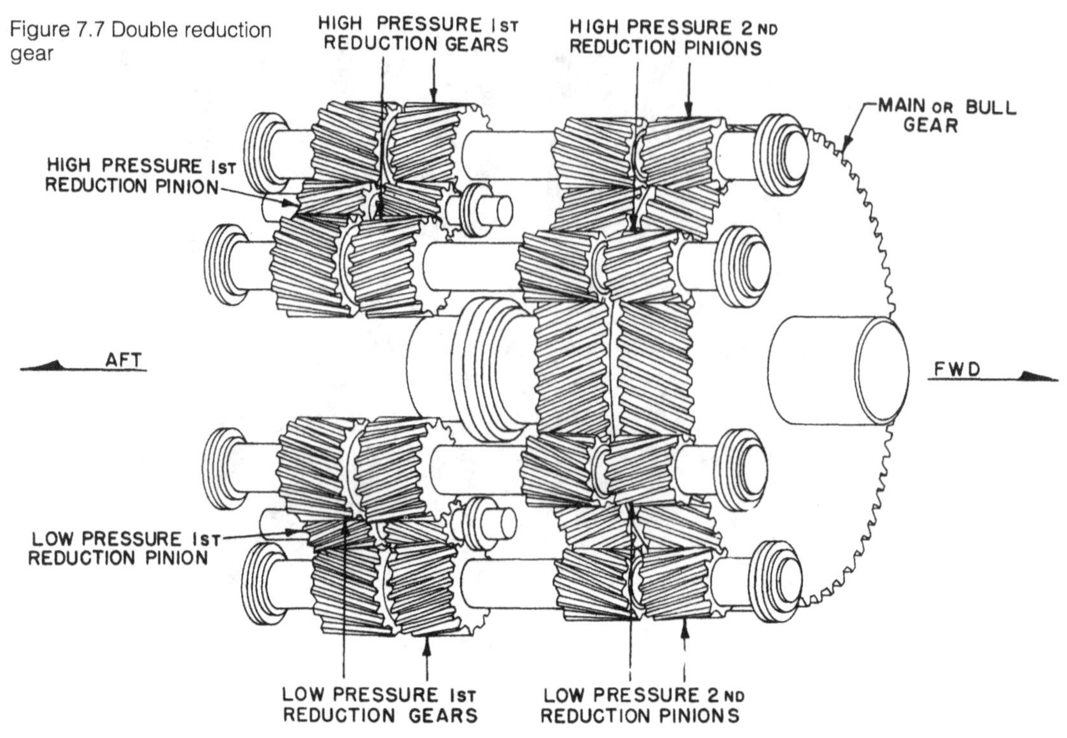

Figure 7.7 Double reduction gear

Figure 7.8 *Spruance* class GT double reduction gear

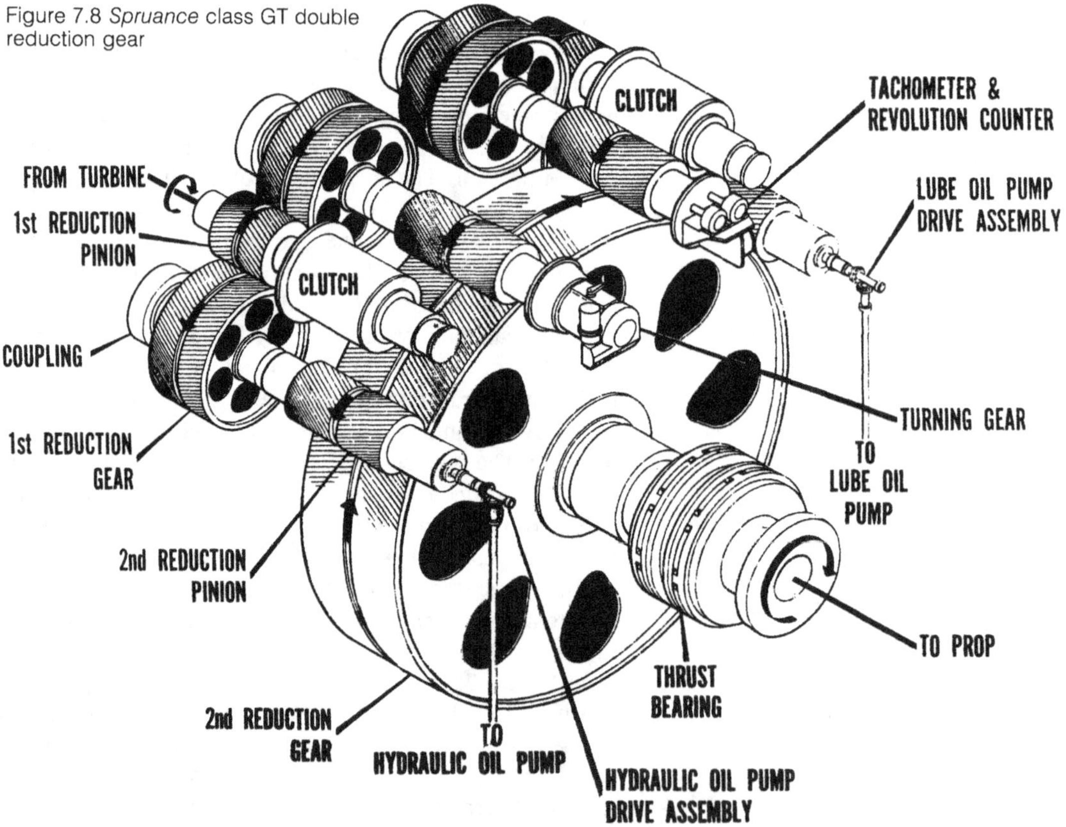

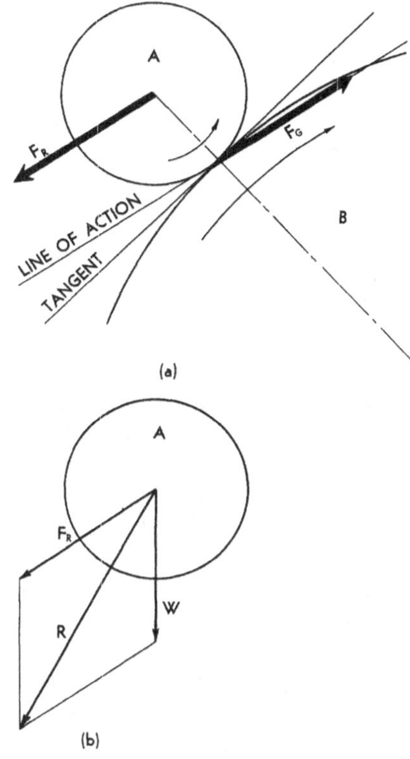

Figure 7.10 Quill shaft arrangement

tant force acting on the bearings. The two halves of the bearing shell are normally located so that their division is perpendicular to the resulting force on the bearing at full load. The load on the bull gear bearings is the vector sum of the reaction forces generated by all four meshing pinions and its weight. When one gear is meshed simultaneously with two other gears diametrically opposite each other, the reaction forces cancel out. This is the case with both the HP and LP turbine pinions in the locked train gear.

Referring to Figures 7.6 and 7.7, note that on each gear the teeth are cut in a double helix. The opposing helices act to balance end thrust on the gears that results from a driving force on a single helix, and to keep meshed gears rigidly positioned with respect to each other axially. However, there is a requirement for axial flexibility between the first and second reductions. The main shaft, connected to the bull gear, has a slight axial freedom, caused by the variable oil clearance in the main Kingsbury thrust bearing. This causes the bull gear to shift axially a few thousandths of an inch under various thrust conditions. This motion is transmitted to the second reduction pinions by the double helical tooth mesh. The motion is absorbed in the quill shaft arrangement that joins the second reduction pinions to the first reduction gears, and it is not transmitted to the other gears.

The quill shaft assembly also provides some flexibility to compensate for slight radial misalignment. The arrangement is shown in detail in Figure 7.10. Note the central shaft, which passes through the hollow gear and pinion. On each end is a spline, or gear type coupling, by which it is attached to the gear and pinion. This coupling permits axial motion of the gear and pinion relative to each other. The length of the shaft provides considerably more torsional flexibility than a short shaft, connecting the inner faces of the two gears, would provide.

The term *locked train* is used to indicate that the power is

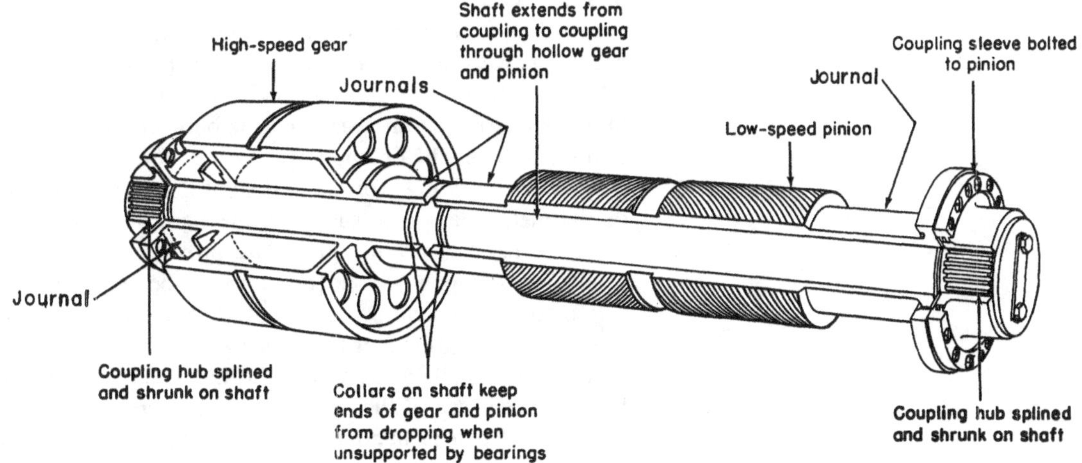

distributed equally between the two quill shafts on each side of the assembly. Consider the HP turbine side of the gear unit. In order for the gear to function properly, it is necessary that the HP pinion deliver its torque equally to the two first reduction gears. It is also necessary that the second reduction pinions deliver equal torques to the bull gear. To ensure this, it is necessary that the rotational position of the second reduction pinions be exactly adjusted so that each delivers the same tooth pressure to the bull gear. This is made possible in assembly of the gear by the vernier action provided by the relationship between the number of teeth on the various gears and on the quill shaft couplings, and by the fact that the circumferential pitches of the teeth in the first reduction, second reduction, and quill shaft couplings are different. Thus it is possible by proper assembly to set the two second reduction pinions in an angular relationship to each other so that each meshes with the bull gear with exactly the same tooth pressure. Automatic equalization of power between the two quill shafts is not assured without such adjustment. The gear train is said to be "locked" when it is adjusted properly.

Some newer ships' reduction gears utilize diaphragm coupling instead of spline tooth couplings on the ends of the quill shafts. Their capability of accommodating misalignment is superior to that of splined couplings, and they require no lubrication.

The gears are mounted in a casing assembly consisting of a weldment of heavy steel plate. Figure 7.11 shows a typical casing arrangement. The lower casing supports the bull gear and serves as a subfoundation for the rest of the assembly. It is bolted directly to the top of the gear foundation built into the ship. The upper casing carries the quill shafts and supports the covers that protect the gear from the entry of foreign matter, and it supports oil lines and spray nozzles arranged to supply a continuous flow of oil to the gear meshes. The whole casing also serves as a support for accessories such as the shaft turning gear, the revolution counter drive, and the attached lube oil pump drive. The lube oil system sump tank, to which all the lube oil from the turbines and reduction gear is led, is built into the foundation immediately below the lower casing. The gear cover is provided with fittings for lube oil sight flow indicators and thermometer leads (not shown in the figure). In addition, several hinged inspection plates are provided for periodic inspection of spray nozzle oil flow to the gear meshes. These covers are kept padlocked to prevent inadvertent admission of foreign matter to the gears.

In some installations, where the bull gear may extend below the surface of the sump oil level, a crescent-shaped sheet metal pan is fitted around the lower portion of the

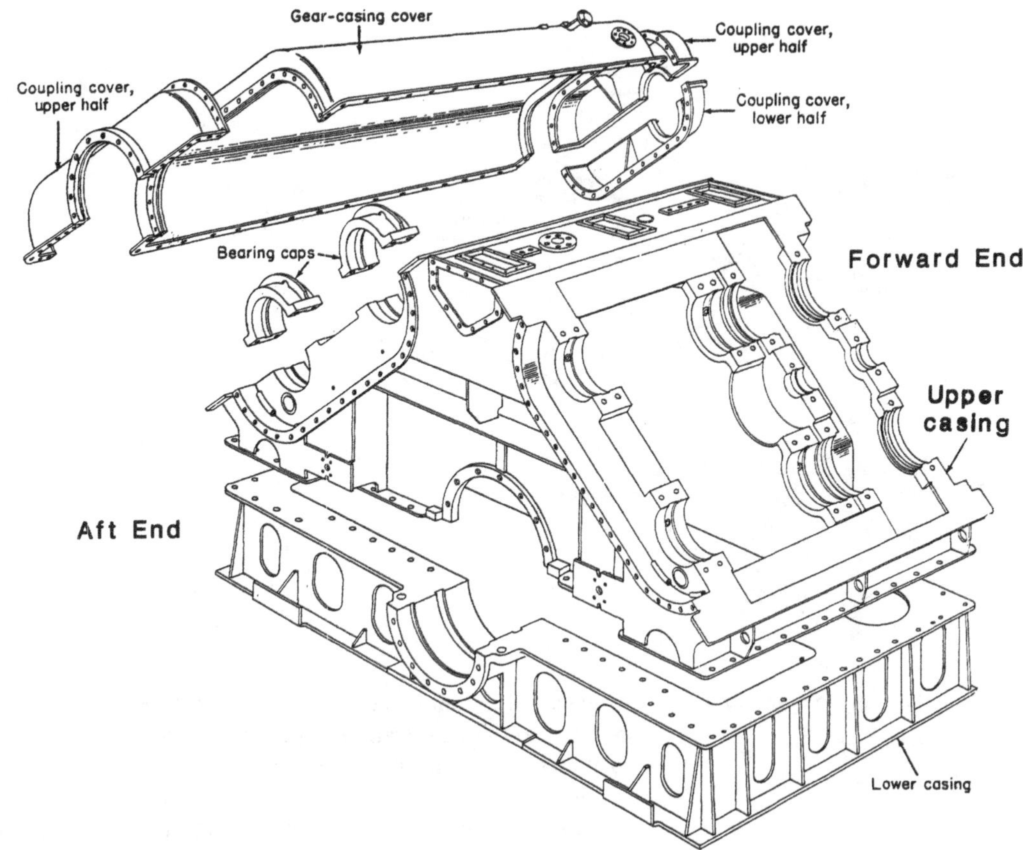

Figure 7.11 Reduction gear casing arrangement

gear. This prevents the gear from turning directly in the body of oil and thus churning and overheating the oil. This fitting is often called the "oil excluding pan."

It is often necessary to turn over the main shaft and the turbines without the use of steam; for example: (1) Immediately after the turbine is secured, the turbine rotor should be turned continually for a period to assure even cooling of the rotor to prevent unequal thermal stresses from developing (this is not required in gas turbines). (2) In port, the shaft is turned over at regular intervals to ensure the maintenance of the oil film in the turbine and gear bearings. (3) When a cold plant is being started up, the turbine is turned over before steam is admitted to the sealing glands and for at least 15 minutes afterward to provide even heating and expansion of the rotor. (4) During maintenance or overhaul, it is often necessary to rotate the shaft to position certain components such as the propeller, gears, or rotor for repair work.

An electric motor–powered gear train is provided for the above purpose. The gear train is normally arranged to engage the HP pinion shaft inside the reduction gear casing. Figure 7.12 shows an isometric sketch of the arrangement.

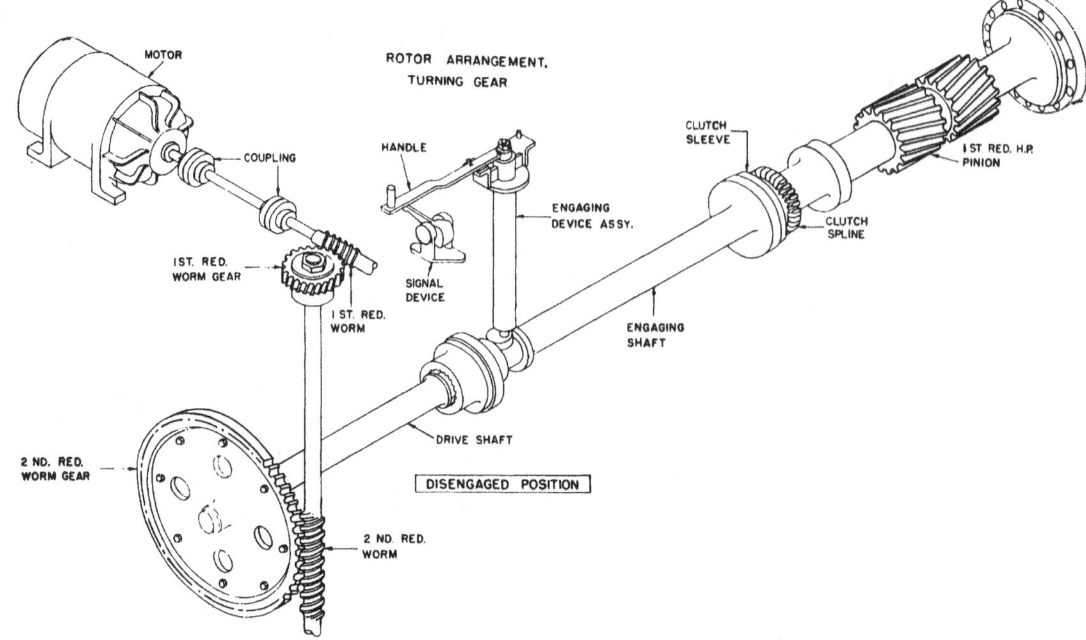

Figure 7.12 Shaft turning (jacking) gear

A typical installation is driven by a 5 HP, 1100 rpm electric motor, and the gear train has a reduction ratio of 283 to 1, so that the HP turbine rotor turns over nearly 4 times a minute.

Another important function of the shaft turning (jacking) gear, provided by the addition of a locking device on the electric motor shaft, is the locking of the propeller shaft while under way. In the event of a propulsion plant casualty while under way, the propeller would otherwise continue to "windmill" the main shaft, and thus the reduction gears and turbines. Generally the ship must be slowed to about 15 knots (depending on the type) and astern steam admitted to the damaged plant to stop it from spinning. Then the jacking gear is engaged and the locking device set. The ship may then resume full power on the remaining shafts. The locking device varies on different installations, but may be in the form of a friction brake or a locked gear arrangement. During the locking operation, the rpm of the other shafts and the astern chest pressure used to stop the disabled shaft are recorded. These conditions are then duplicated in any subsequent unlocking operation.

7.5 Gas Turbine Applications

It has been well established that gas turbines operate most efficiently at full power. For partial load operations, such as low power, long-term cruise operation with good fuel economy, the gas turbine can be combined with a diesel engine, a steam engine, or other equal or smaller-size gas turbines.

Such a combined plant uses the diesel or the small gas turbine for cruise power, and a larger gas turbine unit when full power is required. The basic arrangement of a single gas turbine driving a single propeller is now used only where high power is the normal mode of operation (as in hydrofoil ships, surface effect ships, and so forth); the combined plant is now the rule for surface displacement ships, rather than an exception.

A gas turbine unit can be combined with other power producers in many ways, and a great number of variations already exit, a few of which will be described below. One of the basic arrangements consists of the use of two or more identical units as building blocks to multiply the rated power capability with a minimum increase of weight and space. Figures 7.13–7.15 show two such arrangements. The first of these is a twin LM2500 module with a common reduction gear, used to drive one of the two shafts of the *Spruance* (DD963) class and the *Oliver Hazard Perry* (FFG-7) class, providing a maximum of 40,000 HP per shaft (Figures 7.13 and 7.14). Figure 7.15 shows three TF-35 units (2800 SHP each) used per shaft in a multi-mission patrol ship (PSMM), with primary reduction gear boxes on each engine, which then connect for further speed reduction to a secondary mixer type of gear box. The double gear boxes reduce the turbine speed of 14,000 rpm to 670, which is the full power speed of a four-bladed fully reversible controllable pitch propeller. Continuous rated power is 16,800 SHP, with a maximum of 18,000 SHP

Figure 7.13 *Spruance* class twin engine gear arrangement

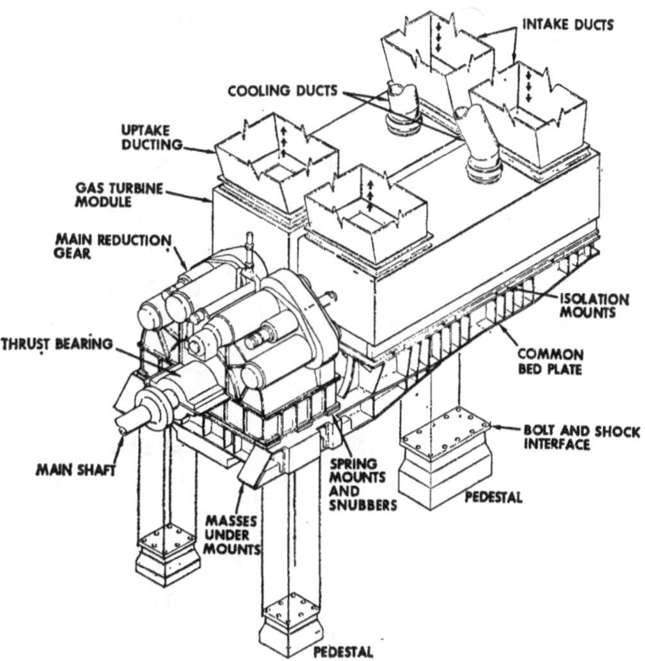

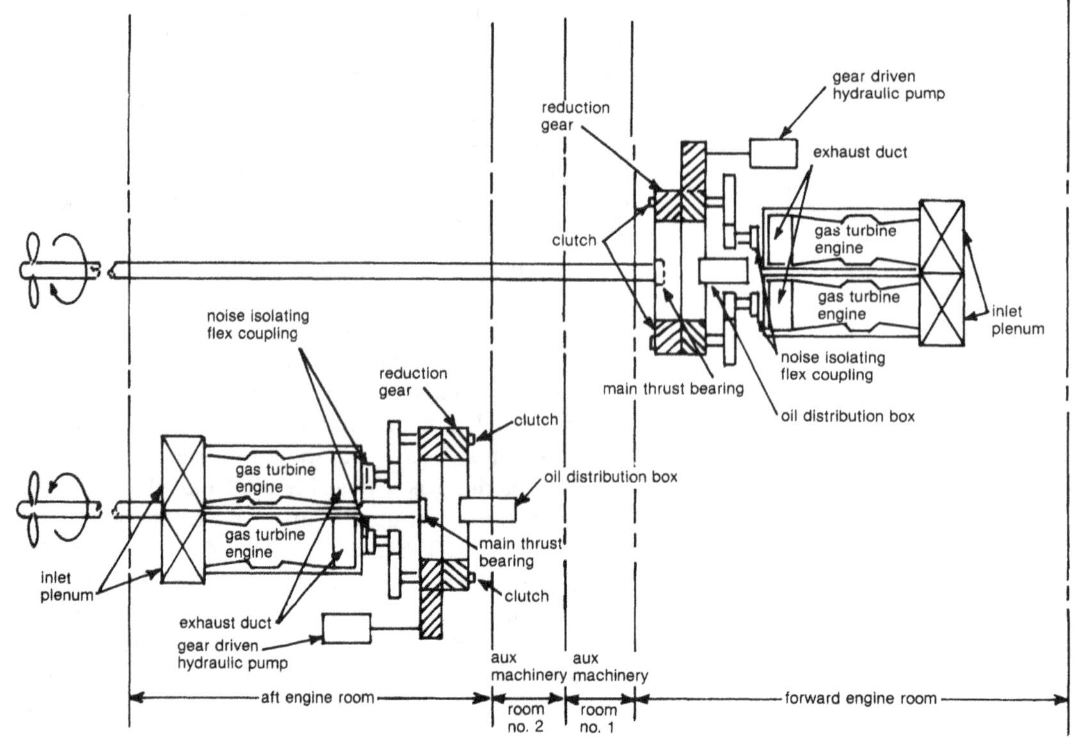

Figure 7.14 *Spruance* class machinery arrangement

available. The advantage of this system is that operation at partial power is possible with some engines operating at full power, while the others are secured entirely.

The type of power plant arrangement described above is known as COGAG (combined gas turbine and gas turbine), which combines a cruising gas turbine *and* a boost gas turbine. The following acronyms have come into common use in describing other various combinations and mixes of gas turbines, diesel engines, and steam turbines:

COGOG   Combined cruise gas turbine *or* boost gas turbine. A smaller gas turbine is used at or near its rated power for cruise and is disengaged after the boost unit takes over.

CODOG   Combined diesel *or* gas turbine. The units are used alternatively. The diesel is used for slow cruising economy and is disengaged after the high power boost gas turbine takes over for higher speeds.

CODAG   Combined diesel *and* gas turbine. Both diesel and gas turbine units are coupled (through hydraulic couplings or clutches) to the shaft. The gas turbine provides boost assist at high powers, and a variable pitch propeller is used.

COSAG   Combined steam turbine *and* gas turbine. Cruising is done with steam turbines; gas is used for boost power. The steam and gas tur-

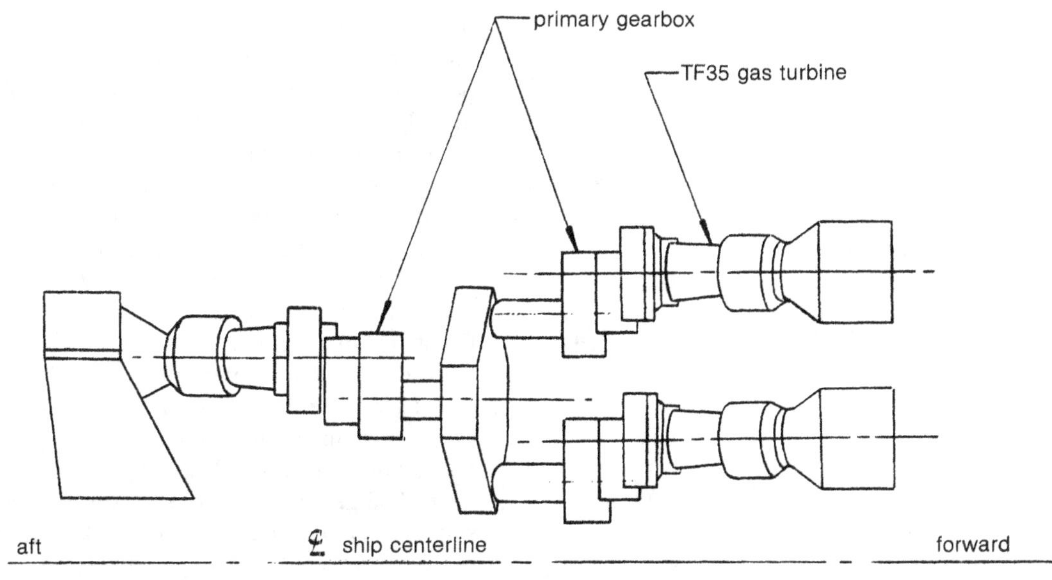

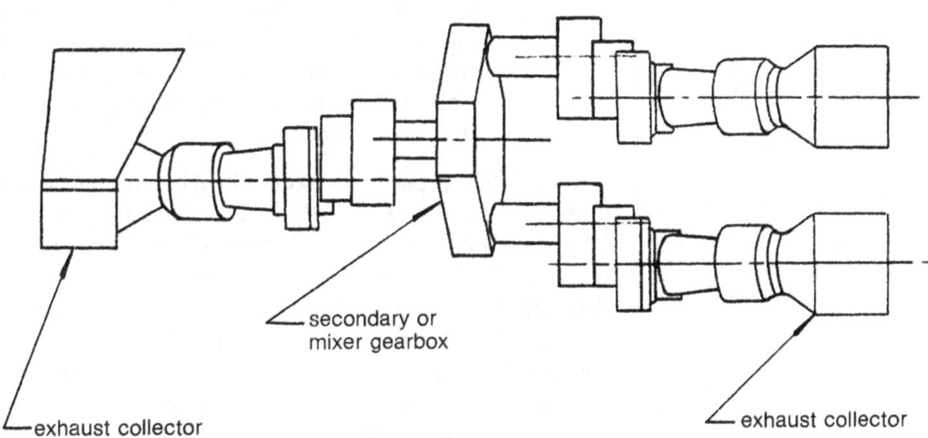

Figure 7.15 Patrol ship multi-mission (PSMM) machinery arrangement. (A) Plan view of four TF-35 units. (B) Machinery and exhaust system (Courtesy AVCO Corp.)

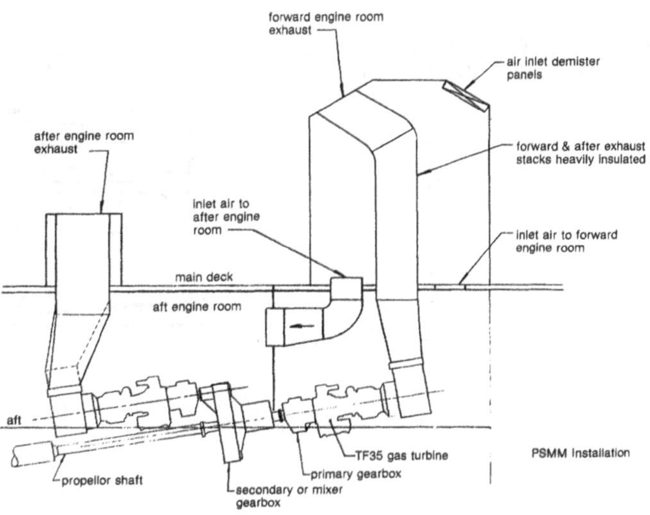

bines are thermodynamically independent, but a common reduction gear is used. Astern power is provided by a steam turbine coupled directly to the gearing and by a gas turbine via astern hydraulic coupling.

COGAS   Combined gas turbine *and* steam turbine. A gas turbine provides basic power, and steam turbine boosts the power. It is a combined cycle, where the gas turbine exhaust provides energy to a waste heat recovery unit.

CODLAG   Combined diesel electric and gas. Direct drive D.C. electric motor, with power supplied by diesel generator through a converter, is used in slow speed mode and for maneuvering. The gas turbine provides high speed power through a clutched gear box. Additional power may be supplied by the electric motor. Intermediate power is provided by the gas turbine. In case of twin screws, the other shaft is declutched and electric power is used only to reduce propeller traveling losses. Reverse power is obtained from the electric drive with gas turbine(s) declutched.

Two COGOG plants are exemplified by propulsion plants of the HMS *Exmouth* and HMS *Sheffield* (first of the British "Type 42" class destroyers) shown in Figures 7.16 and 7.17, respectively. The *Exmouth*'s cruise engines are two 4250 SHP marine Proteus gas turbines, with boost provided by a 28,000 SHP Olympus gas turbine. The Proteus engines are coupled through overrunning clutches to a common reduction gear. Since these engines are geared for maximum speed at cruise conditions, the accelerating boost engine output gear overruns the cruise engines at their maximum speed, and the cruising clutches disengage. These clutches

Figure 7.16 HMS *Exmouth* propulsion units, COGOG (Courtesy Rolls-Royce Ltd.)

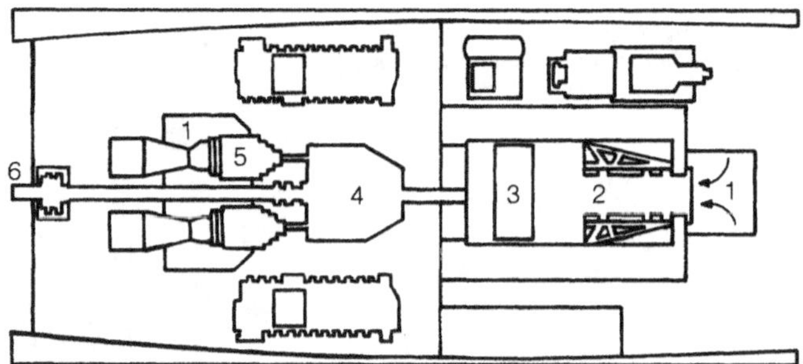

1. plenum chamber
2. Olympus gas turbine
3. exhaust
4. gearbox
5. Proteus gas turbine
6. propeller shaft

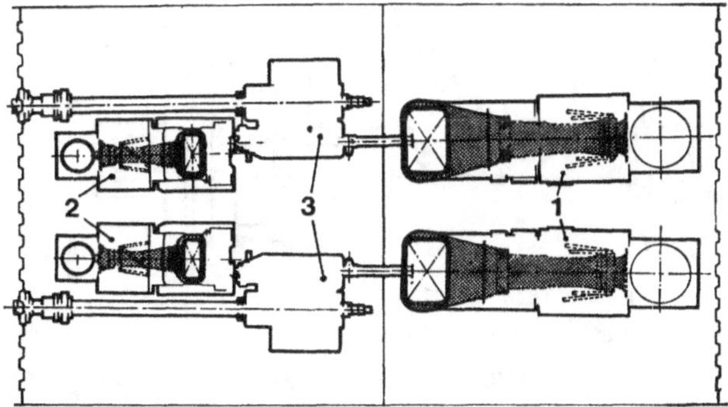

Figure 7.17 HMS *Sheffield* propulsion units, COGOG (Courtesy Rolls-Royce Ltd.)
1. Olympus gas turbine
2. Tyne gas turbine
3. gearbox

also serve to uncouple the cruise engines for service, and allow the use of the boost engine for low-speed, albeit uneconomical, operation.

Two examples of the CODOG type of propulsion plant are shown in Figures 7.18 and 7.19. Figure 7.18 diagrams the plant of the USS *Asheville* (PG-84) which has two 600 HP diesels for cruising and a single GE LM1500 for boost power. The *Asheville*'s first reduction gear provides power to both propellers (controllable pitch) during high speed operation. For cruise, diesels are connected independently to their shafts, and they are clutched out during the high-speed phase. The British *Vosper* Mk 5 destroyer (Figure 7.17) is powered by one 3800 HP diesel and one Olympus per shaft. The *Vosper*'s diesels and gas turbines are connected to a common reduction gear with overrunning clutches and interlocks to prevent simultaneous use of diesels and gas turbines.

The *Lupo* class fast frigate (Italian, Egyptian, Venezuelan, and Peruvian navies) machinery arrangement, shown in Figure 7.20, is a slight variation on the theme. The diesel engine is used here for cruise speeds up to 20 knots, and then two LM2500's are used for high speed boost to beyond 35 knots.

The *PHM-1* (*Pegasus*) class hydrofoil fast patrol boat in Figure 7.21 has another CODOG arrangement. Two diesels are used to drive steerable waterjets for hullborne operation. One LM2500 is used for high speed foilborne operation, and it drives a single water jet. Thus, both rudder and propellers are eliminated.

Figure 7.22 shows the machinery installation arrangement for the Japanese *Ishikari* class destroyer escort CODOG propulsion system. One 5000 HP diesel provides cruise power, with a single Olympus TM3B adding 28,000 HP for boost power to a twin screw CRP arrangement. An example of a COGAG installation is the Japanese DDG-171 class ship of Figure 7.23, where the cruise power (for twin shafts) comes from two 14,000 HP Spey SMIA engines. The boost power comes from two Olympus 28,000 HP TM3B engines, providing a total of 84,000 horsepower.

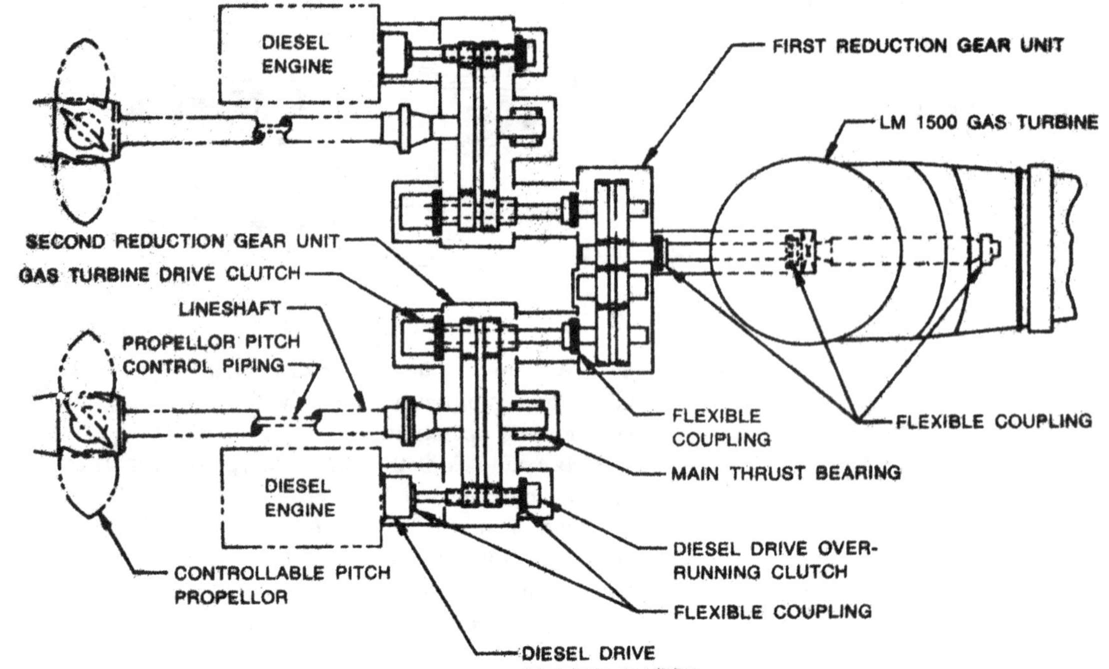

Figure 7.18 USS *Asheville* machinery, CODOG

Figure 7.19 *Vosper* Mk 5 propulsion unit, CODOG (Courtesy Rolls-Royce Ltd.)
1. Olympus gas turbine
2. exhaust
3. gearbox
4. diesel engines

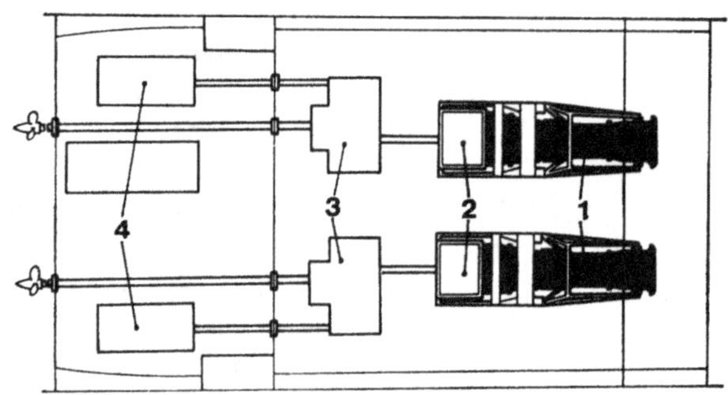

Figure 7.20 *Lupo* class CODOG arrangement

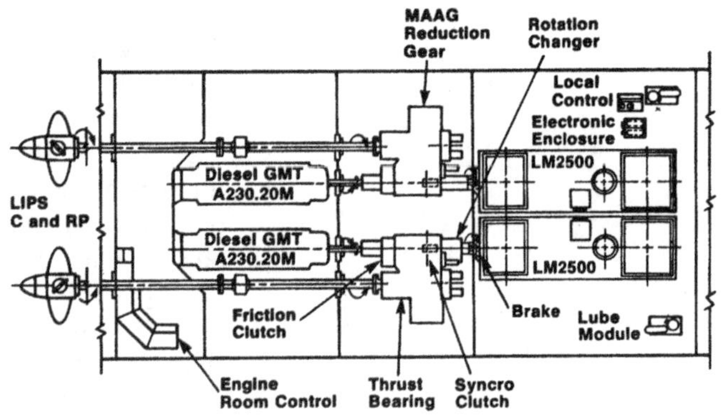

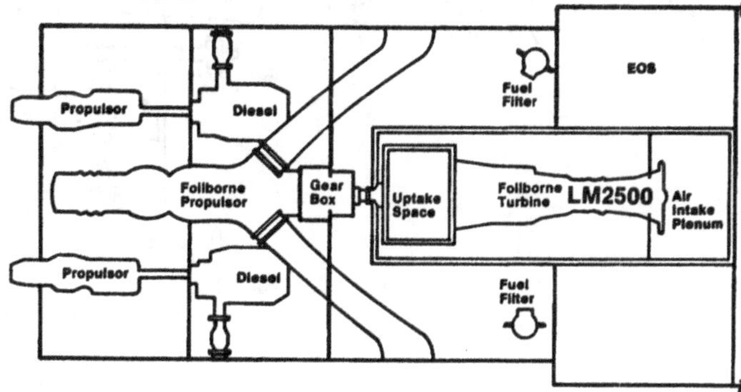

Figure 7.21 PHM-1 CODOG arrangement

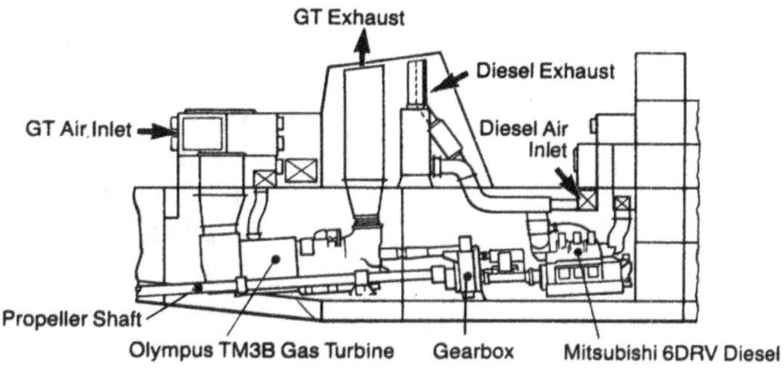

Figure 7.22 DE class CODOG machinery installation

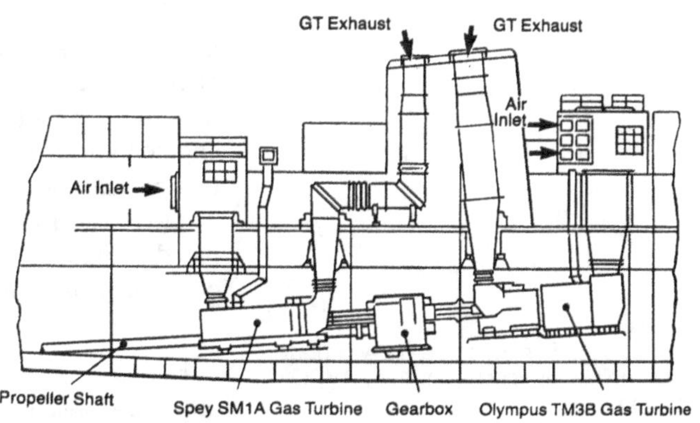

Figure 7.23 DDG class COGAG machinery installation

HMS *Bristol,* a Type 82 destroyer, is an example of a COSAG type of propulsion plant (Figure 7.24). One 15,000 SHP steam turbine and one Olympus gas turbine per shaft provide propulsive power. Gear boxes incorporate a reversing gear train and overrunning clutches for high-speed turbine operation.

In addition to the arrangements described above, there are numerous other ways in which the gas turbine unit can be adapted for marine propulsion. The following examples serve to highlight the versatility of the gas turbine, and may provide some possible directions for future development.

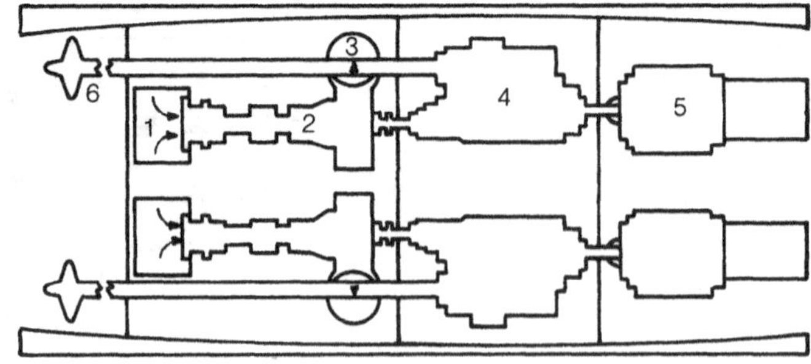

Figure 7.24 HMS *Bristol* propulsion unit, COSAG (Courtesy Rolls-Royce Ltd.)

1. plenum chamber
2. Olympus gas turbine
3. exhaust
4. gearbox
5. steam turbine
6. propeller

Figure 7.25 compares the size of AVCO TF-35 (Figure 6.33) engines (three units) and associated Jacuzzi water jet drives, with an equal shaft horsepower diesel installation of a San Francisco ferry. The gas turbine installation features a typical unmanned engine room, with the controls located in the pilothouse. Two-engine operation is normal, with three engines operating under certain starting conditions or at peak commuting times. The space and weight saving of this arrangement compared to the equal-SHP diesel installation is obvious.

Another example is provided by the Navy LCAC (Landing Craft Air Cushion) craft. The 60-ton payload LCAC is powered by six fully marinized AVCO Lycoming TF40B engines, each rated at 3750 HP. Two TF40B's drive two sets of 4-foot fans that generate the air cushion. The other four engines power four 7.5 foot-diameter shrouded propellers that provide the LCAC's propulsion.

Two basic arrangements in merchant marine use are compared in Figure 7.26. Configurations (a) and (b) are the more commonly used mechanical drive system described earlier.

Figure 7.25 San Francisco ferry propulsion unit comparison (Courtesy AVCO Corp.)

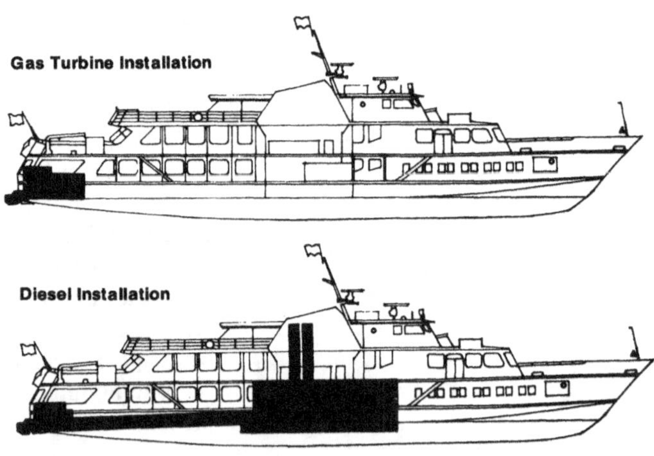

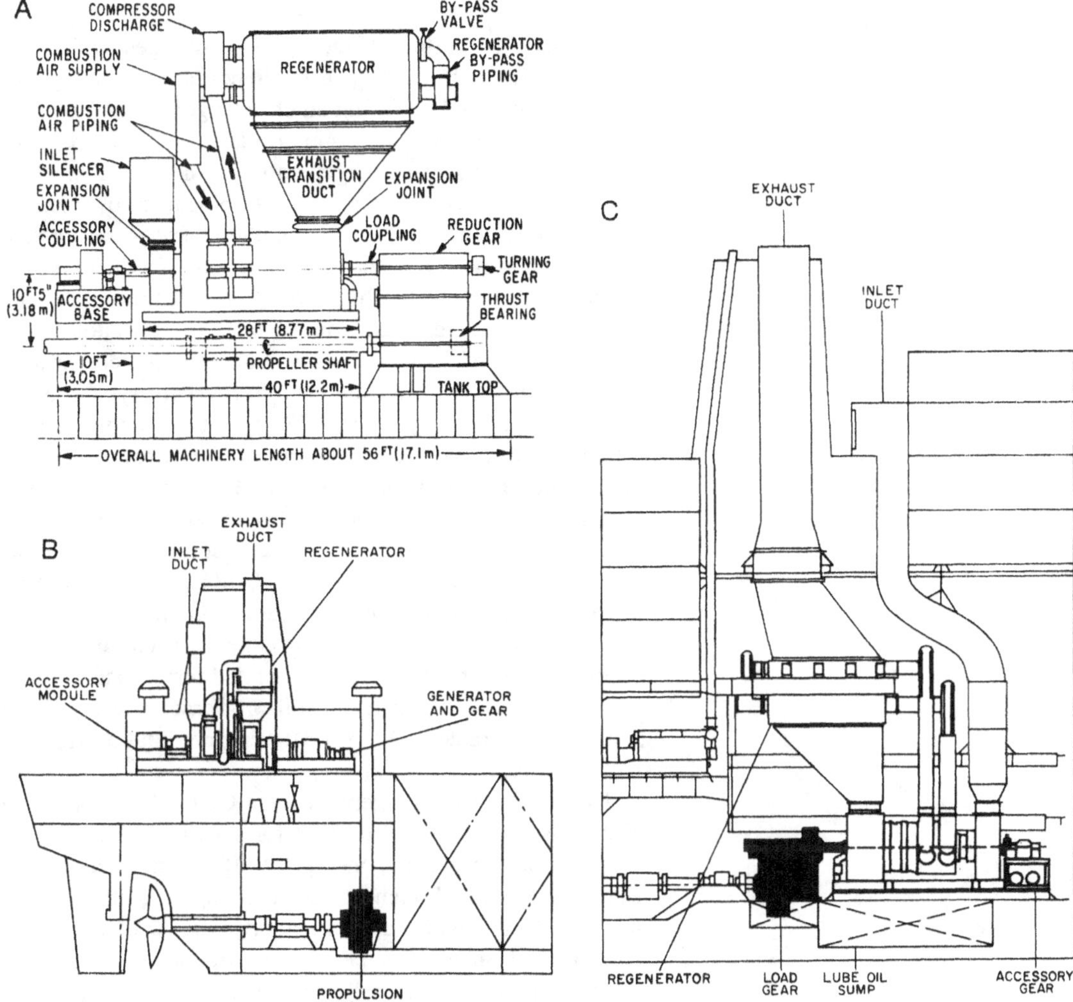

Figure 7.26 Heavy duty gas turbine machinery arrangement. (A) Forward-facing mechanical drive. (B) Mechanical drive installation. (C) Electric drive arrangement (Courtesy General Electric Co.)

It should be noted here, and this is true in general also of the various machinery arrangements described above, that the mechanical drive gas turbine unit is located very low in the ship. Its location is partly determined by the size and swing of the propeller under the stern, and partly by the size of the bull gear. Otherwise it could be mounted directly on the keel with a beneficial effect on stability. A steam plant, on the other hand, is placed higher in the ship, and therefore the shaft must have a higher rake. The placement of the steam turbine is determined essentially by how far the main condenser is under the LP turbine, by the size of the bull gear, and by the additional hotwell height necessary for efficient condensate pump operation. (See also Figure 2.4 for ST ships.)

Figure 7.26c exemplifies the electric drive installation, which was discussed in Chapter 5 (Figure 5.8), and has been used in merchant shipping. This figure depicts the GE heavy duty gas turbine (Figure 6.34) equipped with a regenerator for improved cycle efficiency and economy, located topsides and powering an electric generator. Power is transmitted to a propulsion motor attached to the propeller shaft. Two basic ways of generating-transmitting are feasible. In the AC/AC system the gas turbine turns over an AC synchronous generator, which supplies three-phase current to the synchronous propulsion motor. Here a reduction gear is required between the high speed power turbine and the 60, or 50, hertz generator. In the AC/DC configuration an AC-DC conversion package is required to provide variable power to the propulsion motor. If a thyristor type conversion package is used, variable DC armature voltage is provided to the motor for speed control. Shaft reversal is accomplished by motor field reversal. If a rectifier system is used for AC-DC conversion, then motor armature voltage is varied by controlling the AC generator output voltage. Shaft reversal is accomplished by motor field reversal. However, shaft braking requires either dynamic braking resistors or absorption of this power by other means (pumping into a second power sink as in a double-ended ferry boat). Figure 7.27 shows a 4150T roll on–roll off vessel powered by a John Brown Frame 3, 12,000 SHP engine, with an AC/DC electric drive.

The machinery arrangement for a CODLAG propulsion system is shown in Figure 7.28 for a twin screw Type 23 frigate. For high speeds, one Spey SMC gas turbine on each shaft drives a fixed pitch propeller through a double reduction gear box. Paxman diesels and one GEC D.C. motor per shaft provide power for slow speed operation. The gearboxes and gas turbines are then isolated via SSS clutches.

One gas turbine is used at intermediate speeds with electric motor provided to the other shaft, with gears and the other gas turbine declutched, to minimize the propeller trailing drag.

Last, but by no means least, mention should be made of a more advanced gas turbine engine arrangement whose time seems to have arrived. In Chapters 3 and 6, it was shown how the gas turbine cycle can be improved by intercooling and regeneration (ICR). Adding intercooling increases the net power available for useful work, and incorporating regeneration helps to improve the cycle efficiency (Figure 3.17, Example 6.6). For various reasons, partly developmental, partly operational, until recently the only ICR marine engine in existence was the Rolls Royce RM60 engine installed in HMS *Grey Goose* in the late 1950s. This engine, Figure 7.29, had two intercoolers: one for the eleven-stage low pressure compressor, the other for the two-stage high pressure compressor. An exhaust heat exchanger pro-

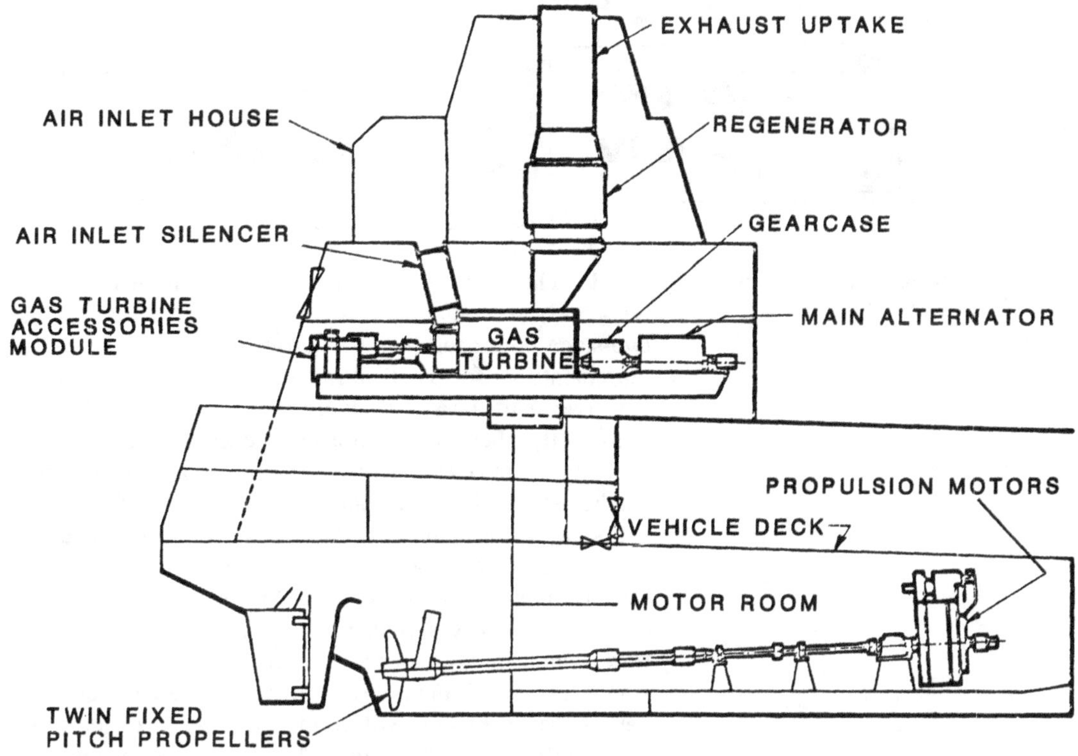

Figure 7.27 Heavy duty roll on–roll off propulsion unit (Courtesy General Electric Co.)

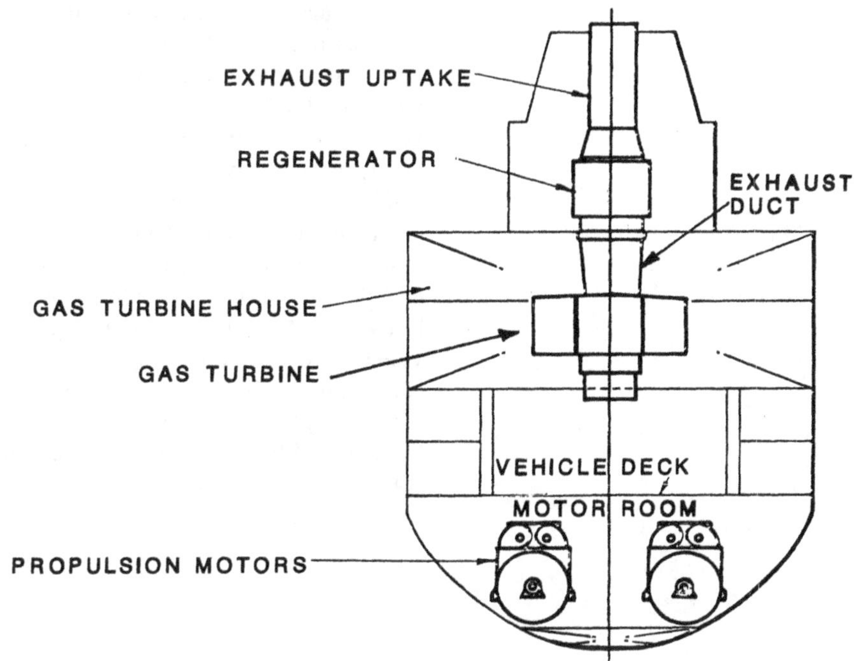

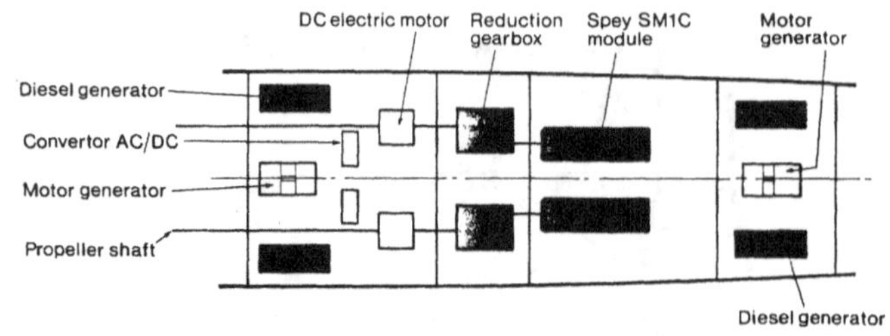

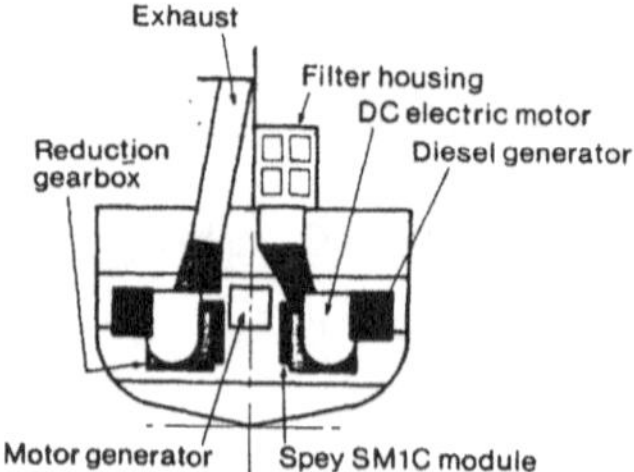

Figure 7.28 Machinery arrangement for CODLAG propulsion system

vided cycle regeneration. Another interesting feature is the high pressure compressor being on the same shaft as the two-stage power turbine. This arrangement typifies the requirement that for practical intercooling, it is necessary to have two shafts or a twin spool engine.

Recently, there has been a resurgence of interest in the ICR engine for several practical reasons. First, as the output can be increased by intercooling to better than 20% over the simple cycle, and since the efficiency will be improved by almost the same amount, the fuel costs should be reduced by almost one-third. Second, the efficiency and specific fuel consumption curves remain flat over about 75% of the power range and do not rise as sharply at low power as the simple cycle GT, Figure 7.5. Thus, the relatively efficient partial load operation will permit elimination of the separate cruise (or boost) turbine. Thus, space and weight will be saved even with the addition of the intercooler and regeneration. Third, due to regeneration the exhaust gas is cooler, thus providing a much lower infrared signature for combatant ships.

For practical (NAVY) applications such an engine will be a successor to the LM2500 which requires about 25,000 HP. Due to power and twinspool requirements, the two engines identified for further marine development are the Spey and LM1600 which are both aircraft derivatives (the TF41 and F404, respectively).

This concludes the brief survey of the essential features and methods of application of gas turbine and steam turbine units and their principal variations. It is readily apparent that the gas turbine power plant is capable of great flexibility, and its variations in size, power, and response provide a wider spectrum of applications than can be economically achieved by steam turbine propulsion. However, it is anticipated that for a number of reasons, not the least being the availability and cost of fuel, gas and steam turbines will continue to coexist for some time to come.

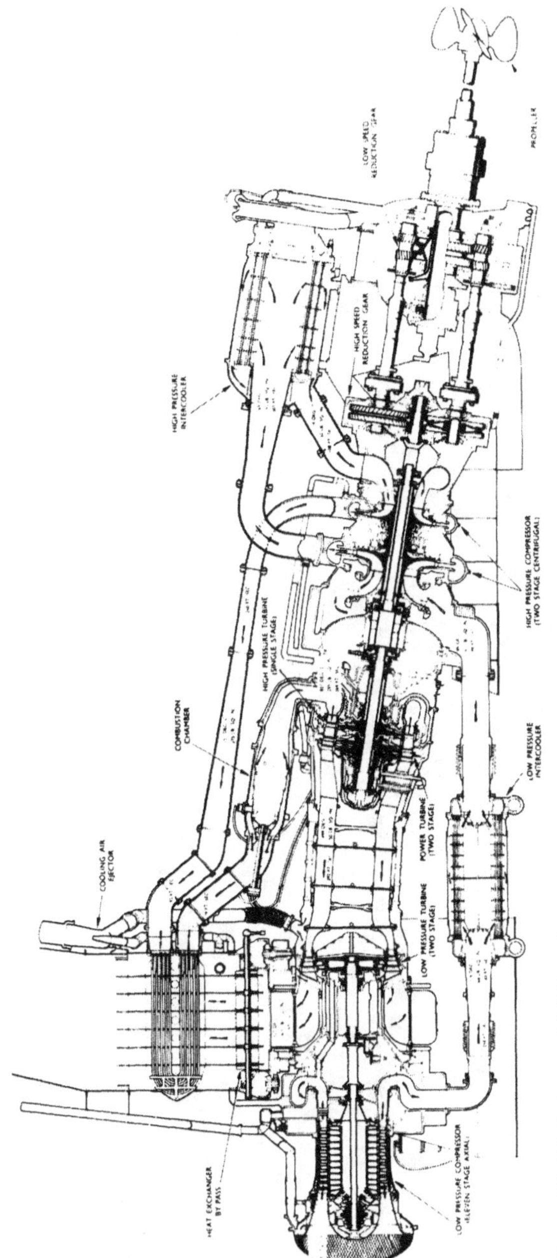

Figure 7.29 Rolls-Royce RM60 marine engine.

Bibliography

Cohen, H., Rogers, G. F. C., and Saravanamuttoo, H. I. H., *Gas Turbine Theory,* New York: John Wiley and Sons, 1973.

Dixon, S. L., *Fluid Mechanics, Thermodynamics of Turbomachinery,* Oxford: Pergamon Press, 1975.

Hunsaker, J. C., and Rightmire, B. G., *Engineering Applications of Fluid Mechanics,* New York: McGraw-Hill Book Company, Inc., 1947.

Lee, J. F., *Theory and Design of Steam and Gas Turbines,* New York: McGraw-Hill Book Company, Inc., 1954.

Pratt and Whitney Aircraft Company, *The Aircraft Gas Turbine and Its Operation,* PWA OI 200, 1970.

Principles of Naval Engineering (NAVPERS 10788-B), Washington, D.C.: U.S. Government Printing Office, 1970.

Sawyer, J. W., Editor, *Sawyer's Gas Turbine Engineering Handbook,* Stamford: Gas Turbine Publications, Inc., 1972.

Shepherd, D. G., *Principles of Turbomachinery,* New York: The Macmillan Company, 1956.

Streeter, V. L., *Fluid Mechanics,* New York: McGraw-Hill Book Company, Inc., 1971.

————, *Steam, Its Generation and Use,* New York: The Babcock and Wilcox Company, 1963.

Index